MONOGRAPHS ON STATISTICS AND APPLIED PROBABILITY

General Editors

**V. Isham, N. Keiding, T. Louis, N. Reid, R. Tibshirani, and H. Tong**

1 Stochastic Population Models in Ecology and Epidemiology *M.S. Barlett* (1960)
2 Queues *D.R. Cox and W.L. Smith* (1961)
3 Monte Carlo Methods *J.M. Hammersley and D.C. Handscomb* (1964)
4 The Statistical Analysis of Series of Events *D.R. Cox and P.A.W. Lewis* (1966)
5 Population Genetics *W.J. Ewens* (1969)
6 Probability, Statistics and Time *M.S. Barlett* (1975)
7 Statistical Inference *S.D. Silvey* (1975)
8 The Analysis of Contingency Tables *B.S. Everitt* (1977)
9 Multivariate Analysis in Behavioural Research *A.E. Maxwell* (1977)
10 Stochastic Abundance Models *S. Engen* (1978)
11 Some Basic Theory for Statistical Inference *E.J.G. Pitman* (1979)
12 Point Processes *D.R. Cox and V. Isham* (1980)
13 Identification of Outliers *D.M. Hawkins* (1980)
14 Optimal Design *S.D. Silvey* (1980)
15 Finite Mixture Distributions *B.S. Everitt and D.J. Hand* (1981)
16 Classification *A.D. Gordon* (1981)
17 Distribution-Free Statistical Methods, 2nd edition *J.S. Maritz* (1995)
18 Residuals and Influence in Regression *R.D. Cook and S. Weisberg* (1982)
19 Applications of Queueing Theory, 2nd edition *G.F. Newell* (1982)
20 Risk Theory, 3rd edition *R.E. Beard, T. Pentikäinen and E. Pesonen* (1984)
21 Analysis of Survival Data *D.R. Cox and D. Oakes* (1984)
22 An Introduction to Latent Variable Models *B.S. Everitt* (1984)
23 Bandit Problems *D.A. Berry and B. Fristedt* (1985)
24 Stochastic Modelling and Control *M.H.A. Davis and R. Vinter* (1985)
25 The Statistical Analysis of Composition Data *J. Aitchison* (1986)
26 Density Estimation for Statistics and Data Analysis *B.W. Silverman* (1986)
27 Regression Analysis with Applications *G.B. Wetherill* (1986)
28 Sequential Methods in Statistics, 3rd edition
*G.B. Wetherill and K.D. Glazebrook* (1986)
29 Tensor Methods in Statistics *P. McCullagh* (1987)
30 Transformation and Weighting in Regression
*R.J. Carroll and D. Ruppert* (1988)
31 Asymptotic Techniques for Use in Statistics
*O.E. Bandorff-Nielsen and D.R. Cox* (1989)
32 Analysis of Binary Data, 2nd edition *D.R. Cox and E.J. Snell* (1989)

33 Analysis of Infectious Disease Data *N.G. Becker* (1989)

34 Design and Analysis of Cross-Over Trials *B. Jones and M.G. Kenward* (1989)

35 Empirical Bayes Methods, 2nd edition *J.S. Maritz and T. Lwin* (1989)

36 Symmetric Multivariate and Related Distributions
*K.T. Fang, S. Kotz and K.W. Ng* (1990)

37 Generalized Linear Models, 2nd edition *P. McCullagh and J.A. Nelder* (1989)

38 Cyclic and Computer Generated Designs, 2nd edition
*J.A. John and E.R. Williams* (1995)

39 Analog Estimation Methods in Econometrics *C.F. Manski* (1988)

40 Subset Selection in Regression *A.J. Miller* (1990)

41 Analysis of Repeated Measures *M.J. Crowder and D.J. Hand* (1990)

42 Statistical Reasoning with Imprecise Probabilities *P. Walley* (1991)

43 Generalized Additive Models *T.J. Hastie and R.J. Tibshirani* (1990)

44 Inspection Errors for Attributes in Quality Control
*N.L. Johnson, S. Kotz and X. Wu* (1991)

45 The Analysis of Contingency Tables, 2nd edition *B.S. Everitt* (1992)

46 The Analysis of Quantal Response Data *B.J.T. Morgan* (1992)

47 Longitudinal Data with Serial Correlation—A state-space approach
*R.H. Jones* (1993)

48 Differential Geometry and Statistics *M.K. Murray and J.W. Rice* (1993)

49 Markov Models and Optimization *M.H.A. Davis* (1993)

50 Networks and Chaos—Statistical and probabilistic aspects
*O.E. Barndorff-Nielsen, J.L. Jensen and W.S. Kendall* (1993)

51 Number-Theoretic Methods in Statistics *K.-T. Fang and Y. Wang* (1994)

52 Inference and Asymptotics *O.E. Barndorff-Nielsen and D.R. Cox* (1994)

53 Practical Risk Theory for Actuaries
*C.D. Daykin, T. Pentikäinen and M. Pesonen* (1994)

54 Biplots *J.C. Gower and D.J. Hand* (1996)

55 Predictive Inference—An introduction *S. Geisser* (1993)

56 Model-Free Curve Estimation *M.E. Tarter and M.D. Lock* (1993)

57 An Introduction to the Bootstrap *B. Efron and R.J. Tibshirani* (1993)

58 Nonparametric Regression and Generalized Linear Models
*P.J. Green and B.W. Silverman* (1994)

59 Multidimensional Scaling *T.F. Cox and M.A.A. Cox* (1994)

60 Kernel Smoothing *M.P. Wand and M.C. Jones* (1995)

61 Statistics for Long Memory Processes *J. Beran* (1995)

62 Nonlinear Models for Repeated Measurement Data
*M. Davidian and D.M. Giltinan* (1995)

63 Measurement Error in Nonlinear Models
*R.J. Carroll, D. Rupert and L.A. Stefanski* (1995)

64 Analyzing and Modeling Rank Data *J.J. Marden* (1995)

65 Time Series Models—In econometrics, finance and other fields
*D.R. Cox, D.V. Hinkley and O.E. Barndorff-Nielsen* (1996)

66 Local Polynomial Modeling and its Applications *J. Fan and I. Gijbels* (1996)
67 Multivariate Dependencies—Models, analysis and interpretation
*D.R. Cox and N. Wermuth* (1996)
68 Statistical Inference—Based on the likelihood *A. Azzalini* (1996)
69 Bayes and Empirical Bayes Methods for Data Analysis
*B.P. Carlin and T.A Louis* (1996)
70 Hidden Markov and Other Models for Discrete-Valued Time Series
*I.L. Macdonald and W. Zucchini* (1997)
71 Statistical Evidence—A likelihood paradigm *R. Royall* (1997)
72 Analysis of Incomplete Multivariate Data *J.L. Schafer* (1997)
73 Multivariate Models and Dependence Concepts *H. Joe* (1997)
74 Theory of Sample Surveys *M.E. Thompson* (1997)
75 Retrial Queues *G. Falin and J.G.C. Templeton* (1997)
76 Theory of Dispersion Models *B. Jørgensen* (1997)
77 Mixed Poisson Processes *J. Grandell* (1997)
78 Variance Components Estimation—Mixed models, methodologies and applications
*P.S.R.S. Rao* (1997)
79 Bayesian Methods for Finite Population Sampling
*G. Meeden and M. Ghosh* (1997)
80 Stochastic Geometry—Likelihood and computation
*O.E. Barndorff-Nielsen, W.S. Kendall and M.N.M. van Lieshout* (1998)
81 Computer-Assisted Analysis of Mixtures and Applications—
Meta-analysis, disease mapping and others *D. Böhning* (1999)
82 Classification, 2nd edition *A.D. Gordon* (1999)
83 Semimartingales and their Statistical Inference *B.L.S. Prakasa Rao* (1999)
84 Statistical Aspects of BSE and vCJD—Models for Epidemics
*C.A. Donnelly and N.M. Ferguson* (1999)
85 Set-Indexed Martingales *G. Ivanoff and E. Merzbach* (2000)
86 The Theory of the Design of Experiments *D.R. Cox and N. Reid* (2000)
87 Complex Stochastic Systems
*O.E. Barndorff-Nielsen, D.R. Cox and C. Klüppelberg* (2001)
88 Multidimensional Scaling, 2nd edition *T.F. Cox and M.A.A. Cox* (2001)
89 Algebraic Statistics—Computational Commutative Algebra in Statistics
*G. Pistone, E. Riccomagno and H.P. Wynn (2001)*
90 Analysis of Time Series Structure—SSA and Related Techniques
*N. Golyandina, V. Nekrutkin and A.A. Zhigljavsky (2001)*
91 Subjective Probability Models for Lifetimes
*Fabio Spizzichino (2001)*
92 Empirical Likelihood *Art B. Owen (2001)*
93 Statistics in the 21st Century *Adrian E. Raftery, Martin A. Tanner, and Martin T. Wells (2001)*
94 Accelerated Life Models: Modeling and Statistical Analysis
*Mikhail Nikulin and Vilijandas Bagdonavičius (2001)*
95 Subset Selection in Regression, Second Edition
*Alan Miller (2002)*

# Subset Selection in Regression

## Second Edition

Alan Miller

CHAPMAN & HALL/CRC

A CRC Press Company

Boca Raton London New York Washington, D.C.

**Library of Congress Cataloging-in-Publication Data**

Miller, Alan J.
Subset selection in regression / Alan Miller.-- 2nd ed.
p. cm. -- (Monographs on statistics and applied probability ; 95)
Includes bibliographical references and index.
ISBN 1-58488-171-2
1. Regression analysis. 2. Least squares. I. Title. II. Series.

QA278.2 .M56 2002
519.5'36--dc21 2002020214



Direct all inquiries to CRC Press LLC, 2000 N.W. Corporate Blvd., Boca Raton, Florida 33431.


International Standard Book Number 1-58488-171-2
Library of Congress Card Number 2002020214
Printed in the United States of America 1 2 3 4 5 6 7 8 9 0
Printed on acid-free paper

# Contents

# Preface to first edition

Nearly all statistical packages, and many scientific computing libraries, contain facilities for the empirical choice of a model, given a set of data and many variables or alternative models from which to select. There is an abundance of advice on how to perform the mechanics of choosing a model, much of which can only be described as folklore and some of which is quite contradictory. There is a dearth of respectable theory, or even of trustworthy advice, such as recommendations based upon adequate simulations. This monograph collects what is known about estimation and presents some new material. This relates almost entirely to multiple linear regression. The same problems apply to nonlinear regression, such as to the fitting of logistic regressions, to the fitting of autoregressive moving average models, or to any situation in which the same data are to be used both to choose a model and to fit it.

This monograph is not a cookbook of recommendations on how to carry out stepwise regression; anyone searching for such advice in its pages will be disappointed. I hope that it will disturb many readers and awaken them to the dangers of using automatic packages that pick a model and then use least squares to estimate regression coefficients using the same data. My own awareness of these problems was brought home to me dramatically when fitting models for the prediction of meteorological variables such as temperature or rainfall. Many years of daily data were available, so we had very large sample sizes. We had the luxury of being able to fit different models for different seasons and using different parts of the data, chosen at random, not systematically, for model selection, for estimation, and for testing the adequacy of the predictions. Selecting only those variables which were very highly 'significant', using '$F$-to-enter' values of 8.0 or greater, it was found that some variables with '$t$-values' as large as 6 or even greater had their regression coefficients reversed in sign from the data subset used for selection to that used for estimation. We were typically picking about 5 variables out of 150 available for selection.

Many statisticians and other scientists have long been aware that the so-called significance levels reported by subset selection packages are totally without foundation, but far fewer are aware of the substantial biases in the (least-squares or other) regression coefficients. This is one aspect of subset selection that is emphasized in this monograph.

The topic of subset selection in regression is one that is viewed by many statisticians as 'unclean' or 'distasteful'. Terms such as 'fishing expeditions', 'torturing the data until they confess', 'data mining', and others are used as descriptions of these practices. However, there are many situations in which it is difficult to recommend any alternative method and in which it is plainly not possible to collect further data to provide an independent estimate of regression coefficients, or to test the adequacy of fit of a prediction formula, yet there is very little theory to handle this very common problem. It is hoped that this monograph will provide the impetus for badly needed research in this area.

It is a regret of mine that I have had to use textbook examples rather than those from my own consulting work within CSIRO. My experience from many seminars at conferences in Australia, North America and the UK, has been that as soon as one attempts to use 'real' examples, the audience complains that they are not 'typical', and secondly, there are always practical problems that are specific to each particular data set and distract attention from the main topic. I am sure that this applies particularly to the textbook examples which I have used, and I am grateful that I do not know of these problems!

This is not in any sense a complete text on regression; there is no attempt to compete with the many hundreds of regression books. For instance, there is almost no mention of methods of examining residuals, of testing for outliers, or of the various diagnostic tests for independence, linearity, normality, etc. Very little is known of the properties of residuals and of other diagnostic statistics after model selection.

Many people must be thanked for their help in producing this monograph, which has taken more than a decade. The original impetus to develop computational algorithms came from John Maindonald and Bill Venables. John Best, John Connell (who provided a real problem with 757 variables and 42 cases), Doug Shaw and Shane Youll tried the software I developed and found the bugs for me. It soon became obvious that the problems of inference and estimation were far more important than the computational ones. Joe Gani, then Chief of CSIRO Division of Mathematics and Statistics, arranged for me to spend a 6-month sabbatical period at the University of Waterloo over the northern hemisphere winter of 1979/1980. I am grateful to Jerry Lawless and others at Waterloo for the help and encouragement they gave me. Hari Iyer is to be thanked for organizing a series of lectures I gave at Colorado State University in early 1984, just prior to reading a paper on this subject to the Royal Statistical Society of London.

The monograph was then almost completed at Griffith University (Brisbane, Queensland) during a further sabbatical spell, which Terry Speed generously allowed me from late 1985 to early 1987. The most important person to thank is Doug Ratcliff, who has been a constant source of encouragement, and has read all but the last version of the manuscript, and who still finds bugs in my software. I of course accept full responsibility for the errors remaining. I would also like to thank Sir David Cox for his support in bringing this monograph to publication.

Alan Miller
Melbourne

# Preface to second edition

What has happened in this field since the first edition was published in 1990?

The short answer is that there has been very little real progress. The increase in the speed of computers has been used to apply subset selection to an increasing range of models, linear, nonlinear, generalized linear models, to regression methods which are more robust against outliers than least squares, but we still know very little about the properties of the parameters of the best-fitting models chosen by these methods. From time-to-time simulation studies have been published, e.g. Adams (1990), Hurvich and Tsai (1990), and Roecker (1991), which have shown, for instance, that prediction errors using ordinary least squares are far too small, or that nominal 95% confidence regions only include the true parameter values in perhaps 50% of cases.

Perhaps the most active area of development during the 1990s has been into Bayesian methods of model selection. Bayesian methods require the assumption of prior probabilities for either individual models or variables, as well as prior distributions for the parameters, that is, the regression coefficients and the residual variance, in addition to the assumptions required by the frequentist approach; these may be that the model is linear with independent residuals sampled from a normal distribution with constant variance. In return, the Bayesian methods give posterior probabilities for the models and their parameters. Rather than make Chapter 5 (the previous Chapter 6) even larger, a new chapter has been added on Bayesian methods.

There is a major divergence between the Bayesian methods that have been developed and those described in the first edition. Most authors in this field have chosen not to select a single model. The underlying Bayesian structure is that of a mixture of models, perhaps many millions of them, each with an associated probability. The result, the posterior model, is not to take a gamble by selecting just one model involving a subset of variables that by chance has come out best using the data at hand and the procedures which we have chosen to use, but it is a weighted average of all of the models with moderately large weights (posterior probabilities) given to some models and very small weights given to most of them. Most authors have then taken the lesser gamble of using a Bayesian Model Average (BMA) of perhaps the top 5% or 10% of these models.

A disadvantage of using a BMA of, say the top 5% of models, is that if we started with 50 candidate predictor models, this BMA may still use 40 of them. An objective of subset selection in many applications is to substantially reduce the number of variables that need be measured in the future.

For those Bayesians who are brave enough to grab hold of the single model with the highest posterior, there is very little advice in the literature. The problems are basically the same as with the frequentist approach. Very little work has been done on the properties of the extreme best-fit model. Some new ideas are given in the last section of Chapter 7.

The bootstrap aggregation (bagging) method of Breiman (1996) bears a similarity to Bayesian model averaging. It uses bootstrapping of the original data set. Instead of picking just one model, the models are averaged. It does not use posterior probabilities to weight the different models found. However, if one model is found several times from different replicates, it is given weight according to the number of times that the same model is selected. This method came to the author's attention too late to be covered adequately in this monograph.

There has been little progress on algorithms to search for the globally optimum subsets of regressors. Problems in global optimization are very difficult. There has been substantial progress in recent years in finding global optima of continuous functions. There is an excellent web site maintained by Arnold Neumaier on this subject at the University of Vienna:

`http:\\www.mat.univie.ac.at/~neum/glopt.html`

However, the problem here is one of combinatorial optimization. In this field, the algorithms seem to be for specific problems such as the travelling salesman problem, or packing problems in one, two, or three dimensions, such as the knapsack problem.

In the statistical literature, the new methods have been Monte Carlo methods, particularly those that have been used by people developing software for Bayesian subset selection. So far, these methods have all been local optimization methods, but with a stochastic element, which gives a chance of breaking away and finding a better local optimum. So far, only an exhaustive search incorporating branch-and-bound (also known as leaps-and-bounds) is available, giving a guarantee that the global optimum has been found. This is only feasible for up to about 30 or so predictors. Despite claims to the contrary, the Monte Carlo methods give absolutely no guarantee that the optimum that has been found is any more than a local optimum. The two-at-a-time algorithms from a random start, which are described in Chapter 3, appear to be the best available at the moment.

The flood of theoretical papers on the asymptotic properties of stopping rules continues unabated. Scant recognition is given to this literature in this monograph, even though an issue of *Statistica Sinica* was almost entirely devoted to this subject. See Shao (1997) for a survey of this vast literature.

The problems in this field are essentially small sample problems. In this regard, there have been a number of advances. Several modifications to Mallows'

$C_p$ have been proposed. These are described in Chapter 5. Also a modification to Akaike's AIC by Hurvich and Tsai (1989) seems to give good results. Perhaps the most important development in this area is the Risk Inflation Criterion (RIC) of Foster and George (1994). In those applications in which there are typically large numbers of available predictor variables, such as meteorology and near-infrared spectroscopy, users of subset selection procedures have typically used $F$-to-enter values of the order of 10 rather than the default value of 4 used in the stepwise regression algorithm of Efroymson or the value of approximately 2 which is implicit in the AIC or Mallows' $C_p$. The RIC appeals to extreme-value theory to argue that when the number of available predictors, $k$, increases then the largest value of the $F$-to-enter statistic from variables that are not in the true model will increase as well. In a way, this is looking at the asymptotics in $k$ rather than those in $n$.

As far as estimation of regression coefficients is concerned, there has been essentially no progress. Chapter 6 outlines some progress that has been made with a similar problem, that of estimating the mean of the population that yielded the largest sample mean. For instance, suppose we plant 40 different varieties of potatoes, and choose the one which produced the largest yield per hectare in our experiment. It is probably fortuitous that this variety came out best, and it is likely that its true yield is lower than in our experiment, but how much lower? (Note: There is a huge literature in this field on the design of such experiments, and on maximizing the PCS, the probability of correct selection, but very little literature on the estimation problem.) The most important contributions in this field known to the author are those of Venter and Steel (1991), and Cohen and Sackrowitz (1989).

For the moment, the best method of estimation that can be recommended for estimating regression coefficients when the same data are used for both model selection and estimation, is to bootstrap the standardized residuals. The model selection procedure is then applied to the bootstrap replicates. The regression coefficients are estimated for the model that has been chosen with the original data, but the difference is noted between the bootstrap replicates in which the same model is selected, and for all cases. The regression coefficients from the original data set are then adjusted by this difference. This method is described in the bootstrap section in Chapter 5. There is no current theory to back up this method. This is not a satisfactory situation, but nobody seems to have found anything better.

In the mathematical derivations in this edition, there is even more emphasis upon $QR$-orthogonalization than in the first edition. Most statisticians learn about principal components (PC), which are closely related to the singular value decomposition (SVD) that is much better known to numerical analysts. These methods (PC and SVD) yield orthogonal directions that usually involve all the predictor variables. The orthogonal directions of the $QR$-orthogonalization involve an increasing number of the predictors. The first column of the $\boldsymbol{Q}$-matrix is in the direction of the first column of the $\boldsymbol{X}$-matrix. The second column is that part of the second column of $\boldsymbol{X}$ that is orthogonal to the first column, after scaling. The third column is that part of the third column of $\boldsymbol{X}$ that is orthogonal to the first two columns, again after scaling. This orthogonalization makes far more sense in the context of subset selection than the PC/SVD decompositions.

$QR$-factorizations are not only accurate methods for least-squares calculations, but they also lead to simple statistical interpretations.

An important property of the $QR$-orthogonalization is that the projections of the dependent variable on these directions all have the same standard deviation, equal to the residual standard deviation for the full model. This means that the projections, which have the same units as the dependent variable whether that be pounds or kilometres per hour or whatever units have a simple interpretation to the layman. In contrast, the analysis of variance, which is often applied to examine the fit of nested models, has units of, say (pounds)$^2$ or (kph)$^2$. How often are the units shown in an analysis of variance?

A little time spent understanding these least-squares projections could be very rewarding. Summing squares of projections leads to much simpler mathematical derivations than the grotesque expressions involving the inverses of parts of $\boldsymbol{X}'\boldsymbol{X}$-matrices multiplied by parts of $\boldsymbol{X}'\boldsymbol{y}$ products. On the negative side, the order of the predictor variables is of vital importance, when the predictors are not orthogonal.

The basic planar rotation algorithm is important as the fast way to change from one subset model to another, simply by changing the order of variables and then omitting the later variables. A deficiency of most of the attempts to use Bayesian methods has been that most of them have used very slow methods to change from one model to another. A notable exception in this regard is the paper by Smith and Kohn (1996).

Much of the fear of ill-conditioning or collinearity seems to stem from a lack of knowledge of the methods of least-squares calculation, such as the Householder reduction, or of Gram-Schmidt-Laplace orthogonalization, or of Jacobi-Givens planar rotation algorithms. Those who are tempted to leave out predictors with correlations of, say 0.8 or 0.9, with other predictors, or to average them or use principal components of the moderately highly correlated variables, may want to consider Bayesian Model Averaging as another alternative. Those who are just nervous because of accuracy problems are probably still using methods based upon the normal equations, and should switch to better computational methods. If high correlations worry you, try looking at near-infrared spectroscopy. I had one set of data for which the majority of correlations between predictors exceeded 0.999; many exceeded 0.9999. Correlations less than 0.999 were sometimes a warning of an error in the data.

I have retained my speed timings performed in the 1980s using an old Cromemco computer. Though ordinary personal computers now run these tests perhaps a thousand times faster, it is the relative times that are of importance.

Software in Fortran for some of the selection procedures in this book can be downloaded from my web site at:

`http://www.ozemail.com.au/~milleraj/`

Finally, I would like to acknowledge my gratitude to CSIRO, which has continued to allow me to use its facilities as an Honorary Research Fellow after my retirement in 1994. I would like to thank Ed George for reading the first draft of Chapter 7, even though he was in the process of getting married and moving home at the time. I would also like to thank Mark Steel for his comments on the same chapter; I have used some of them and disagree with others. Rob Tibshirani persuaded me to pay more attention to bootstrapping and the lasso, and I am grateful to him.

Alan Miller
Melbourne

CHAPTER 1

# Objectives

## 1.1 Prediction, explanation, elimination or what?

There are several fundamentally different situations in which it may be desired to select a subset from a larger number of variables. The situation with which this monograph is concerned is that of predicting the value of one variable, which will be denoted by $Y$, from a number of other variables, which will usually be denoted by $X$'s. It may be necessaryto do this because it is expensive to measure the variable $Y$ and it is hoped to be able to predict it with sufficient accuracy from other variables which can be measured cheaply. A more common situation is that in which the $X$-variables measured at one time can be used to predict $Y$ at some future time. In either case, unless the true form of the relationship between the $X$- and $Y$-variables is known, it will be necessary for the data used to select the variables and to calibrate the relationship to be representative of the conditions in which the relationship will be used for prediction. This last remark particularly applies when the prediction requires extrapolation, e.g. in time, beyond the range over which a relationship between the variables is believed to be an adequate approximation.

Some examples of applications are:

1. The estimation of wool quality, which can be measured accurately using chemical techniques requiring considerable time and expense, from reflectances in the near-infrared region, which can be obtained quickly and relatively inexpensively.
2. The prediction of meteorological variables, e.g. rainfall or temperature, say 24 hours in advance, from current meteorological variables and variables predicted from mathematical models.
3. The prediction of tree heights at some future time from variables such as soil type, topography, tree spacing, rainfall, etc.
4. The fitting of splines or polynomials, often in two or more dimensions, to functions or surfaces.

The emphasis here is upon the task of prediction, not upon the explanation of the effects of the $X$-variables on the $Y$-variable, though the second problem will not be entirely ignored. The distinction between these two tasks is well spelt out by Cox and Snell (1974). However, for those whose objective is not prediction, Chapter 4 is devoted to testing inferences with respect to subsets of regression variables in the situation in which the alternative hypotheses to be tested have not been chosen *a priori*.

Also, we will not be considering what is sometimes called the 'screening' problem; that is the problem of eliminating some variables (e.g. treatments or doses of drugs) so that effort can be concentrated upon the comparison of the effects of a smaller number of variables in future experimentation. The term 'screening' has been used for a variety of different meanings and, to avoid confusion, will not be used again in this monograph.

In prediction, we are usually looking for a small subset of variables which gives adequate prediction accuracy for a reasonable cost of measurement. On the other hand, in trying to understand the effect of one variable on another, particularly when the only data available are observational or survey data rather than experimental data, it may be desirable to include many variables which are either known or believed to have an effect.

Sometimes the data for the predictor variables will be collected for other purposes and there will be no extra cost to include more predictors in the model. This is often the case with meteorological data, or with government-collected statistics in economic predictions. In other situations, there may be substantial extra cost so that the cost of data collection will need to be traded off against improved accuracy of prediction.

In general, we will assume that all predictors are available for inclusion or exclusion from the model, though this is not always the case in practice. In many cases, the original set of measured variables will be augmented with other variables constructed from them. Such variables could include the squares of variables, to allow for curvature in the relationship, or simple products of variables, to allow the gradient of $Y$ on one regressor, say $X_1$, to vary linearly with the value of another variable, say $X_2$. Usually a quadratic term is only included in the model if the corresponding linear term is also included. Similarly, a product (interaction) of two variables is only included if at least one of the separate variables is included. The computational methods which will be discussed in Chapter 3 for finding best-fitting subsets assume that there are no restrictions such as these for the inclusion or exclusion of variables.

In some practical situations we will want to obtain a "point estimate" of the $Y$-variable, that is, a single value for it, given the values of the predictor variables. In other situations we will want to predict a probability distribution for the response variable $Y$. For instance, rather than just predicting that tomorrow's rainfall will be 5 mm we may want to try to assign one probability that it will not rain at all and another probability that the rainfall will exceed say 20 mm. This kind of prediction requires a model for the distribution of the $Y$-variable about the regression line. In the case of rainfall, a log-normal or a gamma distribution is often assumed with parameters which are simple functions of the point estimate for the rainfall, though the distribution could be modelled in more detail. Attention in this monograph will be focussed mainly on the point estimate problem.

All of the models which will be considered in this monograph will be linear; that is they will be linear in the regression coefficients. Though most of the ideas and problems carry over to the fitting of nonlinear models and generalized linear models (particularly the fitting of logistic relationships), the complexity is greatly increased. Also, though there are many ways of fitting regression lines, least squares will be almost exclusively used. Other types of model have been considered by Linhart and Zucchini (1986), while Boyce, Farhi and Weischedel (1974) consider the use of subset selection methods in optimal network algorithms.

There has been some work done on multivariate subset selection. The reader is referred to Seber (1984) and Sparks (1985) for an introduction to this subject. For more recent references, see the paper by Brown et al. (2000).

An entirely different form of empirical modelling is that of classification and regression trees (CART). In this the data are split into two parts based upon the value of one variable, say $X_1$. This variable is chosen as that which minimizes the variation of the $Y$-variable within each part while maximizing the difference between the parts. Various measures of distance or similarity are used in different algorithms. After splitting on one variable, the separate parts of the data are then split again. Variable $X_2$ may be used to split one part, and perhaps $X_3$, or $X_2$ or even $X_1$ again, may be used to split the other part. Such methods are usually employed when the dependent variable is a categorical variable rather than a continuous one. This kind of modelling will not be considered here, but it suffers from the same problems of overfitting and biases in estimation as subset selection in multiple regression. For discussion of some of these clustering methods, see e.g. Everitt (1974), Hartigan (1975), or Breiman, Friedman, Olshen and Stone (1984).

When the noise in the data is sufficiently small, or the quantity of data is sufficiently large, that the detailed shape of the relationship between the dependent variable and the predictors can be explored, the techniques known as projection pursuit may be appropriate. See e.g. Huber (1985), Friedman (1987), Jones and Sibson (1987), or Hall (1989).

## 1.2 How many variables in the prediction formula?

It is tempting to include in a prediction formula all of those variables which are known to affect or are believed to affect the variable to be predicted. Let us look closer at this idea. Suppose that the predictor variable, $Y$, is linearly related to the $k$ predictor variables, $X_1$ , $X_2$ , ... , $X_k$; that is

$$Y = \beta_0 + \sum_{i=1}^{k} \beta_i X_i + \epsilon, \tag{1.1}$$

where the residuals, $\epsilon$, have zero mean and are independently sampled from the same distribution which has a finite variance $\sigma^2$. The coefficients $\beta_0$, $\beta_1$, ... , $\beta_k$ will usually be unknown, so let us estimate them using least squares. The least-squares estimates of the regression coefficients, to be denoted by $b$'s, are given in matrix notation by

$$\boldsymbol{b} = (\boldsymbol{X}'\boldsymbol{X})^{-1}\boldsymbol{X}'\boldsymbol{y},$$

where

$$\boldsymbol{b}' = (b_0, \ b_1, \ ..., \ b_k),$$

$\boldsymbol{X}$ is an $n \times (k+1)$ matrix in which row $i$ consists of a 1 followed by the values of variables $X_1$, $X_2$, ..., $X_k$ for the $i$-th observation, and $\boldsymbol{y}$ is a vector of length $n$ containing the observed values of the variable to be predicted.

Now let us predict $Y$ for a given vector $\boldsymbol{x'} = (1,\ x_1,\ ...,\ x_k)$ of the predictor variables, using

$$\begin{aligned} \hat{Y} &= \boldsymbol{x'b} \\ &= b_0 + b_1 x_1 + ... + b_k x_k . \end{aligned}$$

Then, from standard least-squares theory (see e.g. Seber (1977) page 364), we have that

$$\text{var}(\boldsymbol{x'b}) = \sigma^2 \boldsymbol{x'(X'X)^{-1}x} .$$

If we form the Cholesky factorization of $\boldsymbol{X'X}$, i.e. we find a $(k+1) \times (k+1)$ upper-triangular matrix $R$ such that

$$(\boldsymbol{X'X})^{-1} = \boldsymbol{R^{-1}\ R^{-T}},$$

where the superscript $^{-\boldsymbol{T}}$ denotes the inverse of the transpose, then it follows that

$$\text{var}(\boldsymbol{x'b}) = \sigma^2(\boldsymbol{x'R^{-1}})(\boldsymbol{x'R^{-1}})', \tag{1.2}$$

Now $\boldsymbol{x'R^{-1}}$ is a vector of length $(k+1)$ so that the variance of the predicted value of $Y$ is the sum of squares of its elements. This is a suitable way in which to compute the variance of $\hat{Y}$, though we will recommend later that the Cholesky factorization, or a similar triangular factorization, should be obtained directly from the $\boldsymbol{X}$-matrix without the intermediate step of forming the 'sum of squares and products' matrix $\boldsymbol{X'X}$.

Now let us consider predicting $Y$ using only the first $p$ of the $X$-variables where $p < k$. Write

$$\boldsymbol{X} = (\boldsymbol{X_A},\ \boldsymbol{X_B}),$$

where $\boldsymbol{X_A}$ consists of the first $(p+1)$ columns of $\boldsymbol{X}$, and $\boldsymbol{X_B}$ consists of the remaining $(k-p)$ columns. It is well known that if we form the Cholesky factorization

$$\boldsymbol{X'_A X_A} = \boldsymbol{R'_A R_A},$$

then $\boldsymbol{R_A}$ consists of the first $(p+1)$ rows and columns of $\boldsymbol{R}$, and also that the inverse $\boldsymbol{R_A^{-1}}$ is identical with the same rows and columns of $\boldsymbol{R^{-1}}$. The reader who is unfamiliar with these results can find them in such references as Rushton (1951) or Stewart (1973), though it is obvious to anyone who tries forming a Cholesky factorization and inverting it so that the factorization down to row $p$ and the inverse down to row $p$ are independent of the following rows. The Cholesky factorization of $\boldsymbol{X'X}$ can be shown to exist and to be unique except for signs provided that $\boldsymbol{X'X}$ is a positive-definite matrix.

Then if $\boldsymbol{x_A}$ consists of the first $(p+1)$ elements of $\boldsymbol{x}$ and $\boldsymbol{b_A}$ is the corresponding vector of least-squares regression coefficients for the model with only $p$ variables, we have similarly to (1.2) that

$$\text{var}(\boldsymbol{x'_A b_A}) = \sigma^2(\boldsymbol{x'_A R_A^{-1}})(\boldsymbol{x'_A R_A^{-1}})' ; \tag{1.3}$$

that is, the variance of the predicted values of $Y$ is the sum of squares of the first $(p+1)$ elements that were summed to obtain the variance of $\boldsymbol{x}'\boldsymbol{b}$, and hence

$$\text{var}(\boldsymbol{x}'\boldsymbol{b}) \geq \text{var}(\boldsymbol{x}'_A\boldsymbol{b}_A).$$

Thus the variance of the predicted values increases monotonically with the number of variables used in the prediction - or at least it does for linear models with the parameters fitted using least squares. This fairly well-known result is at first difficult to understand. Taken to its extremes, it could appear that we get the best predictions with no variables in the model. If we always predict $\boldsymbol{Y} = 7$ say, irrespective of the values of the $X$-variables, then our predictions have zero variance but probably have a very large *bias*.

If the true model is as given in (1.1), then

$$\boldsymbol{b}_A = (\boldsymbol{X}'_A\boldsymbol{X}_A)^{-1}\boldsymbol{X}'_A\boldsymbol{y}$$

and hence

$$\begin{aligned} E(\boldsymbol{b}_A) &= (\boldsymbol{X}'_A\boldsymbol{X}_A)^{-1}\boldsymbol{X}'_A\boldsymbol{X}\boldsymbol{\beta} \\ &= (\boldsymbol{X}'_A\boldsymbol{X}_A)^{-1}\boldsymbol{X}'_A(\boldsymbol{X}_A,\ \boldsymbol{X}_B)\boldsymbol{\beta} \\ &= (\boldsymbol{X}'_A\boldsymbol{X}_A)^{-1}(\boldsymbol{X}'_A\boldsymbol{X}_A,\ \boldsymbol{X}'_A\boldsymbol{X}_B)\boldsymbol{\beta} \\ &= \boldsymbol{\beta}_A + (\boldsymbol{X}'_A\boldsymbol{X}_A)^{-1}\boldsymbol{X}'_A\boldsymbol{X}_B\boldsymbol{\beta}_B, \end{aligned}$$

where $\boldsymbol{\beta}_A$, $\boldsymbol{\beta}_B$ consist of the first $(p+1)$ and last $(k-p)$ elements respectively of $\boldsymbol{\beta}$. The second term above is therefore the bias in the first $(p+1)$ regression coefficients arising from the omission of the last $(k-p)$ variables. The bias in estimating $Y$ for a given $\boldsymbol{x}$ is then

$$\begin{aligned} \boldsymbol{x}'\boldsymbol{\beta} - E(\boldsymbol{x}'_A\boldsymbol{b}_A) &= \boldsymbol{x}'_A\boldsymbol{\beta}_A + \boldsymbol{x}'_B\boldsymbol{\beta}_B \\ &\quad - \boldsymbol{x}'_A\boldsymbol{\beta}_A - \boldsymbol{x}'_A(\boldsymbol{X}'_A\boldsymbol{X}_A)^{-1}\boldsymbol{X}'_A\boldsymbol{X}_B\boldsymbol{\beta}_B \\ &= \{\boldsymbol{x}'_B - \boldsymbol{x}'_A(\boldsymbol{X}'_A\boldsymbol{X}_A)^{-1}\boldsymbol{X}'_A\boldsymbol{X}_B\}\boldsymbol{\beta}_B \qquad (1.4) \end{aligned}$$

As more variables are added to a model we are 'trading off' reduced bias against an increased variance. If a variable has no predictive value, then adding that variable merely increases the variance. If the addition of a variable makes little difference to the biases, then the increase in prediction variance may exceed the benefit from bias reduction. The question of how this trade-off should be handled is a central problem in this field, but its answer will not be attempted until Chapter 5 because of the very substantial problems of bias when the model has not been selected independently of the data. Note that the addition of extra variables does not generally reduce the bias for every vector $\boldsymbol{x}$. Also, the best subset for prediction is a function of the range of vectors $\boldsymbol{x}$ for which we want to make predictions.

If the number of observations in the calibrating sample can be increased, then the prediction variance given by (1.3) will usually be reduced. In most practical cases the prediction variance will be of the order $n^{-1}$ while the biases from omitting variables will be of order 1 (that is, independent of $n$). Hence, the number of variables in the best prediction subset will tend to increase with the size of the sample used to calibrate the model.

We note here that Thompson (1978) has discriminated between two prediction situations, one in which the $X$-variables are controllable, as for instance in an experimental situation, and the other in which the $X$-variables are random variables over which there is no control. In the latter case the biases caused by omitting variables can be considered as forming part of the residual variation and then the magnitude of the residual variance, $\sigma^2$, changes with the size of subset.

At this stage we should mention another kind of bias which is usually ignored. The mathematics given above is all for the case in which the subset of variables has been chosen independently of the data being used to estimate the regression coefficients. In practice the subset of variables is usually chosen from the same data as are used to estimate the regression coefficients. This introduces another kind of bias which we will call **selection bias**; the first kind of bias discussed above will be called **omission bias.** It is far more difficult to handle selection bias than omission bias, and for this reason, all of Chapter 6 is devoted to this subject. Apart from a few notable exceptions, e.g. Kennedy and Bancroft (1971), this topic has been almost entirely neglected in the literature.

The question of how many variables to include in the prediction equation, that is of deciding the "stopping rule" in selection, is one which has developed along different lines in the multiple-regression context and in the context of fitting time series, though it is the same problem. In neither case can an answer be given until selection bias is understood, except for the rare situation in which independent data sets are used for the selection of variables (or of the order of the model in fitting time series) and for the estimation of the regression coefficients. This will be attempted in Chapter 5.

## 1.3 Alternatives to using subsets

The main reasons for not using all of the available predictor variables are that, unless we have sufficiently large data sets, some of the regression coefficients will be poorly determined and the predictions may be poor as a consequence, or that we want to reduce the cost of measuring or acquiring the data on many variables in future. Three alternatives that use all of the variables are (i) using 'shrunken' estimators as in ridge regression, (ii) using orthogonal (or nonorthogonal) linear combinations of the predictor variables, and (iii) using Bayesian model averaging.

The usual form in which the expression for the ridge regression coefficients is written is

$$\boldsymbol{b}(\theta) = (\boldsymbol{X}'\boldsymbol{X} + \theta \boldsymbol{I})^{-1}\boldsymbol{X}'\boldsymbol{y}$$

where $\boldsymbol{I}$ is a $k \times k$ identity matrix, and $\theta$ is a scalar. In practice this is usually applied to predictor variables which have first been centered by subtracting the sample average and then scaled so that the diagonal elements of $\boldsymbol{X}'\boldsymbol{X}$ are all equal to one. In this form the $\boldsymbol{X}'\boldsymbol{X}$-matrix is the sample correlation matrix of the original predictor variables. There is a very large literature on ridge regression and on the choice of the value for the ridge parameter $\theta$, including several reviews; see e.g. Draper and van Nostrand (1979), Smith and Campbell (1980) and the discussion which follows this paper, and Draper and Smith (1981, pages 313-325).

The simplest shrunken estimator is that obtained by simply multiplying all of the least-squares regression coefficients by a constant between 0 and 1. Usually this shrinkage is not applied to the constant or intercept in the model (if there is one), and the fitted line is forced to pass through the centroid of the $X$- and $Y$-variables. The best-known of these shrunken estimators is the so-called James-Stein estimator (see Sclove (1968)). There have also been many other types of shrunken estimator proposed, see for instance the 57-varieties considered by Dempster et al. (1977).

Since the publication of the first edition of this book, two new methods have been proposed which essentially give shrunken estimates. These are the garrote and lasso which will be discussed in Chapter 3.

All of these shrunken estimators yield biased estimates of the regression coefficients and hence usually of Y but, with a suitable choice of the parameter(s) controlling the amount of shrinkage, can produce estimates with smaller mean square errors of prediction than the least-squares estimator using all the predictor variables, *over a range of the $X$-variables.*

Broadly speaking, the shrinkage methods err in the direction of caution by leaving in the model too many variables, to make sure that one which should be there is not dropped. At the same time, they give regression coefficients which have been shrunk. Bayesian model averaging (see Chapter 7) gives a similar outcome.

The use of orthogonal linear combinations of some or all of the predictor variables (called principal components in the statistical literature and empirical orthogonal functions in the geophysical sciences) is a way of reducing the number of variables. Usually only the combinations which correspond to the largest eigenvalues of $\boldsymbol{X'X}$ (or, equivalently the largest singular values of $\boldsymbol{X}$) are used. If there are 100 predictor variables there will be 100 eigenvalues, but perhaps only the linear combinations corresponding to the first 5 or 10 eigenvalues will be used. Often the selection of the subset of the new variables, i.e. the linear combinations, is done without reference to the values of the $Y$-variable, thus avoiding the problem of selection bias. It is a common practice to centre and scale the $X$-variables before calculating the eigenvalues or singular values, particularly if the predictor variables are of different orders of magnitude or have different dimensions. In some practical situations the first few eigenvectors may have some sensible interpretation, for instance, if the $X$-variables are the same variable (such as pressure or rainfall) measured at different locations, the first eigenvector may represent a weighted average, the second may be an east-west gradient, and the third a north-south gradient.

The derivation of principal components uses only the values of the predictor variables, not those of the variable to be predicted. In general there is no reason why the predictand should be highly correlated with the vectors corresponding to the largest eigenvalues, and it is quite common in practice to find that the vector which is the most highly correlated is one corresponding to one of the smaller eigenvalues. A valuable example has been given by Fearn (1983).

There is no need for the linear combinations to be orthogonal; any linear combinations could be used to reduce the dimensionality. The advantages are that the coefficients within the linear combinations are taken as known so that only a small number of parameters, that is the regression coefficients, have to be estimated from the data, and that if there is no selection from among the linear combinations based upon the $Y$-values then there is no selection bias. The principal disadvantages are that the linear combinations involve all the predictor variables so that they must still all be measured, and the $Y$-variable may not be well predicted by the chosen linear combinations, though there may be some other linear combination that has been rejected but yields good predictions.

## 1.4 'Black box' use of best-subsets techniques

The ready availability of computer software encourages the blind use of best-subsets methods. The high speed of computers coupled with the use of efficient algorithms means that it may be feasible to find, say the subset of 10 variables out of 150, which gives the closest fit in the least-squares sense to a set of observed $Y$-values. This does not necessarily mean that the subset thus chosen will provide good predictions. Throwing in a few more variables produced using a random number generator or from the pages of a telephone directory could have a very salutary effect!

One of the possible attractions of a subset regression package, or facility in a larger package, is that it relieves the user of the need to think about a sensible model; it comes with a built-in family of linear models; that is, there is an underlying assumption that the family of additive linear models contains one or more models which will fit the user's data adequately. Sometimes this is true. Quite often though there is scientific knowledge already available which will suggest a more sensible family of models to be searched. Quite often this will suggest nonlinear models. These models may have thresholds or asymptotes. There may also be reasons for believing that the rate threshold or slope with respect to one predictor, say $X_1$, will change with the level of another predictor, say $X_2$; that is, there is a need for some kind of interaction term or terms in both $X_1$ and $X_2$.

A number of derogatory phrases have been used in the past to describe the practices of subset selection, such as data grubbing, fishing expeditions, data mining (see Lovell (1983)), and torturing the data until it confesses. Given a sufficiently exhaustive search, some apparent pattern can always be found, even if all the predictors have come from a random number generator. To the author's knowledge, none of the readily available computer packages at the time of writing makes any allowance for the overfitting which undoubtedly occurs in these exercises.

Given a large number of variables and hence a very large number of possible subsets from which to choose, the best subset for prediction may not be the one which gives the best fit to the sample data. In general, a number of the better fitting subsets should be retained and examined in detail. If possible, an independent sample should be obtained to test the adequacy of the prediction equation. If this is not possible, then cross-validation provides a poor substitute which can be used instead. These techniques will be discussed in Chapter 5.

The 'traditional' approach to empirical model building has been a progressive one of plotting and/or correlating the $Y$-variable against the predictor variables one at a time, possibly taking logarithms or using other transformations to obtain approximate linearity and homogeneity of variance, then selecting one variable on some basis, fitting it and repeating the process using the residuals. This type of approach may be feasible when the data are from a well-designed experiment, but can be difficult to apply to observational data when the predictor variables are highly correlated. This type of approach is essentially a form of forward selection, one of the procedures to be discussed in Chapter 3. Forward selection procedures are relatively cheap to apply and easy to use. While forward selection and similar procedures often uncover subsets which fit well, they can fail very badly. The Detroit homicide data used in Chapter 3 provides such a case (see Table 3.15) in which the best-fitting subset of three variables gives a residual sum of squares that is less than a third of that for the subset of three variables found by forward selection. The author has one set of data, not presented here, for which the same ratio is about 90:1. In general, it is gross optimism to hope that an ad hoc procedure of adding one variable, and perhaps plotting residuals against everything which comes to mind, will find the best-fitting subsets.

A sensible compromise between forward selection and the costly extreme of an exhaustive search is often required. If it feasible to carry out an exhaustive search for the best-fitting subsets of, say five variables, then an examination of, say the best 10 subsets each of three, four and five variables may show two variables that appear in all of them. Those two variables can then be forced into all future subsets and an exhaustive search carried out for the best-fitting subsets of up to seven variables including these two.

It will often not be possible to specify in advance a complete set of variables to be searched for best-fitting subsets, though large numbers of polynomial and interaction terms may be included in the initial set. For instance, if possible trends in time have to be considered, then it may be sensible to include interactions between some of the variables and time as a way of allowing for regression coefficients to vary with time. Graphical examination may, for example, show that polynomial terms and interactions are being included in most of the better-fitting subsets because the response variable has an asymptotic level, or a threshold level. In such cases a simple nonlinear model may provide a much better predictor, and by explaining a considerable amount of the noise in the data may demonstrate the relationship with other variables more clearly.

When a small number of promising subsets of variables has been found, the fit of these subsets should be examined in detail. Such methods of examination are often graphical. This may show that there are outliers in the data, that some of the observations have far more influence than others, or that the residuals are highly autocorrelated in time. These methods are not described in detail in this monograph. Useful references on this subject are Barnett and Lewis (1978), Baskerville and Toogood (1982), Belsley, Kuh and Welsch (1980), Cook and Weisberg (1982), Gunst and Mason (1980), Hawkins (1980) and Weisberg (1980).

Because the techniques of empirical data analysis described in these references are for the case in which the model has been chosen independently of the data, the user should be cautious. Nothing appears to be known about the properties of least-squares residuals of best-fitting subsets.

CHAPTER 2

# Least-squares computations

## 2.1 Using sums of squares and products (SSP) matrices

Given a set of $n$ observations of $k$ variables, $x_{ij}$, $i$=1, 2, ..., $n$, $j$=1, 2, ..., $k$, the least-squares coefficients, $b_j$, $j$=1, 2, ..., $k$ and intercept $b_0$, in fitting a linear model can be found by solving the set of $(k+1)$ 'normal' equations:-

$$\begin{aligned}
\sum y_i &= nb_0 + b_1 \sum x_{i1} + \ldots + b_k \sum x_{ik} \\
\sum x_{i1} y_i &= b_0 \sum x_{i1} + b_1 \sum x_{i1}^2 + \ldots + b_k \sum x_{i1} x_{ik} \\
\sum x_{i2} y_i &= b_0 \sum x_{i2} + b_1 \sum x_{i2} x_{i1} + \ldots + b_k \sum x_{i2} x_{ik} \\
\ldots &= \ldots \\
\sum x_{ik} y_i &= b_0 \sum x_{ik} + b_1 \sum x_{ik} x_{i1} + \ldots + b_k \sum x_{ik}^2 \qquad (2.1)
\end{aligned}$$

where the summations are over $i$; that is over the observations.

The oldest method for solving the normal equations is that of Gaussian elimination. The method is well known and is described in many elementary texts on numerical methods or on linear regression. If the normal equations must be solved, it is an efficient method provided that the number of equations, $k+1$, is not large. For $k$ greater than about 15 to 20, an iterative procedure of the Gauss-Seidel or overrelaxation type will usually be faster. Most modern least-squares algorithms do not use the normal equations because of poor accuracy.

It is instructive to look at the first stage of Gaussian elimination. Suppose that we start by eliminating the constant term $b_0$ from all except the first equation. For the $(j+1)$-st equation, we do this by subtracting $(\sum x_{ij}/n)$ times the first equation from it. This leaves the equation as

$$\begin{aligned}
\sum x_{ij} y_i - \sum x_{ij} y_i / n &= b_1 \left( \sum x_{ij} x_{i1} - \sum x_{ij} \sum x_{i1}/n \right) + \ldots \\
&\quad + b_k \left( \sum x_{ij} x_{ik} - \sum x_{ij} \sum x_{ik}/n \right), \qquad (2.2)
\end{aligned}$$

Now it can be readily shown that

$$\sum x_{ij} x_{il} - \sum x_{ij} \sum x_{il}/n = \sum (x_{ij} - \bar{x}_j)(x_{il} - \bar{x}_l), \qquad (2.3)$$

where $\bar{x}_j$, $\bar{x}_l$ are the means of the variables $X_j$ and $X_l$. Hence (2.2) can be written as

$$\sum(x_{ij} - \bar{x}_j)(y_i - \bar{y}) = b_1 \sum(x_{ij} - \bar{x}_j)(x_{i1} - \bar{x}_1) + \ldots + b_k \sum(x_{ij} - \bar{x}_j)(x_{ik} - \bar{x}_k), \qquad (2.4)$$

The set of $k$ equations of the form (2.4) for $j = 1, 2, ..., k$ constitute what are known as the 'centered' or 'corrected' normal equations.

We note that regression programs have sometimes contained code to perform the centering operation on the left-hand side of (2.3) and then called a Gaussian elimination routine to solve the remaining $k$ centered equations when the routine operating on the original $(k+1)$ equations would have performed the same centering operations anyway!

The use of the right-hand side of (2.3) as the computational method for centering requires two passes through the data, the first to calculate the means and the second to calculate the cross-products. An alternative method uses progressively updated means as each new observation is added. If we let $S_r$ be a partial cross-product calculated using the first $r$ observations, then it can be shown that

$$S_{r+1} = \sum_{i-1}^{r+1}(x_i - \bar{x}_{r+1})(y_i - \bar{y}_{r+1}) = S_r + \delta_x \delta_y . r/(r+1), \qquad (2.5)$$

where

$$\delta_x = x_{r+1} - \bar{x}_r,$$
$$\delta_y = y_{r+1} - \bar{y}_r,$$

and $\bar{x}_r$, $\bar{y}_r$, $\bar{x}_{r+1}$, $\bar{y}_{r+1}$ are the means of $x$ and $y$ from the first $r$ or $r+1$ observations. The means are updated using

$$\bar{x}_{r+1} = \bar{x}_r + \delta_x/(r+1).$$

In the form (2.5), two multiplications are required for each observation for each pair of variables. This can be halved by using the deviation from the new mean for one of the two variables and the deviation from the old mean for the other. Thus if

$$\delta_x^* = x_{r+1} - \bar{x}_{r+1},$$

then

$$S_{r+1} = S_r + \delta_x^* \delta_y .$$

This 'trick' seems to have been first discovered by Jennrich (1977).

The use of the right-hand side of (2.3) or of progressive updating (2.5) for calculating the centered equations can give very substantial improvements in accuracy, but only when fitting regressions which include an intercept. A classic example is that given by Longley (1967). He examines the regression of employment against six other variables for the years 1947 to 1962. The table below gives the numbers of accurate digits in the coefficients of the centered

Table 2.1 *Accuracy of formulae for standard deviation calculations*

| Formula used | Range of accurate digits |
|---|---|
| Left-hand side of (2.3) | 3.1 to 7.2 |
| Right-hand side of (2.3) | 6.7 to 7.2 |
| Progressive updating (2.5) | 6.0 to 7.2 |

normal equations as calculated in single precision using a computer which allocates 32 binary bits to floating-point numbers, which are thus stored to an accuracy of about 7 decimal digits. The numbers of accurate digits were taken as $-\log_{10}|(e/x)|$, where $x$ is the true value and $e$ is the error. The number of accurate digits was taken as 7.2 when the error was zero or when the logarithm exceeded 7.2.

In this case, the use of either the right-hand side of (2.3) or of progressive updating of the means (2.4) has resulted in a centered SSP-matrix with very little loss of accuracy. Regression coefficients calculated using the left-hand side of (2.3) yielded no accurate digits in any of the seven coefficients using Gaussian elimination; in fact, five of the seven had the wrong signs. Using either of the other centering methods gave between 2.6 and 4.7 accurate digits for the different coefficients.

Writing the equations (2.1) in matrix form gives

$$\boldsymbol{X'y} = \boldsymbol{X'Xb}, \tag{2.6}$$

where $\boldsymbol{X}$ is an $n \times (k+1)$ matrix with a column of 1's as its first column and such that the element in row $i$ and column $(j+1)$ is the $i$-th value of the variable $X_j$; $\boldsymbol{y}$ is a vector of length $n$ containing the observed values of the variable $Y$, and $\boldsymbol{b}$ is the vector of $(k+1)$ regression coefficients with the intercept $b_0$ as its first element. The matrix $\boldsymbol{X'X}$ is then the SSP-matrix.

This formulation tempts the use of matrix inversion to obtain $b$ using

$$\boldsymbol{b} = (\boldsymbol{X'X})^{-1}\boldsymbol{X'y}. \tag{2.7}$$

If only the regression coefficients are required, this involves considerably more computation than using Gaussian elimination. However, the covariance matrix of $\boldsymbol{b}$ is given by $\sigma^2(\boldsymbol{X'X})^{-1}$ so that the inverse is usually needed.

The SSP-matrix is symmetric and positive definite, or at worst positive semidefinite, and there are extremely efficient ways to invert such matrices. Two of the popular methods are the so-called Gauss-Jordan method (though it was probably unknown to either Gauss or Jordan; it first appeared in a posthumous edition of a book by Jordan), and the Cholesky factorization method. The Cholesky factorization can also be used in a similar way to Gaussian elimination to obtain the regression coefficients by backsubstitution. Both of these methods can be executed using the space required for either the upper or lower triangle of the matrix, with the inverse overwriting the original matrix if desired. Code for the Gauss-Jordan method is given in Wilkinson and Reinsch (1971, pages 45-49), Garside (1971a), and Nash (1979, pages 82-85). Code for the Cholesky factorization is given in Wilkinson and Reinsch (1971, pages 17-21) and Healy (1968a,b), though it can easily be coded from Stewart (1973, algorithm 3.9, page 142). From error analyses, the Cholesky method should be slightly more accurate (see e.g. Wilkinson (1965), pages 244-245), but in an experiment to compare the Gauss-Jordan and Cholesky methods by Berk (1978a), the difference was barely detectable.

The Cholesky method requires the calculation of square roots. Square roots can be calculated very quickly in binary arithmetic using a method that is often taught in schools for calculating square roots in the scale of ten. If we are trying to find the square root of a number $y$ and have so far found a number $x$ containing the first few digits or binary bits, then we attempt to find a $\delta$ such that

$$(x + \delta)^2 \approx y;$$

hence

$$\delta \approx (y - x^2)/(2x + \delta).$$

That is, we divide the current remainder $(y - x^2)$ by twice $x$ plus the next binary bit $\delta$. The particular advantages in the scale of two are that the doubling operation simply requires a shift, and the next bit, $\delta$, can only be 0 or 1 so that the division operation is simply a test of whether $(2x + \delta)$, when shifted the correct number of places, is greater than or equal to the remainder. As the divisor, $2x + \delta$, starts with only one binary bit and averages only half the number in the mantissa, the method is about twice as fast as a floating-point division. Unfortunately most computers use a different method to take advantage of hardware for division. This method usually uses two Newton-Raphson iterations and often gives errors in the last binary bit. This could explain Berk's findings on the relative accuracy of the Gauss-Jordan and Cholesky methods.

The Cholesky method uses the factorization

$$\boldsymbol{X}'\boldsymbol{X} = \boldsymbol{L}\boldsymbol{L}',$$

where $\boldsymbol{L}$ is a lower-triangular matrix. An alternative factorization, which is sometimes credited to Banachiewicz (1938), which avoids the calculation of square roots, is

$$\boldsymbol{X}'\boldsymbol{X} = \boldsymbol{L}\boldsymbol{D}\boldsymbol{L}', \tag{2.8}$$

where $\boldsymbol{L}$ is lower-triangular with 1's on its diagonal, and $\boldsymbol{D}$ is a diagonal matrix. In computations, the elements on the diagonal of $\boldsymbol{D}$ can be stored overwriting the diagonal elements of $\boldsymbol{L}$. This can be expected to be slightly more accurate than the Gauss-Jordan method and as efficient in terms of both speed and storage requirements. Code for forming the factorization (2.8) is given in Wilkinson and Reinsch (1971, pages 21-24).

An advantage of the Cholesky or Banachiewicz methods over the Gauss-Jordan method in some situations is that the triangular factorizations can easily be updated when more data become available. A method for doing this will be presented in section 2.2. If the Gauss-Jordan method has been used, the inverse matrix $(\boldsymbol{X}'\boldsymbol{X})^{-1}$ can be updated using

$$\boldsymbol{A}^* = \boldsymbol{A} - (\boldsymbol{A}\boldsymbol{x}\boldsymbol{x}'\boldsymbol{A})/(1 + \boldsymbol{x}'\boldsymbol{A}\boldsymbol{x}), \tag{2.9}$$

where $\boldsymbol{A}$ is the old inverse and $\boldsymbol{A}^*$ is the updated inverse after an extra observation, $x$, has been added. Unless the inverse matrix is required after every new observation is added, this method is slow and can yield inaccurate results. If only the regression coefficients are required after each observation is added, then the regression coefficients can be obtained quickly from a triangular factorization using back-substitution without the need for matrix inversion. The update formula (2.9) is usually credited to Plackett (1950) or Bartlett (1951), though it has been suggested by Kailath (1974) that the method was known to Gauss.

So far we have looked at some aspects of least-squares computations when all of the variables are to be included in the model. In selecting a subset of variables, we will want to perform calculations for a number of different subsets, sometimes a very large number of them, to find those which give good fits to the data. Several ways of doing this are described in Chapter 3. In choosing the computational procedure, we will need both speed and accuracy. Accuracy is particularly important for two reasons: (i) subset selection procedures are often used when there is a choice among several fairly highly correlated variables that are attempting to measure the same attribute, and in such cases the normal equations can be badly ill-conditioned, and (ii) some of the procedures for searching for best-fitting subsets require a very large number of arithmetic operations, and we need to be sure that rounding errors accumulate as slowly as possible. The emphasis will be upon calculating the residual sum of squares for each subset investigated, not upon regression coefficients which can be found later for the small number of subsets singled out for closer scrutiny. We will want to obtain the residual sum of squares for a subset with the smallest amount of additional computation from calculations already carried out for previous subsets. We shall see later that methods operating on the SSP-matrix (or equivalently on the correlation matrix) and its inverse are sometimes slightly faster than methods based around triangular factorizations, which will be described in section 2.2.

However, these latter methods have a very definite advantage in accuracy, particularly if the starting triangularization is computed accurately using either orthogonal reduction or extra precision. On many computers, two levels of precision are available, a single precision which often represents floating-point numbers to about 7 significant decimal digits, and a double precision which represents them to about 16 decimal digits. The accuracy advantage is such that it is often feasible to perform the search for best-fitting subsets in single precision using the methods based around triangular factorizations when double precision is necessary using SSP-matrices and Gauss-Jordan methods. When a floating-point processor is not available, double-precision calculations are usually several times slower.

## 2.2 Orthogonal reduction methods

The basic idea here is to find an orthogonal basis on which to express both the $X$- and $Y$-variables, to perform regression calculations on this basis, and then to transform back to obtain regression coefficients in the dimensions of the real problem. In most of the orthogonal reduction methods, the matrix $\boldsymbol{X}$ of $n$ observations of each of $k$ variables, $X_1$, $X_2$, ..., $X_k$ (where $X_1$ will be identically equal to one in most practical cases), is factored as

$$\boldsymbol{X} = \boldsymbol{QR} \tag{2.10}$$

where either

- $\boldsymbol{Q}$ is an $n \times k$ matrix and $\boldsymbol{R}$ is a $k \times k$ upper-triangular matrix, or
- $\boldsymbol{Q}$ is an $n \times n$ matrix and $\boldsymbol{R}$ is an $n \times k$ matrix containing an upper-triangular matrix in its first $k$ rows and zeroes elsewhere.

In either case, the columns of $\boldsymbol{Q}$ are orthogonal and usually normalized so that

$$\boldsymbol{Q'Q} = \boldsymbol{I}$$

where $\boldsymbol{I}$ is either a $k \times k$ or an $n \times n$ identity matrix. The principal methods which use the factorization (2.10) are the modified Gram-Schmidt, Householder reduction, and various methods using planar rotations (also known as Jacobi or Givens rotations). General introductions to these alternative methods are contained in Seber (1977, Chapter 11) and Späth (1992). For a more detailed analysis of least-squares methods, Bjorck's (1996) book has become the standard reference.

Code for the modified Gram-Schmidt method has been given by Farebrother (1974), Wampler (1979a,b), and Longley (1981), and for the Householder method by Lawson and Hanson (1974, algorithm HFTI, pages 290-291), Lawson et al. (1979) and in the LINPACK package (Dongarra et al., 1979).

An alternative type of orthogonal reduction method uses the singular-value decomposition (SVD) in which $\boldsymbol{X}$ is factored as

$$\boldsymbol{X} = \boldsymbol{U\Lambda V'} \tag{2.11}$$

where $\boldsymbol{U}$ is $n \times k$, $\boldsymbol{\Lambda}$ is a $k \times k$ diagonal matrix with the singular values along its diagonal, $\boldsymbol{V}$ is $k \times k$, and $\boldsymbol{U'U} = \boldsymbol{V'V} = \boldsymbol{VV'} = \boldsymbol{I}$. Then

$$\boldsymbol{X'X} = \boldsymbol{V\Lambda^2V'},$$

which provides a quick and accurate way of calculating principal components without first forming the SSP-matrix. In most statistical work the matrix $\boldsymbol{U}$ is not required, and in such cases $\boldsymbol{\Lambda}$ and $\boldsymbol{V}$ can conveniently be obtained starting from the factorization (2.10).

Principal components are usually formed after subtracting the means from each variable. If the orthogonal reduction (2.10) has been formed from a matrix $\boldsymbol{X}$, which had a column of 1's as its first column, then the means are conveniently removed simply by omitting the top row of $\boldsymbol{R}$. Similarly, the

$\boldsymbol{X}$-variables can be scaled to have unit sum of squares about the mean by scaling each column of $\boldsymbol{R}$, after removing the top row, so that the sum of squares of elements in each column is one. The modified version of $R$ is then used in place of $\boldsymbol{X}$ in (2.11) to obtain the same principal components which are usually obtained (with poor accuracy) from the correlation matrix. As the matrix $\boldsymbol{R}$ has only $(k-1)$ rows, which is usually far fewer than the $n$ rows of $\boldsymbol{X}$, the SVD calculation is fairly fast; the bulk of the computational effort goes into calculating the orthogonal reduction (2.10).

The SVD can be used for multiple regression work but is usually somewhat slower than the other orthogonal reductions. It is often used for determining the rank of $\boldsymbol{X}$, though the matrix $\boldsymbol{X}$ must be scaled first. Code for the SVD is given in Lawson and Hanson (1974, pages 295-297), Nash (1979, pages 30-31), Wilkinson and Reinsch (1971, pages 134-151), Späth (1992), and the popular Numerical Recipes books (Press et al., 1986).

If we write $\boldsymbol{R} = \{r_{ij}\}$, then from (2.10) we have

$$\begin{aligned} \boldsymbol{X_1} &= r_{11}\boldsymbol{Q_1} \\ \boldsymbol{X_2} &= r_{12}\boldsymbol{Q_1} + r_{22}\boldsymbol{Q_2} \\ \boldsymbol{X_3} &= r_{13}\boldsymbol{Q_1} + r_{23}\boldsymbol{Q_2} + r_{33}\boldsymbol{Q_3}, \text{ etc.} \end{aligned}$$

Thus $\boldsymbol{Q_1}$ spans the space of $\boldsymbol{X_1}$; $\boldsymbol{Q_1}$ and $\boldsymbol{Q_2}$ span the space of $\boldsymbol{X_2}$, where $r_{22}\boldsymbol{Q_2}$ is that component of $\boldsymbol{X_2}$ which is orthogonal to $\boldsymbol{X_1}$, etc. We notice also that

$$\boldsymbol{X}'\boldsymbol{X} = \boldsymbol{R}'\boldsymbol{R},$$

that is, that $\boldsymbol{R}$ is the upper triangle of the Cholesky factorization. SSP-matrices can thus be constructed from the $R$-matrix if needed. If we omit the first row and column of $\boldsymbol{R}$, the remaining coefficients give the components of $\boldsymbol{X_2}$, $\boldsymbol{X_3}$, etc., which are orthogonal to $\boldsymbol{X_1}$, in terms of the direction vectors $\boldsymbol{Q_2}$, $\boldsymbol{Q_3}$, etc. If $\boldsymbol{X_1}$ is a column of 1's, then $\boldsymbol{R}'_{-1}\boldsymbol{R}_{-1}$ gives the centered SSP-matrix, where $\boldsymbol{R}_{-1}$ is $\boldsymbol{R}$ after removal of the first row and column. Correlations between variables can then be calculated from the SSP-matrix. Similarly, by removing, say the first three rows and columns of $\boldsymbol{R}$, the matrix of partial correlations among $X_4$, $X_5$, ..., $X_k$ can be obtained after regressing out $X_1$, $X_2$, and $X_3$.

The orthogonal reduction (2.10) can be achieved by multiplying $X$ on the left by a series of orthonormal matrices, each one chosen to reduce one or more elements of $X$ to zero. One such type of matrix is the planar rotation matrix

$$\begin{pmatrix} c & s \\ -s & c \end{pmatrix},$$

where $c$ and $s$ can be thought of as cosine and sine, and for the matrix to be orthonormal we require that $c^2 + s^2 = 1$. For instance, if we want to reduce the element $y$ to zero by operating on the two rows of the matrix below, we choose $c = w/(w^2 + y^2)^{\frac{1}{2}}$ and $s = y/(w^2 + y^2)^{\frac{1}{2}}$, then

$$\begin{pmatrix} c & s \\ -s & c \end{pmatrix} \begin{pmatrix} w & x & \cdots \\ y & z & \cdots \end{pmatrix} = \begin{pmatrix} (w^2 + y_2)^{\frac{1}{2}} & cx + sz & \cdots \\ 0 & -sx + cz & \cdots \end{pmatrix}.$$

The full planar rotation matrix looks like

$$\begin{pmatrix} 1 & & & & & \\ & 1 & & & & \\ & & c & & s & \\ & & & 1 & & \\ & & -s & & c & \\ & & & & & 1 \end{pmatrix},$$

where the blanks denote zero values, for the rotation of rows 3 and 5 of a $6 \times 6$ matrix.

By applying planar rotations to one additional row of $\boldsymbol{X}$ at a time, the factorization (2.10) can be achieved requiring the storage of only one row of $\boldsymbol{X}$ and the $k$ rows of $\boldsymbol{R}$ at any time. It is possible to eliminate the calculation of square roots by producing the Banachiewicz factorization instead of the Cholesky factorization. Efficient algorithms for this have been given by Gentleman (1973, 1974), Hammarling (1974), Buckley (1981), Reichel and Gragg (1990), and Miller (1992). Using this type of method, $\boldsymbol{Q}'$ is the product of the planar rotations as the product is such that $\boldsymbol{Q}'\boldsymbol{X} = \boldsymbol{R}$. The matrix $\boldsymbol{Q}$ is not usually formed explicitly, though the $c$'s and $s$'s can be stored if needed.

The linear transformations applied to the $X$-variables are simultaneously applied to the $Y$-variable giving a vector $\boldsymbol{Q}'\boldsymbol{y}$. The vector of values of $Y$ can be added to $\boldsymbol{X}$ as an extra column if desired, though that method will not be used here. If the orthogonal reduction is of the kind for which $\boldsymbol{Q}$ is an $n \times k$ matrix, e.g. the modified Gram-Schmidt method, then $\boldsymbol{Q}'\boldsymbol{y}$ is of length $k$, and there is an associated residual vector which is orthogonal to the columns of $\boldsymbol{Q}$ and hence orthogonal to the $X$-space. If the reduction method is of the kind for which $\boldsymbol{Q}$ is an $n \times n$ matrix, e.g. Householder reduction or planar rotation methods, then $\boldsymbol{Q}'\boldsymbol{y}$ is of length $n$. In either case, the first $k$ elements of $Q'y$ are the projections of $\boldsymbol{y}$ in the directions $Q_1$, $Q_2$, ..., $Q_k$.

The residuals in the last $(n - k)$ elements of $\boldsymbol{Q}'\boldsymbol{y}$ from either the Householder reduction method or the planar rotation method can be shown to be uncorrelated and to be homogeneous, i.e. to have the same variance, if the true but unknown residuals from the model also have these properties. This is a property which least-squares residuals do not have, and it can be useful in model testing. The residuals from using Householder reduction are known as LUSH (linear unbiased with scalar covariance Householder) and are discussed by Grossman and Styan (1972), Ward (1973) and Savin and White (1978). The residuals from using planar rotations have been shown by Farebrother (1976) to be identical to the 'recursive' residuals of Brown, Durbin and Evans (1975) and are much more readily calculated using planar rotations than by using the elaborate method given by them.

If we let $r_{iy}$ denote the $i^{th}$ element of $\boldsymbol{Q'y}$, then

$$\boldsymbol{y} = r_{1y}\boldsymbol{Q_1} + r_{2y}\boldsymbol{Q_2} + \ldots + r_{ky}\boldsymbol{Q_k} + \boldsymbol{e}, \qquad (2.12)$$

where $\boldsymbol{e}$ is the vector of residuals orthogonal to the directions $\boldsymbol{Q_1}$, $\boldsymbol{Q_2}$, ..., $\boldsymbol{Q_k}$. The $r_{iy}$'s are therefore the least-squares regression coefficients of $\boldsymbol{y}$ upon the $\boldsymbol{Q}$'s. Using (2.10), we can substitute for the $\boldsymbol{Q}$'s in terms of the variables of interest, that is, the $X$'s. In matrix notation (2.12) is

$$\boldsymbol{Y} = \boldsymbol{Q}(\boldsymbol{Q'y}) + \boldsymbol{e}$$

when $\boldsymbol{Q}$ is an $n \times k$ matrix. Substituting for $\boldsymbol{Q}$ from (2.10) then gives

$$\begin{aligned} \boldsymbol{Y} &= \boldsymbol{XR^{-1}Q'y} + \boldsymbol{e} \\ &= \boldsymbol{Xb} + \boldsymbol{e}, \end{aligned}$$

where $\boldsymbol{b}$ is the vector of regression coefficients of $Y$ upon $X$. Hence $\boldsymbol{b}$ can be calculated from

$$\boldsymbol{b} = \boldsymbol{R^{-1}Q'y}$$

or more usually by back-substitution in

$$\boldsymbol{Rb} = \boldsymbol{Q'y}. \qquad (2.13)$$

The formula (2.13) can also be obtained by substitution in the 'usual' formula (2.7). By using only the first $p$ equations, we can quickly obtain the regression coefficients of $Y$ upon the subset $X_1$, $X_2$, ..., $X_p$.

The breakup of the sum of squares is readily obtained from (2.12). The total sum of squares of $Y$ is

$$\boldsymbol{y'y} = r_{1y}^2 + r_{2y}^2 + \ldots + r_{ky}^2 + \boldsymbol{e'e}.$$

If the first variable, $X_1$, is just a column of 1's then the total sum of squares of $Y$ about its mean is

$$r_{2y}^2 + r_{3y}^2 + \ldots + r_{ky}^2 + \boldsymbol{e'e}.$$

The residual sum of squares after regressing $Y$ against $X_1$, $X_2$, ..., $X_p$ is

$$r_{p+1,y}^2 + \ldots + r_{ky}^2 + \boldsymbol{e'e}.$$

Thus, from the factorization (2.10), and the associated vector $\boldsymbol{Q'y}$, the set of $k$ sequential regressions $Y$ upon $X_1$; $Y$ upon $X_1$ and $X_2$; ...; $Y$ upon $X_1$, $X_2$, ..., $X_k$ can be carried out very quickly. In the situation in which there is a hierarchical order for carrying out a set of regressions, such as when fitting polynomials, the various methods of orthogonal reduction provide a fast computational method.

Now suppose that we want to regress $Y$ against a subset of the predictor variables such as $X_1$, $X_3$ and $X_4$ but excluding $X_2$. We can rearrange the columns of $\boldsymbol{X}$ and $\boldsymbol{Q}$ in (2.10), and both the rows and columns of $\boldsymbol{R}$. Unfortunately the rearranged matrix $\boldsymbol{R}$, let us call it $\boldsymbol{R_T}$, is no longer triangular. We have then

$$\boldsymbol{X_1X_3X_4X_2} = \boldsymbol{Q_1Q_3Q_4Q_2R_T},$$

where $\boldsymbol{R_T}$ looks like

$$\begin{pmatrix} \times & \times & \times & \times \\ & \times & \times & \\ & & \times & \\ & \times & \times & \times \end{pmatrix}$$

where an '$\times$' denotes a nonzero element. A more serious problem is that the orthogonal directions $\boldsymbol{Q_1}$, $\boldsymbol{Q_3}$ and $\boldsymbol{Q_4}$ do not form a basis for $\boldsymbol{X_1}$, $\boldsymbol{X_3}$ and $\boldsymbol{X_4}$ as both $\boldsymbol{X_3}$ and $\boldsymbol{X_4}$ have components in the direction $\boldsymbol{Q_2}$, which is orthogonal to this basis. A solution to this problem is to use planar rotations to restore $\boldsymbol{R_T}$ to upper-triangular form. By operating upon rows 2 and 4, the element in position (4,2) can be reduced to zero. Other elements in these two rows will be changed, in particular a nonzero element will be introduced into position (2,4). Another planar rotation applied to rows 3 and 4 removes the element in position (4,3) and the matrix is in upper-triangular form. Let $\boldsymbol{P}$ be the product of these two planar rotations, so that $\boldsymbol{PR_T}$ is the new triangular matrix. Then

$$\boldsymbol{X_1X_3X_4X_2} \;=\; (\boldsymbol{Q_1Q_3Q_4Q_2P'})(\boldsymbol{PR_T}),$$

so that the first three columns of $(\boldsymbol{Q_1Q_3Q_4Q_2P'})$ form the new orthogonal basis. This method appears to have been used first by Elden (1972); it is described in more detail by Clarke (1980). The first publication of the method appears to be in Osborne (1976).

Software for changing the order of variables can be found in Osborne (1976) in the LINPACK package (Dongarra et al., 1979), in Reichel and Gragg (1990), and Miller (1992).

It is possible to generate subset regressions from the SVD as well as from triangular factorizations, but the process of adding or deleting a variable changes the entire factorization, not just one or two rows.

## 2.3 Gauss-Jordan v. orthogonal reduction methods

We have said that the orthogonal reduction methods, which lead to operations on triangular matrices, are much more accurate than methods which require the inversion of parts of an SSP-matrix. Why is this? Let $e_i$ be used to denote the $i^{th}$ least-squares residual, then

$$e_i \;=\; (y_i - \bar{y}) \;-\; \sum_{j=1}^{p} b_j\,(x_{ij} - \bar{x}_j), \tag{2.14}$$

where the $b_j$'s are the least-squares estimates of the regression coefficients and it is assumed that we are fitting models containing a constant. $\bar{x}_i$, $\bar{y}$ denote the sample means of the appropriate variables. Suppose that the $e_i$'s are calculated using (2.14) with each quantity correctly rounded to $t$ decimal digits. Now let us suppose that the $e_i$'s are of a smaller order of magnitude than the $(y_i - \bar{y})$'s, which will be the case if the model fits the data closely. The order of magnitude could be defined as the average absolute value of the quantities, or as their root-mean square, or as some such measure of their spread. Then if the $e_i$'s are of order $10^{-d}$ times the order of the $(y_i - \bar{y})$'s, the $e_i$'s will be accurate to about $(t - d)$ decimal digits, and the sum $\sum e_i^2$ will have a similar accuracy. This loss of $d$ decimal digits is because of the cancellation errors which occur in performing the subtraction in (2.14). Suppose, for instance, that we are working to 7 decimal digits, then the right-hand side of (2.14) for one observation might be

$$3333333. \quad - \quad 3311111.$$

giving a difference $e_i = 22222$. If the two quantities were correctly rounded, then the maximum error in each is 0.5. Hence the maximum error in $e_i$ is 1.0 or a possible error of one in the fifth decimal digit. This is the principal source of error in orthogonal reduction methods in which we work in the scale of the original variables not in the scale of their squares as with SSP-matrices.

In using SSP-matrices, residual sums of squares are usually calculated as

$$\sum_i^n e_i^2 = \sum_i^n (y_i - \bar{y})^2 - \sum_j^n b_j \sum_i (x_{ij} - \bar{x}_j)(y_i - \bar{y}). \tag{2.15}$$

Now if $\sum e_i^2$ is of the order of $10^{-2d}$ times the order of $\sum(y_i - \bar{y})^2$, and all of the terms on the right-hand side of (2.15) are correctly rounded to $t$ decimal digits, then $2d$ digits in each of the terms on the right-hand side must cancel giving an accuracy of $(t - 2d)$ digits for the residual sum of squares. Thus we can expect to lose about twice as many digits of accuracy using methods based upon SSP-matrices as we lose using orthogonal reduction methods.

The notation $(1 - r^2)$ is commonly used for the ratio of $\sum e_i^2$ to $\sum(y_i - \bar{y})^2$, that is, of the residual sum of squares to the total sum of squares about the mean. Then if $r^2 = 0.9$ we can expect to lose one decimal digit due to cancellation errors when using SSP-matrices. We can expect to lose two digits if $r^2 = 0.99$, three if $r^2 = 0.999$, etc. This assumes that $\sum(y_i - \bar{y})^2$ was calculated accurately in the first place and, as was shown in section 2.1, this can also be a source of serious error. These are lower limits to the loss in accuracy as they ignore errors in the $b_j$'s, which can be substantial if there are high correlations among the $X$-variables.

Now let us look at how this applies to regressions involving subsets of variables. Let us denote by $a_{ij}$ the element in row $i$ and column $j$ of the SSP-matrix. In adding variable number $r$ into the regression, we calculate new elements, $a_{ij}^*$, usually overwriting the old ones, using

$$\begin{aligned} a_{ij}^* &= a_{ij} - a_{ir}a_{rj}/a_{rr} && \text{for all } i, j \neq r, \\ a_{ir}^* &= -a_i r/a_{rr} && \text{for } i \neq r, \text{ and} \\ a_{rr}^* &= 1/a_{rr}. \end{aligned}$$

The elements, $a_{ii}^*$, along the diagonal are of particular interest. Initially $a_{ii}$ is the sum of squares for variable $X_i$, which may have been centered if a constant is being fitted. $(a_{ir}/a_{rr})$ is the regression coefficient for the regression of variable $X_i$ on variable $X_r$ which we will denote by $b_{ir}$. Then

$$a_{ii}^* = a_{ii} - b_{ir}a_{ir} \quad \text{for } i \neq r. \tag{2.16}$$

This has the same form as (2.15) as $b_{ir}$ and $a_{ir}$ have the same sign. Hence if the correlation coefficient between $X_i$ and $X_r$ is $\pm 0.95$ (i.e. $r^2 \approx 0.9$), one decimal digit will be lost in cancellation errors in calculating $a_{ii}^*$, two digits will be lost if the correlation is $\pm 0.995$, etc. This means that high correlations among the predictor variables can lead to losses of one or more digits in calculating the second term on the right-hand side of (2.15), so that the residual sum of squares may be of low accuracy even when the $Y$-variable is not well predicted by the $X$-variables. Again, these losses are roughly halved when working in the scale of the $X$-variables, e.g. with triangular factorizations, rather than with sums of squares and products.

Planar rotations are stable transformations which lead to little buildup in error when used repeatedly. Suppose, for instance, that we need to calculate

$$x = uc + vs,$$

where $c$, $s$ are such that $c^2 + s^2 = 1$, and that the values of $u$ and $v$ have been calculated with errors $\delta u$ and $\delta v$, respectively. Then, neglecting errors in $c$ and $s$, the error in the calculated $x$ is

$$\delta x = c.\delta u + s.\delta v,$$

plus any new rounding error in this calculation. The maximum value of $\delta x$ is then $\{(\delta u)^2 + (\delta v)^2\}^{\frac{1}{2}}$ and occurs when $c = \pm\delta u / \{(\delta u)^2 + (\delta v)^2\}^{\frac{1}{2}}$, though in most cases it will be much smaller and often the two components of $\delta x$ will have opposite signs and partially cancel each other. Thus the absolute size of the errors will remain small, though the relative errors may be large when the $x$ resulting from calculations such as this is very small. Thus we can anticipate only little buildup of error when planar rotations are used repeatedly to change the order of variables in triangular factorizations. There is no such result limiting the size of buildup of errors in the inversion and reinversion of parts of matrices in the Gauss-Jordan method.

The above is a heuristic discussion of the errors in least-squares calculations. Detailed error analyses have been given for one-off least-squares calculations by Golub (1969, particularly pages 382-385), Jennings and Osborne (1974), Gentleman (1975), Stewart (1977), and Bjorck (1996) among others. The basic building blocks for these analyses were given by Wilkinson (1965) on pages 209-217 for Gaussian elimination, and on pages 131-139 for planar rotations.

To investigate the buildup of errors in a subset selection procedure, the Garside (1971b,c) algorithm for generating all subsets of variables using a Gauss-Jordan method, and an equivalent routine using planar rotations were compared with respect to both speed and accuracy in single precision on machines which use 32 binary bits for floating-point numbers and hence store numbers with about seven significant decimal digits. In each case the appropriate starting matrix, the upper triangle of the SSP-matrix for the Garside algorithm, and the upper triangular matrix from an orthogonal reduction for the planar rotation algorithm, were calculated in double precision then rounded to single precision. Generating all possible subsets is usually not very sensible in practice; it was chosen here as an extreme example of a subset selection procedure requiring the maximum amount of computational effort and hence giving the slowest speeds and poorest accuracy.

The Garside algorithm was programmed in Fortran with the upper triangle of the SSP-matrix stored as a one-dimensioned array to minimize the effort required in evaluating storage addresses. The tests for singularities and the facility for the grouping of variables were omitted.

Many alternative ways of coding the planar rotation algorithm were considered. Two alternative orders for producing the subsets were examined, namely the Hamiltonian cycle, as used by Garside (1965) and Schatzoff et al. (1968), and the binary sequence as used by Garside (1971b,c). Unless otherwise stated, the 'Garside algorithm' will be used to refer to his 1971 algorithm.

The Hamiltonian cycle can be thought of as a path linking the corners of a hypercube so that each step is to a neighbouring corner. If the corners of a three-dimensional cube with side of unit length are at (0,0,0), (0,0,1), ..., (1,1,1), then there are many Hamiltonian cycles, of which one is

000 001 011 010 110 111 101 100.

In applying this to subset selection, a '1' or '0' indicates whether a variable is in or out of the subset. The attractive feature of the Hamiltonian cycle is that only one variable is operated upon at each step. In the binary sequence, the string of 0's and 1's is treated as a binary number to which one is added at each step. Thus a sequence for three variables is

000 001 010 011 100 101 110 111.

This sequence often requires several variables to be added or removed from the subset in one step. However, as the right-hand end digits change rapidly, these can be made to correspond to the bottom corner of the triangular matrix, in either the Garside or the planar rotation algorithm, where there are fewer elements to be operated on than in the upper rows. More details of these alternatives are contained in Appendix 2A, where it is shown that for a moderate number, $k$, of $X$-variables, the number of multiplications or divisions per subset is about $(10 + k)$ for the Hamiltonian cycle and 15.75 per subset for the binary sequence. In contrast, Garside's algorithm requires 14 operations (multiplications or divisions) per subset (not 8 as stated by Garside). Both the Hamiltonian cycle and the binary sequence have been used with the planar rotation algorithm in the comparisons which follow.

The other major alternative considered was in the treatment of the response variable $Y$. It could either be left as the last variable or it could be allowed to move in the triangular matrix being placed immediately after the last variable included in the current subset. It was found to be faster to leave its position as the last variable fixed using the Hamiltonian cycle, otherwise two rows of the

Table 2.2 *Summary of data sets used*

| Data set name | Source | k predictors | n cases |
|---|---|---|---|
| WAMPLER | Wampler (1970) using his Y3 | 5 | 21 |
| LONGLEY | Longley (1967), Y = total derived employment | 6 | 16 |
| STEAM | Draper and Smith (1981) page 616 | 9 | 25 |
| DETROIT | Gunst and Mason (1980) Set A3 on page 360 Origin, Fisher (1976) | 11 | 13 |
| POLLUTE | Gunst and Mason (1980) Set b1 on pages 370-371 Origin, McDonald and Schwing (1973) | 15 | 60 |

triangular matrix had to be swapped at each step instead of one. The column corresponding to the $Y$-variable, $\boldsymbol{Q'y}$ in our earlier notation, was stored as a separate vector as were the progressive residual sums of squares, that is, the quantities $\boldsymbol{e'e}$, $\boldsymbol{e'e} + r_{ky}^2$, $\boldsymbol{e'e} + r_{ky}^2 + r_{k-1,y}^2$, etc.

Another alternative considered was whether to use the Gentleman (1973) algorithm or the Hammarling (1974) algorithm for the planar rotations. The Gentleman algorithm is usually slower but can be applied so that only small subtriangles within the main triangular matrix need to be operated upon when a binary sequence is used to generate the subsets. The Hammarling algorithm has a disadvantage in that it requires occasional rescaling to prevent its row multipliers becoming too small. In the following comparisons, both the Hammarling and Gentleman algorithms have been used, the first with the Hamiltonian cycle and the second with the binary sequence.

For the comparisons of accuracy, five data sets were used. These are briefly summarized in Table 2.2. The WAMPLER data set is an artificial set that was deliberately constructed to be very ill-conditioned; the other data sets are all real and were chosen to give a range of numbers of variables and to give a range of ill-conditioning such as is often experienced in real problems.

Table 2.3 shows measures of the ill-conditioning of the SSP-matrices. If we denote the eigenvalues of the correlation matrix by $\lambda_i$'s, then the ratio of the largest to the smallest, $\lambda_{max}/\lambda_{min}$, is often used as a measure of the ill-conditioning of the SSP-matrix. Berk (1978) compared a number of measures of ill-conditioning and found that for matrix inversion the accuracy of the inverse matrix was most highly correlated with the trace of the inverse matrix, which is $\sum(1/\lambda_i)$. Both of these measures are shown in the table, first for the matrix of correlations among the $X$-variables only and then with the $Y$-variable added.

Table 2.4 shows the lowest and average numbers of accurate decimal digits in the residual sums of squares for all subsets of variables for the Gauss-Jordan

Table 2.3 *Measures of ill-conditioning of the test data sets*

| Data set | $\lambda_{max}/\lambda_{min}$ | | $\sum(1/\lambda_i)$ | |
|---|---|---|---|---|
| | X only | X and Y | X only | X and Y |
| WAMPLER | 600,000 | 700,000 | 130,000 | 130,000 |
| LONGLEY | 12,000 | 6,000 | 3,100 | 2,200 |
| STEAM | 800 | 1,100 | 290 | 350 |
| DETROIT | 7,000 | 4,200 | 1,500 | 1,500 |
| POLLUTE | 900 | 1,000 | 260 | 280 |

Table 2.4 *Lowest and average numbers of accurate decimal digits in the calculation of residual sums of squares for all subsets*

| Data | Gauss-Jordan | | Hamiltonian cycle | | Binary sequence | |
|---|---|---|---|---|---|---|
| set | Lowest | Average | Lowest | Average | Lowest | Average |
| WAMPLER | 1.6 | 3.5 | 4.9 | 6.6 | 5.0 | 6.4 |
| LONGLEY | 3.6 | 5.5 | 6.1 | 6.9 | 6.3 | 7.2 |
| STEAM | 4.4 | 5.5 | 5.4 | 6.6 | 5.6 | 6.6 |
| DETROIT | 2.0 | 5.0 | 4.9 | 6.1 | 4.5 | 6.2 |
| POLLUTE | 3.6 | 4.3 | 5.0 | 6.0 | 4.5 | 5.5 |

(GJ) and the two planar rotation algorithms. The calculations were performed using a Cromemco Z2-D microcomputer using Microsoft Fortran version 3.37 and the CDOS operating system. As mentioned earlier, the centered SSP-matrix and orthogonal reduction were calculated in double precision (equivalent to about 17 decimal digits) and then rounded to single precision. Some calculations were repeated on a PDP11-34 with identical results.

The performance of the planar rotation algorithms was very impressive in terms of accuracy. The LONGLEY data set is often used as a test of regression programs and we see that in the worst case only about one decimal digit of accuracy was lost; in fact the accuracy was calculated, using the logarithm to base 10 of the relative accuracy, to be at least 7.0 for 22 out of the 63 residual sums of squares using the Hamiltonian cycle and the Hammarling rotations, and even better for the binary sequence and Gentleman rotations. The POLLUTE data set required a large amount of computation, yet there was very little sign of buildup of errors using planar rotations; for the last 100 subsets generated (out of 32767), the lowest accuracy was 5.3 decimal digits and the average was 5.9 using the Hamiltonian cycle. In contrast, the Gauss-Jordan algorithm performed poorly on the very ill-conditioned WAMPLER data set, and on the moderately ill-conditioned DETROIT data set, and on the well-conditioned POLLUTE data set. On the POLLUTE data set a steady build-up of errors was apparent. The last 100 subsets generated contained none with an accuracy greater than 4.0 decimal digits.

How much accuracy is needed? A subset selection procedure is used to pick a

Table 2.5 *Lowest and average numbers of accurate decimal digits in the residual sums of squares using planar rotations and an initial orthogonal reduction calculated in single precision*

| Data set | Lowest accuracy | Average accuracy |
|---|---|---|
| WAMPLER | 4.6 | 5.4 |
| LONGLEY | 3.5 | 4.9 |
| STEAM | 5.2 | 6.0 |
| DETROIT | 4.3 | 6.0 |
| POLLUTE | 5.0 | 5.6 |

small number of subsets of variables that give small residual sums of squares. Regression coefficients and other quantities can be calculated later using a saved, accurate copy of the appropriate matrix. In most practical cases, an accuracy of two decimal digits, equivalent to errors of 1%, will be quite adequate as this is usually less than the difference needed between two or more subsets to be statistically significant. Hence, provided that the initial SSP-matrix is calculated accurately, the Gauss-Jordan algorithm is adequate for all of these subsets except the artificial WAMPLER set, using a machine, such as the Cromemco or PDP11-34, which allocates 32 binary bits to floating-point numbers and performs correctly rounded arithmetic. However, some computers perform truncated arithmetic. Limited experience with one such computer showed an alarming buildup of errors for the Gauss-Jordan algorithm but only slightly worse performance for the planar rotation algorithm.

The Gauss-Jordan method will usually yield barely acceptable accuracy in single precision using 32-bit arithmetic, provided that the initial SSP-matrix is calculated using greater precision, and using a computer which performs rounded arithmetic; either planar rotation method will almost always be adequate in such cases. How well do the planar rotation methods perform if the initial orthogonal reduction is also carried out in single precision? Table 2.5 shows the results obtained using a Hamiltonian cycle. Except for the LONGLEY data set, the accuracy is only about half a decimal digit worse than before. The poorer performance with the LONGLEY data is associated with one variable, the year. Subsets which included the year gave accuracies between 3.5 and 5.3 decimal digits, averaging 4.2 digits; the other subsets gave 5.1 to 6.7 accurate digits with an average of 5.6 digits. The value of this variable ranged from 1947 to 1962 so that the first two decimal digits were the same for each line of data, and there was not much variation in the third digit. This one variable causes a loss of about 2.5 decimal digits using orthogonal reduction and about 5 decimal digits using SSP-matrices with crude centering.

To compare the speeds of the planar rotation and Gauss-Jordan algorithms, artificial data were generated. Table 2.6 gives times in seconds taken to calculate the residual sums of squares for all subsets on a PDP11-34 in Fortran IV under a Unix operating system. No accuracy calculations were carried out

Table 2.6 *Times in seconds, and their ratios, for calculating residual sums of squares for all subsets using planar rotation and Gauss-Jordan algorithms*

| No. of vars. | Garside (GJ) | Hamiltonian + Hammarling | Binary + Gentleman | Ratios Ham/GJ | Bin/GJ |
|---|---|---|---|---|---|
| 10 | 3.8 | 5.5 | 4.7 | 1.42 | 1.23 |
| 11 | 7.7 | 12.1 | 9.8 | 1.56 | 1.27 |
| 12 | 15.4 | 25.3 | 19.1 | 1.64 | 1.24 |
| 13 | 31. | 54. | 38. | 1.75 | 1.23 |
| 14 | 63. | 113. | 80. | 1.80 | 1.27 |
| 15 | 129. | 235. | 152. | 1.82 | 1.18 |
| 16 | 241. | 494. | 300. | 2.05 | 1.25 |
| 17 | 484. | 1037. | 602. | 2.14 | 1.24 |

during these speed tests and no use was made of the residual sums of squares which were not even stored. Averages of three runs were recorded for $k$ up to 14; averages of two runs are recorded for larger $k$. We see that the Gauss-Jordan algorithm is faster for all values of $k$ in the table, though its speed advantage over planar rotations using the binary sequence is not large. It is possible to increase the speed of this planar rotation algorithm by not operating upon the first row when a variable is deleted. This makes the algorithm fairly complex but increases both speed and accuracy. This has not been done by the author.

## 2.4 Interpretation of projections

The projections, $r_{iy}$, of the $Y$-variable on each of the orthogonal directions $\boldsymbol{Q_1}$, $\boldsymbol{Q_2}$, ..., $\boldsymbol{Q_k}$ are simple linear combinations of the values of the $Y$-variable, and hence have the same dimensions (e.g. length, mass, time, temperature, etc.) as the $Y$-variable. Similarly, the elements $r_{ij}$ of matrix $\boldsymbol{R}$ in the orthogonal reduction (2.10) have the same units as variable $X_j$, i.e. the column variable. $r_{ij}$ is simply the projection of variable $X_j$ upon direction $\boldsymbol{Q_i}$, where $j \geq i$.

The size, $r_{iy}$, of a projection is dependent upon the ordering of the predictor variables, unless they are orthogonal. When the $X$-variables are correlated among themselves, changing their order often produces substantial changes in the projection of the $Y$-variable on the direction associated with a particular $X$-variable. Usually, the earlier that a variable occurs in the ordering, the larger will be the projection associated with it, but this is not a mathematical law and it is possible to construct examples for which the opposite applies. To illustrate the effect of ordering, consider the artificial data in Table 2.7.

If we construct two further variables, $X_2$ and $X_3$, equal to the square and cube of $X_1$, respectively, then the LS regression equation relating $Y$ to $X_1$, $X_2$ and $X_3$, including a constant, is

$$Y = 1 + X_1 + X_2 + X_3,$$

Table 2.7 *Artificial data for fitting a cubic polynomial*

| | | | | | | |
|---|---|---|---|---|---|---|
| $X_1$ | 40 | 41 | 42 | 43 | 44 | 45 |
| $Y$ | 65647 | 70638 | 75889 | 81399 | 87169 | 93202 |
| $X_1$ | 46 | 47 | 48 | 49 | 50 | |
| $Y$ | 99503 | 106079 | 112939 | 120094 | 127557 | |

Table 2.8 *Projections in natural and reverse order for the cubic example*

| Natural order | | Reverse order | |
|---|---|---|---|
| Constant | 313607 | Cubic | 320267 |
| Linear | 64856 | Quadratic | –477 |
| Quadratic | 3984 | Linear | 0.69 |
| Cubic | 78.6 | Constant | 0.00087 |

that is, all the regression coefficients should equal 1.0 exactly. This is a useful test of how well a regression package handles ill-conditioning. The projections of $Y$ in the 'natural' and reverse order are shown in Table 2.8.

The residual sum of squares is exactly 286. From the first ordering, we see that if the cubic is omitted, the residual sum of squares is

$$286 + (78.6)^2 = 6464,$$

whereas, from the second ordering, if both the constant and linear term are omitted, the residual sum of squares is only

$$286 + (0.00087)^2 + (0.69)^2 = 286.48.$$

A useful statistical property of the projections is that if the true relationship between $Y$ and the $X$-variables is linear with uncorrelated residuals which are normally distributed with homogeneous variance, $\sigma^2$, then the projections are also uncorrelated and normally distributed with variance $\sigma^2$. This can be demonstrated as follows. Let

$$Y = X\beta + \epsilon,$$

where the $\epsilon$ are independent and $N(0,\ \sigma^2)$, then the projections are given by

$$\begin{aligned} \boldsymbol{Q}'\boldsymbol{y} &= \boldsymbol{Q}'\boldsymbol{X}\boldsymbol{\beta} + \boldsymbol{Q}'\boldsymbol{\epsilon} \\ &= \begin{pmatrix} \boldsymbol{R} \\ 0 \end{pmatrix} \boldsymbol{\beta} + \boldsymbol{Q}'\boldsymbol{\epsilon}. \end{aligned}$$

The vector $\boldsymbol{R\beta}$ contains the expected values of the first $k$ projections, where $k$ is the number of columns in $\boldsymbol{X}$. The remaining $(n-k)$ projections have zero expected values. The stochastic part of the projections is $\boldsymbol{Q}'\boldsymbol{\epsilon}$. This vector has covariance matrix equal to

$$\begin{aligned} E(\boldsymbol{Q}'\boldsymbol{\epsilon}\boldsymbol{\epsilon}'\boldsymbol{Q}) &= \boldsymbol{Q}'.E(\boldsymbol{\epsilon}\boldsymbol{\epsilon}').\boldsymbol{Q} \\ &= \sigma^2\, \boldsymbol{Q}'\boldsymbol{Q} \end{aligned}$$

$$= \sigma^2 \boldsymbol{I}.$$

That is, the elements of $\boldsymbol{Q}'\boldsymbol{\epsilon}$ are uncorrelated and all have variance $\sigma^2$. This part of the result is distribution-free. If the $\epsilon$ are normally distributed, then the elements of $\boldsymbol{Q}'\boldsymbol{\epsilon}$, being linear combinations of normal variables, will also be normally distributed.

Notice that the size of the elements of $\boldsymbol{Q}'\boldsymbol{\epsilon}$ is independent of both the sample size, $n$, and the number of predictors, $k$. As the sample size increases, if the $X$-predictors and $Y$ continue to span roughly the same ranges of values, the elements of $\boldsymbol{R}$ will increase roughly in proportion to $\sqrt{n}$ so that the stochastic element in the projections decreases relative to $\boldsymbol{R\beta}$.

To summarize, it can be seen that the $QR$-factorization has a number of properties that are very useful in subset selection.

1. The matrix $\boldsymbol{R}$ is the Cholesky factor of $\boldsymbol{X}'\boldsymbol{X}$, that is, $\boldsymbol{X}'\boldsymbol{X} = \boldsymbol{R}'\boldsymbol{R}$.
2. The first $p$ rows and columns of $\boldsymbol{R}$ are the Cholesky factor of the first $p$ rows and columns of $\boldsymbol{X}'\boldsymbol{X}$. This means that for the regression of $Y$ against the first $p$ columns of $\boldsymbol{X}$, we can use the same calculations as for the full regression but just 'forget' the last $(k - p)$ rows and columns.
3. The order of the variables in the Cholesky factor, $\boldsymbol{R}$, can easily be changed to derive regressions of $Y$ upon other subsets of $X$-variables.
4. If the data satisfy $Y = X\beta + \epsilon$, where the true residuals, $\epsilon$, are uncorrelated and identically distributed with variance $\sigma^2$, then the elements of $\boldsymbol{t} = \boldsymbol{Q}'\boldsymbol{y}$ are uncorrelated and have variance equal to $\sigma^2$. This result can be found in Grossman and Styan (1972). By appealing to the Central Limit Theorem, the elements of $\boldsymbol{t}$ will usually be approximately normally distributed.
5. If only the first $p$ elements of $\boldsymbol{\beta}$ are nonzero, then the expected values of the last $(n - p)$ elements of $\boldsymbol{t}$ are equal to zero.
6. The square of the $i^{th}$ projection, $t_i$, is the reduction in the residual sum of squares when the variable in position $i$ is added to the linear model containing the first $(i - 1)$ variables.
7. In general, rearranging the order of variables, between positions $i$ and $j$, will change all elements in those rows, and interchange elements in the same columns.
8. To calculate the regression or residual sums of squares for a new model, given those for a previous model, may require some planar rotations, followed by the appropriate addition and subtraction of the squares of only those projections which have changed, or are associated with variables being added to or deleted from the model.

# Appendix A
# Operations counts for all-subsets regression

In the following derivations for Garside's Gauss-Jordan algorithm and for the planar rotations algorithms, the notation differs slightly from Garside's in that $k$ is defined as the number of $X$-variables (excluding any variable representing a constant in the model). In Garside (1971b), the value of $k$ is one larger than our value as it includes the $Y$-variable. The word 'operation' will be used to mean a multiplication or a division.

The derivations require the following two sums with various limits of summation and with $\alpha = 2$.

$$\sum_{r=1}^{k-1} r\alpha^{r-1} = \{(k-1)\alpha^k - k\alpha^{k-1} + 1\}/(1-\alpha)^2 \tag{A.1}$$

$$\begin{aligned}\sum_{r=1}^{k} r(r-1)\alpha^{r-2} = {} & \{-k(k-1)\alpha^{k+1} + 2(k-1)(k+1)\alpha^k \\ & - k(k+1)\alpha^{k-1} + 2\}/(1-\alpha)^3\end{aligned} \tag{A.2}$$

## A.1 Garside's Gauss-Jordan algorithm

Using a binary order for subset generation, the variable in position 1 is deleted only once, the variable in position 2 is deleted twice and reinstated once, and the variable in position i is deleted $2^{i-1}$ times and reinstated one less time. That is, variable number i is pivotted in or out a total of $(2^i - 1)$ times. In pivotting on row i, that row is left unchanged as variable i will be reinstated before any variable in a lower numbered position is used as a pivot. Each higher numbered row requires one operation to set up the calculations and then one operation per element in the row on and to the right of the diagonal. The setup operation also uses the reciprocal of the diagonal element in the pivot row. These setup operations, which were omitted from Garside's operations counts, are exactly those which would be calculated in operating upon the pivot row. In pivotting on row i, $(k+3-j)$ operations are performed on row j, including the set-up operation for that row, where j ranges from $(i+1)$ to $(k+1)$. Thus the total number of operations required to pivot on row i is $(k+2-i)(k+3-i)/2$. Hence, the total operations count for the Garside algorithm is

$$\sum_{i-1}^{k} (2^i - 1)(k+2-i)(k+3-i)/2 = 14(2^k) - (k+4)(k^2+8k+21)/6.$$

As there are $(2^k - 1)$ subsets, the operations count is approximately 14 per subset for moderately large k. For small $k$, the numbers of operations per subset are as shown in Table A.1.

Table A.1 *Operations per subset using Garside's algorithm*

| k | 2 | 3 | 5 | 10 | 20 |
|---|---|---|---|---|---|
| Opns. per subset | 5. | 7. | 10.29 | 13.56 | 13.998 |

Table A.2 *Operations per subset using Hammarling's algorithm*

| k | 2 | 3 | 5 | 10 | 20 |
|---|---|---|---|---|---|
| Opns. per subset | 4.67 | 8. | 12.84 | 19.82 | 29.9995 |

## A.2 Planar rotations and a Hamiltonian cycle

The following simple algorithm generates a Hamiltonian cycle, which is suitable for generating all subsets of $k$ variables. Each time step 3 is executed, a new subset of $p$ variables is generated. The subsets of variables 1, 1 2, 1 2 3, ..., 1 2 3 ... k, are obtained from the initial ordering without any row swaps; the calculation of residual sums of squares for these subsets requires $2(k-1)$ operations.

1. For $i = 1$ to $k-1$, set $index(i) = i$.
2. Set $p = k - 1$.
3. Swap rows $p$ and $p+1$.
4. Add 1 to $index(p)$.
5. If $index(p) \leq k-1$, set $index(p+1) = index(p)$, add 1 to $p$, go to step 3.
   Otherwise, subtract 1 from $p$,
   If $p > 0$, go to step 3.
   Otherwise the end has been reached.

A new subset is generated each time two rows are swapped. Hence, rows $i$ and $(i+1)$ are swapped $({}^kC_i - 1)$ times. Using the Hammarling algorithm, it requires $10 + 2(k-i)$ operations to perform the swap. This count comprises 8 operations to set up the rotation and calculate the new elements in columns $i$ and $(i+1)$, 2 operations on the $\boldsymbol{Q'y}$-vector, 2 operations to calculate the residual sum of squares for the new subset of i variables, and 1 operation for each remaining element in rows $i$ and $(i+1)$. Hence, the total count of operations is

$$\sum_{i-1}^{k-1}({}^kC_i \;-\; 1)(10 + 2^k - 2^i) \;+\; 2(k-1) \;=\; (10+k)2^k - k^2 - 9k - 12$$

or about $(10+k)$ operations per subset for moderately large $k$. For small $k$, the numbers of operations per subset are as shown in Table A.2.

At the end of this algorithm, the variables are in a scrambled order.

Table A.3 *Elements which need to be recalculated when variable 5 is deleted then reinstated*

| Variable represented in the row | Variable represented in the column | | | | | |
|---|---|---|---|---|---|---|
| | 1 | 3 | 4 | 5 | 6 | 2 |
| 1 | X | X | X | X | X | X |
| 3 | | X | X | X | X | X |
| 4 | | | X | X | X | X |
| 5 | | | | * | * | X |
| 6 | | | | | * | X |
| 2 | | | | | | X |

## A.3 Planar rotations and a binary sequence

To see how the algorithm works in this case, consider a case with six $X$-variables. Let us suppose that variable number 2 has just been deleted. The current order of the variables is 1 3 4 5 6 2. The binary code for this situation is 1 0 1 1 1 1, i.e. the only variable with a zero index is number 2. Subtracting 1 from the binary value gives 1 0 1 1 1 0, i.e. we drop variable number 6. This requires no change of order as the subset 1 3 4 5 is already in order. Subtracting another 1 gives 1 0 1 1 0 1. This requires the interchange of rows 4 and 5 which contain variables 5 and 6. After variable number 2 was deleted, the triangular factorization was as in Table A.3 where X's and *'s denote non-zero elements. As variables are reinstated in exactly the reverse order in which they were introduced in the binary sequence, though not in the Hamiltonian cycle, we prefer to avoid operating on elements other than those marked with an * in deleting variable number 5, and to omit the column swaps which are a feature of the planar rotation algorithm. Unfortunately, if we use the Hammarling algorithm, when variable 5 is later reinstated, its row multiplier is not the same as it was immediately before the variable was deleted. This means that the new row multiplier is not that which applied to the last elements in the rows for variables 5 and 6. If the Gentleman algorithm is used, this problem does not arise, nor does it arise with the so-called fast planar rotation given by the updating formula (9') in Gentleman (1973). However the latter is liable to give severe accuracy problems, which can be far worse than those associated with Gauss-Jordan methods.

In developing code to use the binary sequence with a planar rotation algorithm, it is necessary to find the appropriate variable to add to or delete from the current subset. Fortunately the variables in the subset are always represented in increasing order in the triangular factorization while those deleted are always in reverse numerical order. In the algorithm which follows, $nout(i)$ stores the number of variables with number less than $i$ which are currently out of the subset. This means that if variable number $i$ is in the subset, its position is $i - nout(i)$. As the next-to-last variable is always reinstated immediately after being deleted, it is only necessary to calculate the value of the residual sum of squares when it is deleted; there is no need to calculate the new values for the three elements in the bottom corner of the triangular factorization. The array ibin() stores the 0-1 codes indicating whether a variable is in (1) or out (0) of the subset, $k$ = the total number of variables, and $ifirst$ is the number of the first variable which may be deleted (e.g. if there is a constant in the model which is to be in all subsets, then $ifirst = 2$). The algorithm is

1. Calculate the initial residual sums of squares.
   Simulate the deletion of variable number $(k - 1)$.
2. For $i = 1$ to $k - 2$, set $ibin(i) = 1$ and $nout(i) = 0$.
   Set $last = k$.
3. Subtract 1 from position $k - 2$ in the binary value of the sequence.
   Set $p = k - 2$.
4. If $ibin(p) = 1$, go to step 6.
   Otherwise set $ibin(p) = 1$.
   Set $ipos = p - nout(p)$.
   Raise variable from position $(last + 1)$ to position $ipos$.
   Set $last = last + 1$.
5. Set $p = p - 1$.
   If $p > 0$, go to step 4.
   Otherwise end has been reached.
6. Delete variable number p.
   Set $ibin(p) = 0$.
   Set $ipos = p - nout(p)$.
   Lower variable from row $ipos$ to row $last$.
   Set $last = last - 1$.
   Calculate new residual sums of squares for rows $ipos$ to $last$.
   For $i = p + 1$ to $k - 2$, set $nout(i) = nout(p) + 1$.
   Simulate the deletion of variable number $(k - 1)$ which is in row $(last - 1)$.
   Go to step 3.

As for the Garside algorithm, variable number $i$ is operated upon $(2^i - 1)$ times, except that no calculations are required when $i = k$. In general, when variable number $i$ is deleted, all of the higher numbered variables are in the subset. Hence, the variable must be rotated past variables numbered $i+1, i+2, ..., k$. Using the Gentleman algorithm (formula (9) in Gentleman (1973)), the number of operations required to swap variables $i$ and $(i+1)$ is $10+3(k-1+i)$. This is made up of 5 operations to set up the rotation, 3 operations on the $\boldsymbol{Q'y}$-vector, 2 operations to calculate the new residual sum of squares for the new subset of $i$ variables, and 3 operations for each pair of remaining elements in the two rows up to and including column $last$. In the case of variable number $(k - 1)$, the residual sum of squares when it is deleted can be calculated in 7

Table A.4 *Number of operations per subset – Gentleman's algorithm*

| k | 2 | 3 | 5 | 10 | 20 |
|---|---|---|---|---|---|
| Opns. per subset | 3. | 4.86 | 9.29 | 14.84 | 15.745 |

operations. Hence the total count of operations is

$$\sum_{i-1}^{k-1}(2^i - 1)\{7 + 3(k-i)(k+1-i)/2\} \;+\; 7(2^k - 2) \;+\; 2(k-1)$$

$$= \;(14 + 7/4).2^k \;-\; (k^3 + 6k^2 + 27k + 22)/2.$$

For moderately large $k$, this is about 15.75 operations per subset, or one-eighth more than for the Garside algorithm. For small $k$, the numbers of operations per subset are as follows:

## A.4 Fast planar rotations

As described so far, planar rotations are applied to a pair or rows at a time, and have the form

$$\begin{pmatrix} c & s \\ -s & c \end{pmatrix}\begin{pmatrix} w & x & \ldots \\ y & z & \ldots \end{pmatrix} = \begin{pmatrix} (w^2+y^2)^{\frac{1}{2}} & cx + sz & \ldots \\ 0 & -sx + cz & \ldots \end{pmatrix}. \qquad \text{(A.3)}$$

For the off-diagonal elements, we must calculate the quantities $(cx + sz)$ and $(-sx + cz)$ for each pair. This means that there are 4 multiplications and 2 additions for each pair.

By using row multipliers, the number of multiplications per pair can be reduced to 3 or 2. The equations equivalent to (A.3) are

$$\begin{pmatrix} c & s \\ -s & c \end{pmatrix}\begin{pmatrix} \sqrt{d_1} & \\ & \sqrt{d_2} \end{pmatrix}\begin{pmatrix} w & x & \ldots \\ y & z & \ldots \end{pmatrix}$$

$$= \begin{pmatrix} \sqrt{d_1^*} & \\ & \sqrt{d_2^*} \end{pmatrix}\begin{pmatrix} w^* & x^* & \ldots \\ 0 & z^* & \ldots \end{pmatrix},$$

where

$$\sqrt{d_1^*}x^* \;=\; c\sqrt{d_1}x \;+\; s\sqrt{d_2}z \qquad \text{(A.4)}$$

$$\sqrt{d_2^*}z^* \;=\; -s\sqrt{d_1}x \;+\; c\sqrt{d_2}z, \qquad \text{(A.5)}$$

and the condition upon $s$ and $c$ for the value of $y$ to be reduced to zero is that

$$0 \;=\; -\,s\sqrt{d_1}w \;+\; c\sqrt{d_2}y,$$

that is

$$c = \frac{\sqrt{d_1}w}{\sqrt{d_1 w^2 + d_2 y^2}}$$
$$s = \frac{\sqrt{d_2}y}{\sqrt{d_1 w^2 + d_2 y^2}},$$

which means that

$$\sqrt{d_1^*}w^* = (d_1 w^2 + d_2 y^2)^{\frac{1}{2}} \tag{A.6}$$
$$\sqrt{d_1^*}x^* = (d_1 wx + d_2 yz)/(d_1 w^2 + d_2 y^2)^{\frac{1}{2}} \tag{A.7}$$
$$\sqrt{d_2^*}z^* = (d_1 d_2)^{\frac{1}{2}}(-xy + wz)/(d_1 w^2 + d_2 y^2)^{\frac{1}{2}}, \tag{A.8}$$

In Gentleman's (1973, 1974, 1975) algorithm, the multiplier $d_1^*$ is chosen so that the diagonal element $w^* = 1$, and $d_2$ is chosen to give a '3-multiplication' algorithm. This means that (as $w = 1$ for this algorithm)

$$d_1^* = d_1 + d_2 y^2$$
$$d_2^* = d_1 d_2/(d_1 + d_2 y^2)$$
$$x^* = \bar{c}x + \bar{s}z$$
$$z^* = z - yx,$$

where $\bar{c} = d_1/(d_1 w^2 + d_2 y^2)$ and $\bar{s} = d_2 y/(d_1 w^2 + d_2 y^2)$.

The '3-multiplications' applies to the calculation of the remaining pairs of elements along each pair of rows, where $x$ and $z$ is the first such pair.

Notice that by storing the $d$'s rather than their square roots, the calculation of square roots is not required.

Hammarling (1974) showed 6 basic ways of modifying the Gentleman algorithm to give '2-multiplication' algorithms. From (A.4) and (A.5), by choosing $d_1^*$ and $d_2^*$, these are

$$d_1^* = c^2 d_1, \quad d_2^* = c^2 d_2$$
$$d_1^* = c^2 d_1, \quad d_2^* = s^2 d_1$$
$$d_1^* = c^2 d_1, \quad d_2^* = c^4 d_1/s^2$$
$$d_1^* = s^2 d_2, \quad d_2^* = c^2 d_2$$
$$d_1^* = s^2 d_2, \quad d_2^* = s^2 d_1$$
$$d_1^* = s^2 d_2, \quad d_2^* = s^4 d_2/c^2$$

There are several more ways of constructing '2-multiplication' algorithms. As the new values for the multipliers $d_i$ and the row elements, $w$, $x$ and $z$, are usually stored in the same locations as the old values, the new values for the upper row of each pair must be stored elsewhere for 4 of these 6 methods while the second row value is calculated. This would appear to make the third and sixth methods more attractive; these are the two ways which do not require temporary storage elsewhere as the calculation of $z^*$ uses the new $x^*$. Unfortunately, these cases can sometimes give rise to large cancellation errors and are not recommended.

A disadvantage of these '2-multiplication' algorithms is that the row multipliers tend to becoming either very small or very large and must occasionally be rescaled to prevent either underflow or overflow occurring. A number of algorithms have been proposed addressing this problem. One which is particularly directed toward least-squares calculations is that of Anda and Park (1994). This seems to be particularly fast for producing the initial $QR$-factorization from the input file containing the $X$ and $Y$ data.

CHAPTER 3

# Finding subsets which fit well

## 3.1 Objectives and limitations of this chapter

In this chapter we look at the problem of finding one or more subsets of variables whose models fit a set of data fairly well. Though we will only be looking at models that fit well in the least-squares sense, similar ideas can be applied with other measures of goodness-of-fit. For instance, there have been considerable developments in the fitting of models to categorical data, see e.g. Goodman (1971), Brown (1976), and Benedetti and Brown (1978), in which the measure of goodness-of-fit is either a log-likelihood or a chi-squared quantity. Other measures used in subset selection have included that of minimizing the maximum deviation from the model, known simply as minimax fitting or as $L_\infty$ fitting (e.g. Gentle and Kennedy (1978)), and fitting by maximizing the sum of absolute deviations or $L_1$ fitting (e.g. Roodman (1974), Gentle and Hanson (1977), Narula and Wellington (1979), Wellington and Narula (1981)).

Logistic regression is an important area of application of stepwise regression. The model to be fitted in this case is

$$\text{Prob.}(Y = 1) = 1 - \frac{1}{1 + e^{\Sigma_j x_j \beta_j}}$$

where $Y$ is a variable which can only take the values 0 or 1. In medical applications, the variable $Y$ could be the survival of a patient or the success of a treatment; in meteorology $Y$ could be 1 if rain occurs on a particular day and 0 otherwise. The conceptual problems are the same as those in linear regression. The computations are slower as they are iterative, and linear approximations are often used in heuristic algorithms in the attempt to find models that fit well. Useful references are Gabriel and Pun (1979), Nordberg (1981, 1982), Cox and Snell (1989), Hosmer and Lemeshow (1989), and Freedman et al. (1992)

We will only be considering models involving linear combinations of variables, though some of the variables may be functions of other variables. For instance, there may be four basic variables $X_1$, $X_2$, $X_3$ and $X_4$ from which other variables are derived such as $X_5 = X_1^2$, $X_6 = X_1 X_2$, $X_7 = X_4/X_3$, $X_8 = X_1 \log X_3$, etc.

The problem we will be looking at in this chapter is that, given a set of variables $X_1, X_2, ..., X_k$, we want to find a subset of $p < k$ variables $X_{(1)}$, $X_{(2)}$,

Table 3.1 *Use of dummy variables to represent a categorical variable*

| Age of subject (years) | $X_{17}$ | $X_{18}$ | $X_{19}$ | $X_{20}$ |
|---|---|---|---|---|
| 0-16 | 0 | 0 | 0 | 0 |
| 17-25 | 1 | 0 | 0 | 0 |
| 26-40 | 0 | 1 | 0 | 0 |
| 41-65 | 0 | 0 | 1 | 0 |
| >65 | 0 | 0 | 0 | 1 |

..., $X_{(p)}$ which minimizes or gives a suitably small value for

$$S = \sum_{i=1}^{n}(y_i - \sum_{j=1}^{p} b_{(j)}x_{i,(j)})^2, \tag{3.1}$$

where $x_{i,(j)}$, $y_i$ are the $i^{th}$ observations, $i = 1, 2, ..., n$ of variables $X_{(j)}$ and $Y$, and $b_{(j)}$ is a least-squares regression coefficient. In most practical cases, the value of $p$ is not predetermined and we will want to find good subsets for a range of values of $p$. To overcome the problem of bias in the regression coefficients, we should find not just one, but perhaps the best 10 or more subsets of each size $p$. Problems of selection bias will be considered in Chapter 5, and the problem of deciding the best value of $p$ to use, i.e. the so-called 'stopping rule' problem, will be considered in Chapter 6. This chapter is concerned with the mechanics of selecting subsets, not with their statistical properties.

In many practical cases, the minimization of (3.1) will be subject to constraints. One of these is that we may wish to force one or more variables to be in all subsets selected. For instance, most regression models include a constant which can be accommodated by making one of the variables, usually $X_1$, a dummy variable, which always takes the value 1 and which is forced into all subsets. There may also be constraints that one variable may only be included in a selected subset if another variable(s) is also included. For instance, it is often considered unreasonable to include a variable such as $X_1^2$ unless $X_1$ is also included. Dummy variables are often used to represent categorical variables. Thus if we have five age groups, 0-16, 17-25, 26-40, 41-65, and over 65 years, we may introduce four dummy variables $X_{17}$, $X_{18}$, $X_{19}$, $X_{20}$, with values assigned as in Table 3.1.

In such cases, it is often required that either all or none of the variables in such a group should be in the model.

It may be arguable whether such constraints should be applied. Here a distinction has to be made between a model that is intended to be *explanatory* and meant to be an approximation to the real but unknown relationship between the variables, and a model thatis intended to be used for *prediction*. The latter is our objective in this monograph, though if the selected subset is to be used for prediction outside the range of the data used to select and calibrate the model, an explanatory model may be safer to use.

In general, we will expect that the number, $p$, of variables in the subset will be less than the number of observations, $n$, though the number of available predictors, $k$, often exceeds $n$. The $X$-variables need not be linearly independent; for instance, it is often useful to include a difference $(X_1 - X_2)$ as another variable in addition to both $X_1$ and $X_2$.

## 3.2 Forward selection

In this procedure, the first variable selected is that variable $X_j$ for which

$$S = \sum_1^n (y_i - b_j x_{ij})^2$$

is minimized, where $b_j$ minimizes $S$ for variable $X_j$. As the value of $b_j$ is given by

$$b_j = \sum_1^n x_{ij} y_i / \sum_1^n x_{ij}^2,$$

it follows that

$$S = \sum_1^n y_i^2 - (\sum_1^n x_{ij} y_i)^2 / \sum_1^n x_{ij}^2.$$

Hence, the variable selected is that which maximizes

$$(\sum_1^n x_{ij} y_i)^2 / \sum_1^n x_{ij}^2. \tag{3.2}$$

If this expression is divided by $\sum_1^n y_i^2$, then we have the square of the cosine of the angle between vectors $X_j$ and $Y$. If the mean has been subtracted from each variable, then the cosine is the correlation between variables $X_j$ and $Y$.

Let the first variable selected be denoted by $X_{(1)}$; this variable is then forced into all further subsets. The residuals $Y - X_{(1)} b_{(1)}$ are orthogonal to $X_{(1)}$, and so to reduce the sum of squares by adding further variables, we must search the space orthogonal to $X_{(1)}$. From each variable $X_j$, other than the one already selected, we could form

$$\boldsymbol{X_{j.(1)}} = \boldsymbol{X_j} - b_{j,(1)} \boldsymbol{X_{(1)}},$$

where $b_{j,(1)}$ is the least-squares regression coefficient of $X_j$ upon $X_{(1)}$. Now we find that variable, $X_{j.(1)}$, which maximizes expression (3.2) when $Y$ is replaced with $Y - X_{(1)} b_{(1)}$ and $X_j$ is replaced with $X_{j.(1)}$. The required sums of squares and products can be calculated directly from previous sums of squares and products without calculating these orthogonal components for each of the $n$ observations; in fact, the calculations are precisely those of a Gauss-Jordan pivotting out of the selected variable. If the mean had first been subtracted from each variable, then the new variable selected is that which has the largest partial correlation in absolute value with $Y$ after variable $X_{(1)}$ has been fitted.

Thus, variables $X_{(1)}$, $X_{(2)}$, ..., $X_{(p)}$ are progressively added to the prediction equation, each variable being chosen because it minimizes the residual sum of squares when added to those already selected.

A computational method for forward selection using an orthogonal reduction is as follows. Let us suppose that we have reached the stage where $r$ variables have been selected or forced into the subset, where $r$ may be zero. Planar rotations are used to make these variables the first ones in the triangular factorization; that is, they occupy the top $r$ rows. Then the orthogonal reduction can be written as

$$\begin{aligned}(\boldsymbol{X_A}, \boldsymbol{X_B}) &= (\boldsymbol{Q_A}, \boldsymbol{Q_B}) \begin{pmatrix} \boldsymbol{R_A} \\ \boldsymbol{R_B} \end{pmatrix} \\ \boldsymbol{y} &= (\boldsymbol{Q_A}, \boldsymbol{Q_B}) \begin{pmatrix} \boldsymbol{r_{yA}} \\ \boldsymbol{r_{yB}} \end{pmatrix} + \boldsymbol{e},\end{aligned}$$

where $\boldsymbol{X_A}$ is an $n \times r$ matrix consisting of the values of the variables selected so far; $\boldsymbol{X_B}$ is an $n \times (k-r)$ matrix of the remaining variables; $\boldsymbol{Q_A}$ and $\boldsymbol{Q_B}$ have $r$ and $(k-r)$ columns, respectively, all of which are orthogonal, $\boldsymbol{R_A}$ consists of the top $r$ rows of a $k \times k$ upper-triangular matrix; $\boldsymbol{R_B}$ consists of the last $(k-r)$ rows and hence is triangular; $\boldsymbol{r_{yA}}$ and $\boldsymbol{r_{yB}}$ consist of the first $r$ and last $(k-r)$ projections of $\boldsymbol{y}$ on the directions given by the columns of $\boldsymbol{Q_A}$ and $\boldsymbol{Q_B}$; and $\boldsymbol{e}$ is a vector of $n$ residuals. The information on the components of $\boldsymbol{y}$ and of the remaining unselected $X$-variables, which are orthogonal to the selected variables, is then contained in $\boldsymbol{r_{yB}}$ and $\boldsymbol{R_B}$.

Let us write

$$\boldsymbol{R_B} = \begin{pmatrix} r_{11} & r_{12} & r_{13} & \cdots \\ & r_{22} & r_{23} & \cdots \\ & & r_{33} & \cdots \\ & & & \cdots \end{pmatrix} \quad \boldsymbol{r_{yB}} = \begin{pmatrix} r_{1y} \\ r_{2y} \\ r_{3y} \\ \cdots \end{pmatrix};$$

then if the variable in the top row of the submatrix $\boldsymbol{R_B}$ is added next, the reduction in the residual sum of squares (RSS) is $r_{1y}^2$. To find the reduction in RSS if the variable in the second row is added next instead, a planar rotation can be used to bring this variable into the top row and then the reduction in RSS is equal to the square of the value which is then in the top position of $\boldsymbol{r_{yB}}$. There is no need to perform the full planar rotation of whole rows to calculate the effect of bringing a variable from row $i$ to row 1 of $\boldsymbol{R_B}$. In swapping a pair of rows, only the elements on the diagonal and immediately next to it are needed to calculate the required planar rotations which are then applied only to calculate the new diagonal element for the upper row and the new element in $\boldsymbol{r_{yB}}$ for the upper row. Using Hammarling rotations, the reduction in RSS for the $i^{th}$ row of $\boldsymbol{R_B}$ can be calculated in $7(i-1)+2$ operations, and hence the total number of operations for all $(k-r)$ variables

Table 3.2 *Example in which two variables together have high predictive value*

| Observation number | $X_1$ | $X_2$ | $X_3$ | $Y$ |
|---|---|---|---|---|
| 1 | 1000 | 1002 | 0 | –2 |
| 2 | –1000 | –999 | –1 | –1 |
| 3 | –1000 | –1001 | 1 | 1 |
| 4 | 1000 | 998 | 0 | 2 |

is $(7/2)(k-r)(k-r-1) + 2(k-r)$. At the end, the variable selected to add next is rotated to the top row, if it is not already there, using the full planar rotations, adding a few more operations to the count.

An alternative and quicker way to calculate the reductions in RSS is to use (3.2), but with the $x$'s and $y$'s replaced by their components orthogonal to the already selected variables. The sums of squares of the $x$'s are obtained from the diagonal elements of $\boldsymbol{R}'_{\boldsymbol{B}}\boldsymbol{R}_{\boldsymbol{B}}$, and the cross-products of $x$'s and $y$'s from $\boldsymbol{R}'_{\boldsymbol{B}}\boldsymbol{r}_{\boldsymbol{yB}}$. Thus in the notation above, the sum of squares for the $X$-variable in the third row is $r_{13}^2 + r_{23}^2 + r_{33}^2$ and the cross-product is $r_{13}r_{1y} + r_{23}r_{2y} + r_{33}r_{3y}$. To calculate all $(k-r)$ reductions in RSS requires $(k-r)(k-r+3)$ operations if $\boldsymbol{R_B}$ is actually stored without the row multipliers used in the Hammarling and Gentleman algorithms, or about 50% more operations if row multipliers are being used. The selected variable is then rotated into the top row of $\boldsymbol{R_B}$, and the process is repeated to find the next variable with one row and column less in $\boldsymbol{R_B}$.

In general, there is no reason why the subset of $p$ variables that gives the smallest RSS should contain the subset of $(p-1)$ variables that gives the smallest RSS for $(p-1)$ variables. Table 3.16 provides an example in which the best-fitting subsets of 3 and 2 variables have no variables in common, though this is a rare situation in practice. Hence, there is no guarantee that forward selection will find the best-fitting subsets of any size except for $p = 1$ and $p = k$.

Consider the artificial example in Table 3.2. The correlations (cosines of angles) between $Y$ and $X_1$, $X_2$, $X_3$ are 0.0, –0.0016, and 0.4472, respectively. Forward selection picks $X_3$ as the first variable. The partial correlations of $Y$ upon $X_1$ and $X_2$ after being made orthogonal to $X_3$ are 0.0 and –0.0014 respectively. With $X_3$ selected, the subset of $X_1$ and $X_2$, which gives a perfect fit, $Y = X_1 - X_2$, cannot be obtained.

Examples similar to that above do occur in real life. The difference $(X_1 - X_2)$ may be a proxy for a rate of change in time or space, and in many situations such rates of change may be good predictors even when the values of the separate variables have little or no predictive value. In such cases, any subset-finding method which adds, deletes, or replaces only one variable at a time may find very inferior subsets. If there is good reason to believe that differences may provide good predictors, then they can be added to the subset of variables available for selection, though care must be taken to make sure that the software used can handle linear dependencies among the predictor variables.

## 3.3 Efroymson's algorithm

The term 'stepwise regression' is often used to mean an algorithm proposed by Efroymson (1960). This is a variation on forward selection. After each variable (other than the first) is added to the set of selected variables, a test is made to see if any of the previously selected variables can be deleted without appreciably increasing the residual sum of squares. Efroymson's algorithm incorporates criteria for the addition and deletion of variables as follows.

*(a) Addition*

Let $RSS_p$ denote the residual sum of squares with $p$ variables and a constant in the model. Suppose the smallest RSS, which can be obtained by adding another variable to the present set, is $RSS_{p+1}$. The ratio

$$R = \frac{RSS_p - RSS_{p+1}}{RSS_{p+1}/(n-p-2)} \tag{3.3}$$

is calculated and compared with an '$F$-to-enter' value, $F_e$. If $R$ is greater than $F_e$, the variable is added to the selected set.

*(b) Deletion*

With $p$ variables and a constant in the selected subset, let $RSS_{p-1}$ be the smallest RSS that can be obtained after deleting any variable from the previously selected variables. The ratio

$$R = \frac{RSS_{p-1} - RSS_p}{RSS_p/(n-p-1)} \tag{3.4}$$

is calculated and compared with an '$F$-to-delete (or drop)' value, $F_d$. If $R$ is less than $F_d$, the variable is deleted from the selected set.

*(c) Convergence of algorithm*

From (3.3), it follows that when the criterion for adding a variable is satisfied

$$RSS_{p+1} \leq RSS_p/\{1 + F_e/(n-p-2)\},$$

while from (3.4), it follows that when the criterion for deletion of a variable is satisfied

$$RSS_p \leq RSS_{p+1}\{1 + F_d/(n-p-2)\}.$$

Hence, when an addition is followed by a deletion, the new RSS, $RSS_p^*$ say, is such that

$$RSS_p^* \leq RSS_p . \frac{1 + F_d/(n-p-2)}{1 + F_e/(n-p-2)}, \tag{3.5}$$

The procedure stops when no further additions or deletions are possible which satisfy the criteria. As each $RSS_p$ is bounded below by the smallest RSS for any subset of $p$ variables, by ensuring that the RSS is reduced each time that a new subset of $p$ variables is found, convergence is guaranteed. From (3.5), it follows that a sufficient condition for convergence is that $F_d < F_e$.

*(d) True significance level*

The use of the terms '$F$-to-enter' and '$F$-to-delete' suggests that the ratios $R$ have an $F$-distribution under the null hypothesis, i.e. that the model is the true model, and subject to the true residuals being independently, identically, and normally distributed. This is not so. Suppose that after $p$ variables have been entered, these conditions are satisfied. If the value of $R$ is calculated using (3.3) but using the value of $RSS_{p+1}$ for one of the remaining variables *chosen at random*, then the distribution of $R$ is the $F$-distribution. However, if we choose that variable that maximizes $R$, then the distribution is not an $F$-distribution or anything remotely like an $F$-distribution. This was pointed out by Draper et al. (1971) and by Pope and Webster (1972); both papers contain derivations of the distribution of the $R$-statistic for entering variables. Evaluation of the distribution requires multidimensional numerical integration. This has been done by Derflinger and Stappler (1976).

A rough approximation to the percentage points can be obtained by treating $R$ as if it were the maximum of $(k - p)$ independent $F$-ratios. The $R$-value corresponding to a significance level of $\alpha$ is then the $F$-value which gives a significance level of $\alpha^*$ where

$$(1 - \alpha^*)^{k-p} = 1 - \alpha, \tag{3.6}$$

Pope and Webster suggest that it would be better to replace $RSS_{p+1}/\,(n - p - 2)$ in the denominator of (3.3) with $RSS_k/(n - k - 1)$, i.e. to use all of the available predictor variables in estimating the residual variance. Limited tables for the distribution of $R$ have been given by Draper et al. (1979); these show that the nominal 5% points for the maximum $F$-to-enter as incorrectly obtained from the $F$-distribution will often give true significance levels in excess of 50%. An early use of (3.6) to calculate $F$-to-enter values was by Miller (1962).

From (3.6), we can derive the rough Bonferroni bound that $\alpha < (k - p)\alpha^*$ for the true significance level, $\alpha$, in terms of the value, $\alpha^*$, read from tables of the $F$-distribution. Butler (1984) has given a fairly tight lower bound for the selection of one more variable out of the remaining $(k - p)$ variables, provided that the information that none of them had been selected earlier can be neglected. This lower bound is

$$(k - p)\alpha^* \;-\; \sum_{i<j} p_{ij},$$

where $p_{ij}$ is the probability that two of the remaining variables, $X_i$ and $X_j$, satisfy the *a priori* condition for significance at the $\alpha^*$ level. Butler's derivation is in terms of partial correlations of the dependent variable with the predictors, though this could be translated into $F$-values. The joint probabilities, $p_{ij}$, can be expressed as bivariate integrals which must be evaluated numerically. An algorithm for doing this is described in Butler (1982).

As with forward selection, there is no guarantee that the Efroymson algorithm will find the best-fitting subsets, though it often performs better than forward selection when some of the predictors are highly correlated. The algorithm incorporates its own built-in stopping rule; recommendations with respect to suitable values for $F_e$ and $F_d$ will be given in Chapter 6.

## 3.4 Backward elimination

In this procedure, we start with all $k$ variables, including a constant if there is one, in the selected set. Let $RSS_k$ be the corresponding residual sum of squares. That variable is chosen for deletion which yields the smallest value of $RSS_{k-1}$ after deletion. Then that variable from the remaining $(k-1)$, which yields the smallest $RSS_{k-2}$, is deleted. The process continues until there is only one variable left, or until some stopping criterion is satisfied.

Backward elimination can be carried out fairly easily starting from an orthogonal reduction. If the variable to be deleted is in the last row of the triangular reduction, then the increase in RSS when it is deleted is simply $r_{ky}^2$. Hence, each variable in turn can be moved to the bottom row to find which gives the smallest increase in RSS when deleted. The actual movement of rows can be simulated without altering any values in the triangular factorization. The values in the row to be moved are copied into a separate storage location and the effect of rotation past each lower row is calculated. Numbering rows *from the bottom*, the number of operations necessary to lower the variable from row $i$ to row 1, using Hammarling rotations, is $(i-1)(i+12)/2$ plus 2 operations to calculate the increase in RSS. Hence, the total for all $k$ variables is $k(k^2+18k-7)/6$. The selected variable is then moved to the last row and the process repeated on the $(k-1)\times(k-1)$ matrix obtained by omitting the last row and column.

There is an alternative method using SSP-matrices. It can be shown that the increase in RSS when variable $i$ is deleted is $b_i^2/c^{ii}$ where $b_i$ is the least-squares regression coefficient with all variables in the model and $c^{ii}$ is the $i^{th}$ diagonal element of $(\boldsymbol{X}'\boldsymbol{X})^{-1}$. This formula can be verified very easily from the orthogonal reduction. Consider the case of the last variable. By back-substitution,

$$b_k = r_{ky}/r_{kk}.$$

Now as

$$(\boldsymbol{X}'\boldsymbol{X})^{-1} = \boldsymbol{R}^{-1}\boldsymbol{R}^{-T},$$

the element in the bottom right-hand corner of the inverse of the SSP-matrix is simply

$$c^{kk} = 1/r_{kk}^2.$$

Hence,

$$b_k^2/c^{kk} = r_{ky}^2,$$

which is the correct increase. Now any other variable can be moved into the last row of the triangular decomposition without changing the value of $b_i^2/c^{ii}$ for that variable, and hence the formula holds for all variables. Alternative derivations of this well-known result are given in many books on linear regression.

Using this alternative method requires $k^2(k+1)/2$ operations for the inversion of the SSP-matrix. After the first variable has been deleted, the inverse matrix for the remaining variables can be obtained by pivoting out the selected variable using the usual Gauss-Jordan formulae. This only requires $k(k+1)/2$ operations. The method using SSP-matrices is usually much faster but the accuracy can be very poor if the SSP-matrix is ill-conditioned, and calculated RSS's or increases in RSS can be negative unless great care is exercised.

Backward elimination is usually not feasible when there are more variables than observations. If we have 100 variables but only 50 observations then, provided that the 100 columns of the predictor variables have rank 50, the residual sum of squares will be zero. In most cases 51 variables will have to be deleted before a nonzero RSS is obtained. The number of ways of selecting 51 out of 100 variables is approximately $9.9 \times 10^{28}$.

It has been argued by Mantel (1970) that in situations similar to that in the example in section 3.2 in which the variable $Y$ is highly correlated with some linear combination of the X-variables, such as $X_1 - X_2$ or a second difference $X_1 - 2X_2 + X_3$, but where the correlations with the individual variables are small, backward elimination will tend to leave such groups of variables in the subset, whereas they would not enter in a forward selection until almost all variables have been included.

Beale (1970) countered with the common situation in which the X-variables are percentages, e.g. of chemical constituents, which sum to 100% or perhaps slightly less if there are small amounts of unidentified compounds present. Any one variable is then given either exactly or approximately by subtracting the others from 100%. It is then a matter of chance which variable is deleted first in the backward elimination; the first one could be the only one which is of any value for prediction from small subsets, and once a variable has been deleted, it cannot be reintroduced in backward elimination. Example 3 in section 3.10 is a case in which the first variable deleted in backward elimination is the first one inserted in forward selection. Of course a backward analogue of the Efroymson procedure is possible.

Backward elimination always requires far more computation than forward selection. If, for instance, we have 50 variables available and expect to select a subset of less than 10 of them, in forward selection we would only proceed until about 10 or so variables have been included. In backward elimination we start with 50 variables, then 49, then 48, until eventually we reach the size of interest.

Both forward selection and backward elimination can fare arbitrarily badly in finding the best-fitting subsets. Berk (1978b) has shown that even when forward selection and backward elimination find exactly the same subsets of all sizes, it gives no guarantee that there are not much better fitting subsets of some sizes.

A variation of the Efroymson algorithm has been proposed by Broersen (1986). After forward selection has been used until all variables have been included, backward elimination is commenced. At each stage of both forward selection and backward elimination, the Mallows' $C_p$-statistic is calculated (see Chapter 5 for details of this statistic). If the elimination of a variable causes an increase in $C_p$, then the variable is reinstated and the algorithm backtracks to the previous step where a different variable is selected and a new path is tried. A record is kept of the values of $C_p$ that have been found, and when no further progress can be made, the subset giving the smallest value is selected. This subset is not necessarily the final one. After the forward selection phase, the algorithm is like the Efroymson algorithm, but using backward elimination instead of forward selection, and using Mallows' $C_p$ as the criterion rather than $F$-to-enter. We shall see later that using Mallows' $C_p$ is equivalent to using an $F$-to-enter value of about 2.0. Broersen calls his algorithm Stepwise Directed Search.

## 3.5 Sequential replacement algorithms

The basic idea here is that once two or more variables have been selected, we see whether any of those variables can be replaced with another that gives a smaller residual sum of squares. For instance, if we have 26 variables, which are conveniently denoted by the letters of the alphabet, we may at some stage have selected the subset of four variables

$$ABCD.$$

Let us try replacing variable $A$. There may be several variables from the remaining 22 that give a smaller $RSS$ when in a subset with $B$, $C$ and $D$. Suppose that of these, variable $M$ yields the smallest $RSS$. We can replace $A$ with $M$ giving the subset

$$MBCD.$$

Now we try replacing variable $B$, then variable $C$, then $D$, then back to $M$, etc. Some of these attempts will not find a variable which yields a reduction in the $RSS$, in which case we move on to the next variable. Sometimes variables we have replaced will return. The process continues until no further reduction is possible by replacing any variable.

The procedure must converge as each replacement reduces the $RSS$ that is bounded below. In practice the procedure usually converges very rapidly.

Unfortunately, this type of replacement algorithm does not guarantee convergence upon the best-fitting subset of the size being considered. In the above

Table 3.3 *Illustrating variable replacement*

| | Variables | RSS |
|---|---|---|
| Initial subset | ABCD | 100 |
| Best replacement for A | MBCD | 93 |
| Best replacement for B | AMCD | 91 |
| Best replacement for C | ABXD | 96 |
| Best replacement for D | ABCQ | 94 |

example, if we had started by trying to replace variable B instead of variable $A$, then the procedure might have converged upon a different subset. Let us label these final subsets as *stationary subsets.*

Suppose that in our hypothetical example, the subset of four variables which gives the smallest $RSS$ is

$$BEST.$$

If we start with the subset $PEST$, we are certain to reach the stationary subset $BEST$ only if variable $P$ is the first one to be replaced. Thus, we are certain to reach the absolute minimum from only 23 out of the 14,950 possible starting subsets of four variables, though in practice the best subset will usually be found from many more starting subsets. Even if we are lucky and find the best-fitting subset of this size, we have no way of determining that it is the best one, other than by doing an exhaustive search perhaps using branch-and-bound (Section 3.8).

The following modification improves our chances of finding the best-fitting subset. Suppose that we start with the subset $ABCD$ and that this gives an $RSS = 100$. We now look for the best replacement for variable $A$, but we do not make the replacement yet. Similarly, we try replacing variable $B$ but with variable $A$ still in the subset. Suppose we obtain the results in Table 3.3 for the best replacements of each variable.

Replacing variable $B$ gives the smallest $RSS$, so we make the replacement and then repeat the process. Notice that at the next stage we know that we cannot replace variable $M$ with any variable which gives a smaller $RSS$; therefore we only consider replacing the other three at the next step.

Using this replacement algorithm, there are now 92 out of the 14,950 possible starting subsets from which we are guaranteed to find the best-fitting subset.

Either of the above replacement algorithms can be used in conjunction with any of the algorithms described in earlier sections of this chapter. Thus a sequential replacement algorithm can be obtained by taking the forward selection algorithm and applying a replacement procedure after each new variable is added.

Sequential replacement requires more computation than forward selection or the Efroymson algorithm, but it is still feasible to apply to problems with several hundred variables when subsets of, say up to 20-30 variables, are required.

An alternative technique which can be used when there are large numbers of variables is to randomly choose starting subsets and then apply a replacement procedure. However, on one problem on which the author was involved there were 757 variables; 74 different stationary subsets of 6 variables were obtained from 100 random starts! Even then a subset that gave only about two-thirds of the $RSS$ of the best of the 74 was later found by chance.

## 3.6 Replacing two variables at a time

Replacing *two* variables at a time substantially reduces the maximum number of stationary subsets and means that there is a much better chance of finding good subsets when, for instance, a difference between two variables is a good predictor but was not included in the available set of predictor variables. However, for subsets of $p$ variables there are $p(p-1)/2$ pairs of variables to be considered for replacement and hence much more computation is required. Similarly, forward selection and backward elimination are possible, two variables at a time.

An algorithm which has been very successful for the author in problems with hundreds of predictors, as occurs in meteorological forecasting or in near-infrared (NIR) spectroscopy is as follows. We suppose that a subset of maximum size $p$ variables is being sought.

1. Generate a random subset of about $p/2$ variables.
2. Consider all possible replacements of pairs of variables in the currently selected subset. This includes replacements of only one of each pair of variables, as well as replacing both.
3. Find the best replacement. If it reduces the $RSS$ for the current size of subset, then make the replacement and go back to step 2.
4. If the maximum subset size of interest has not been reached, increase the current size by one by adding any variable, then go back to step 2.
5. Repeat the exercise from step 1 until satisfied.

## 3.7 Generating all subsets

We saw earlier in Chapter 2 that it is feasible to generate all subsets of variables provided that the number of predictor variables is not too large, say less than about 20, if only the residual sum of squares is calculated for each subset. After the complete search has been carried out, a small number of the more promising subsets can be examined in more detail. The obvious disadvantage of generating all subsets is cost. The number of possible subsets of one or more variables out of $k$ is $(2^k - 1)$. Thus, the computational cost roughly doubles with each additional variable.

In most cases, we are not interested in all subsets of all sizes. For instance,

Table 3.4 *Lexicographic order of generation of all subsets of 3 variables out of 7*

| | | | | |
|---|---|---|---|---|
| 123 | 136 | 167 | 247 | 356 |
| 124 | 137 | 234 | 256 | 357 |
| 125 | 145 | 235 | 257 | 367 |
| 126 | 146 | 236 | 267 | 456 |
| 127 | 147 | 237 | 345 | 457 |
| 134 | 156 | 245 | 346 | 467 |
| 135 | 157 | 246 | 347 | 567 |

someone with 100 variables (and possibly only 50 observations) is unlikely to be interested in subsets of 45 variables; subsets of 10 variables may be large enough. There are $(2^{100} - 1) = 1.3 \times 10^{30}$ subsets of one or more variables out of 100, but only $1.9 \times 10^{13}$ subsets of 10 or less. Even this last number is too large for an exhaustive evaluation of all subsets to be feasible. However, the device to be described in the next section will usually render this case feasible.

Many algorithms have been published for performing exhaustive evaluations, including those of Garside (1965, 1971b,c), Schatzoff et al. (1968), Furnival (1971), and Morgan and Tatar (1972). These evaluate all subsets of all sizes, and all require that the number of observations is at least as great as the number of predictor variables. Kudo and Tarumi (1974) have published an algorithm for searching for all subsets of $p$ or less variables out of $k$. These algorithms use matrix inversion and SSP-matrices and present problems when the SSP-matrix is ill-conditioned or of less than full rank. In a series of papers, Narula and Wellington (1977a, 1977b, 1979) and Wellington (1981) have presented an algorithm for finding the best-fitting subsets of $p$ variables out of $k$ variables when the criterion used is that of minimizing the sum of absolute deviations, i.e. minimizing the $L_1$-norm.

Let us consider algorithms for generating all subsets of $p$ variables out of $k$. Table 3.4 illustrates the generation of all subsets of 3 variables out of 7 in what is known as lexicographic order. It should be read by columns.

This would appear to be a good order of generation for software using SSP-matrices as the variable at the right-hand end is changing most rapidly and it is possible to perform the calculation of the $RSS$'s by operating only on submatrices of the SSP-matrix. In the case of the variable in position $p$, the calculation of the new $RSS$ when this variable is changed only requires one multiplication and one division. However the changes to variables in other positions require operations on almost every element in the SSP-matrix. An algorithm for generating the lexicographic order has been given by Gentleman (1975).

In using SSP-matrices, the order of the variables remains unchanged and the operations consist of inverting and reinverting part of the matrix so that the variables to be pivoted in or out are easily found. A lexicographic order

Table 3.5 *Start of a possible sequence to generate the lexicographic order*

| | |
|---|---|
| 123 | 4567 |
| 124 | 3567 |
| 125 | 4367 |
| 126 | 5437 |
| 127 | 6543 |
| 134 | 2765 |
| 135 | 4276 |
| 136 | 5427 |
| 137 | 6542 |
| 145 | 3762 |

of generation is not well suited to a planar rotation algorithm. Table 3.4 only shows the order of the first 3 variables; below we see what the full order of variables could look like. One of many possible orderings is shown in Table 3.5 for the first part of the lexicographic order above.

In the algorithm used to generate the start of the sequence, the next variable or variables needed were moved up from wherever they were in the complete ordering and the other variables kept their previous order but moved down the appropriate number of places. As can be seen, the above ordering requires a large number of interchanges of variables. In particular, the move from 127 6543 to 134 2765 seems very wasteful. At this stage, all of the subsets of 3 variables including 12 have been exhausted so that either 1 or 2 must be dropped, but it would be much more efficient to now introduce the 6 from the next position. This can be done in the following algorithm, which operates on whichever variables happen to be in the position from which a variable is to be moved irrespective of the number of that variable.

The basic idea of this combinatoric algorithm is that for each of the $p$ positions to be filled there is a 'last' position to which the next variable taken from that position is moved. Initially it is position $k$ for all of the $p$ positions. The first variable moved from position $i$ goes to position $last(i)$ amongst the last $(k-p)$. To make sure that the variable is not moved again with the current set of variables in positions 1, 2, ..., $i-1$, one is subtracted from $last(i)$. Also, to make sure that all subsets of the variables in positions $p+1$, ..., (new) $last(i)$ are used without moving the variable just moved from position $i$, the values of $last(j)$ for $j = i+1$, ..., $p$ are set equal to the new value of $last(i)$. The algorithm is as follows:

For $i$=1 to $p$, set $last(i) = k$.
Set the current pointer, $ipt = p$.
Then loop as follows:
DO
  If ($ipt < last(ipt)$ AND $last(ipt) > p$) then
    Move variable from position $ipt$ to position $last(ipt)$

Table 3.6 *Ordering of generated subsets of 3 variables out of 7, indicating also the positions of excluded variables*

| | | | |
|---|---|---|---|
| 123 4567 | 134 5672 | 452 7361 | 735 2641 |
| 124 5673 | 135 6472 | 427 3651 | 752 6341 |
| 125 6743 | 136 5472 | 423 6751 | 756 2341 |
| 126 7543 | 165 4372 | 426 3751 | 762 5341 |
| 127 6543 | 164 5372 | 463 7251 | 625 3741 |
| 176 5432 | 145 6372 | 467 3251 | 623 5741 |
| 175 4362 | 456 3721 | 473 6251 | 635 2741 |
| 174 3562 | 453 7261 | 736 2541 | 352 6741 |
| 173 4562 | 457 2361 | 732 5641 | |

```
    Decrease last(ipt) by 1.
    For i = ipt + 1 to p, set last(i) = last(ipt).
    Set ipt = p.
  Otherwise
    Decrease ipt by 1.
    If (ipt < 1) then END.
  End If
END DO
```

The above algorithm is the same as the Gentleman (1975) algorithm, except that instead of the index of the simulated nested DO-loops being incremented, the upper limit, i.e. $last(ipt)$, is decremented. It could equally well have been written in the same way as for the Gentleman algorithm; in fact the first exhaustive search algorithm written by the author was in this form. The algorithm does not generate a lexicographic order as the index of the DO-loops (or the upper limits here) are not used to determine the *number* but the *position* of a variable. The algorithm generates the sequence in Table 3.6 of subsets of 3 variables out of 7.

The algorithm above generates all subsets of $p$ variables out of $k$; it does not in general generate all subsets of less than $p$ variables. For instance, in Table 3.6, the pairs 23, 25, 27, 34 and 56 do not occur in the first two positions, and the single variables 2 and 5 do not appear in the first position. The algorithm can easily be modified to generate all subsets of $p$ variables or less out of $k$, by simply removing the condition 'AND $last(ipt) > p$' from the top of the main loop. With this modification to the algorithm, the generated sequence is as given in Table 3.7.

We noted earlier that the inner cycle of the lexicographic algorithm can be performed extremely rapidly using SSP-matrices. This gives it a major speed advantage over algorithms using planar rotations. However, the outer cycles are much slower using SSP-matrices as operations must be performed on all except a few rows at the top of the matrix, whereas planar rotation algorithms usually require operations on only a very small number of adjacent rows near

Table 3.7 *Ordering of generated subsets of 3 or less variables out of 7*

| | | | |
|---|---|---|---|
| 123 4567 | 136 5472 | 526 3741 | 362 4751 |
| 124 5673 | 165 4372 | 563 7241 | 326 4751 |
| 125 6743 | 164 5372 | 567 3241 | 264 7351 |
| 126 7543 | 145 6372 | 573 6241 | 267 4351 |
| 127 6543 | 154 6372 | 537 6241 | 274 6351 |
| 176 5432 | 546 3721 | 376 2451 | 247 6351 |
| 175 4362 | 543 7261 | 372 4651 | 476 2351 |
| 174 3562 | 547 2361 | 374 2651 | 467 2351 |
| 173 4562 | 542 7361 | 342 6751 | 674 2351 |
| 134 5672 | 527 3641 | 346 2751 | 764 2351 |
| 135 6472 | 523 6741 | | |

the top of the matrix. The planar rotation algorithm can be made faster by using the trick employed in section 2.3 on forward selection. Instead of the slow inner cycle, an SSP-matrix is formed of those components of the variables in rows $p$, $p+1$, ..., $last(p)$, which are orthogonal to the variables in rows 1, 2, ..., $p-1$. This simply means that we form

$$\boldsymbol{X}'\boldsymbol{X} = \boldsymbol{R}'\boldsymbol{R},$$

using as $\boldsymbol{R}$ the triangular submatrix between rows and columns $p$ and $last(p)$ inclusive of the triangular factorization which is current at that time. Further details will be given at the end of the next section.

## 3.8 Using branch-and-bound techniques

Suppose that we are looking for the subset of 5 variables out of 26 that gives the smallest $RSS$. Let the variables again be denoted by the letters $A$ to $Z$. We could proceed by dividing all the possible subsets into two 'branches', those which contain variable $A$ and those which do not. Within each branch we can have subbranches including and excluding variable $B$, etc. Now suppose that at some stage we have found a subset of 5 variables containing $A$ or $B$ or both, which gives $RSS = 100$. Let us suppose that we are about to start examining that subbranch which excludes both $A$ and $B$. A lower bound on the smallest $RSS$ which can be obtained from this subbranch is the $RSS$ for all of the 24 variables $C - Z$. If this is, say 108, then no subset of 5 variables from this subbranch can do better than this, and as we have already found a smaller $RSS$, this whole subbranch can be skipped.

This simple device appears to have been used first in subset selection by Beale et al. (1967), and by Hocking and Leslie (1967). It is further exploited by LaMotte and Hocking (1970). Using this device gives us the advantage of exhaustive searches that are guaranteed to find the best-fitting subsets, and at the same time the amount of computation is often reduced substantially.

The branch-and-bound device can similarly be applied with advantage with

Table 3.8 *Hypothetical residual sums of squares*

| No. of variables | 1 | 2 | 3 | 4 | 5 | 6 |
|---|---|---|---|---|---|---|
| RSS | 503 | 368 | 251 | 148 | 93 | 71 |

most other criteria of goodness-of-fit such as minimum sum of absolute deviations or maximum likelihood. One such application has been made by Edwards and Havranek (1987) to derive so-called minimal adequate sets, and this will be described briefly in section 4.2.

Branch-and-bound can be applied in a number of ways. For instance, if we want to find the 10 best subsets of 5 variables, then the $RSS$ for all the variables in the subbranch is compared against the 10th best subset of 5 variables that has been found up to that stage.

Alternatively, consider the task of finding the best-fitting subsets of all sizes up to and including 6 variables. Suppose that we are about to start examining a subbranch and that the smallest $RSS$'s found up to this stage have been as shown in Table 3.8.

Suppose that the $RSS$, including all of the variables of this subbranch is 105, then we cannot possibly find a better-fitting subset of 5 or 6 variables from this subbranch, though there could be better-fitting subsets of 1, 2, 3 or 4 variables. Hence, until we complete searching this subbranch, we look only for subsets of 4 or fewer variables.

Branch-and-bound is particularly useful when there are 'dominant' variables such that good-fitting subsets must include these variables. The device is of almost no value when there are more variables than observations as the lower bounds are nearly always zero.

An algorithmic trick which can be used in conjunction with branch-and-bound is that of saving the current state of matrices and arrays immediately before a variable is deleted. For instance, suppose we are looking for the best subset of 10 variables out of 100 and have reached the stage where the variable in position 7 is about to be deleted for the first time. Using orthogonal reduction methods, the variable is then moved from position 7 to position 100, an operation which requires a very large amount of computation. The $RSS$ with the other 99 variables in the regression is then calculated. Suppose that this shows that with the variable from position 7 deleted we cannot possibly improve upon the best subset of 4 variables found so far. The deleted variable must then be reinstated, that is, moved all the way back to position 7. If a copy of the matrices and arrays had been kept, then a considerable amount of computation is avoided.

The calculation of the bounds requires a similar amount of extra computation using methods based upon SSP-matrices. At the start of any branch, e.g. that excluding variables A and B, the SSP-matrix will usually be available with none of the variables included in the model. All of the variables except

A and B must then be pivoted on to obtain the bound. A way around this problem is to work with two matrices available, the SSP-matrix and the other its complete inverse. One of these matrices is then used to calculate the $RSS$'s for small subsets while the other is used to obtain bounds. Furnival and Wilson (1974) have published an algorithm based upon SSP-matrices that uses this trick. It finds the best subsets of all sizes though and can be very slow. One important feature noted by Furnival and Wilson is that the amount of work needed to find, say the 10 best-fitting subsets of each size, is usually not much greater than that needed to find the best ones.

If model building, rather than prediction, is the objective, then the algorithm of Edwards and Havranek (1987) may be used. This is similar to the Furnival and Wilson algorithm but attempts to find only what Aitkin (1974) has called minimal adequate (sub)sets. More details are given in section 4.2.

An equivalent algorithm is used by Narula and Wellington (1977b, 1979), Armstrong and Kung (1982), and Armstrong et al. (1984) but using the weighted sum of absolute errors as the criterion instead of least squares.

The equivalence of keeping an inverse matrix when using orthogonal reductions is to keep a second triangular factorization with the variables in reverse order. The bounds are then calculated from operations on variables that are in the lower rows of the triangular factorization. If the use of bounds results in variables frequently being reinstated immediately after deletion, then this idea obviously has considerable merit. However, the author has usually used a replacement algorithm before calling on an exhaustive search algorithm, and this usually establishes fairly good bounds and orders variables so that the 'best' variables are at the start and so are rarely deleted. Also, many of the author's problems have involved more variables than observations, in which case branch-and-bound is of no value until the final stages of the search.

An exhaustive search can be very time consuming if a large number of possible subsets have to be examined. Experience suggests that the maximum feasible size of problem is one for which the number of possible subsets, i.e. $\sum {}^kC_p$ where $k$ is the number of available predictor variables and $p$ ranges from 1 up to the maximum size of subset of interest, is of the order of $10^7$. It is recommended that any software for exhaustive search should periodically write out to disk the best subsets found up to that stage, so that all is not lost if the computer run exceeds its time limit. If this information can be read back in later, it can provide good bounds to save much of the computation during a future run.

## 3.9 Grouping variables

Gabriel and Pun (1979) have suggested that when there are too many variables for an exhaustive search to be feasible, it may be possible to break them down into groups within which an exhaustive search for best-fitting subsets is feasible. The grouping of variables would be such that if two variables, $X_i$ and $X_j$, are in different groups, then their regression sums of squares are additive;

that is, that if the reduction in $RSS$ due to adding variable $X_i$ to a previously selected subset of variables is $S_i$, and the corresponding reduction if variable $X_j$ is added instead is $S_j$, then the reduction when both are added is $S_i + S_j$.

If we have a complete set of say 100 variables for which we want to find good subsets of 5 variables, we may be able to break the variables into groups containing, say 21, 2, 31, 29 and 17 variables. Within each group, except the one with only 2 variables, the best subsets of 1, 2, 3, 4 and 5 variables would be found. All combinations of 5 variables made up from those selected from the separate groups would then be searched to find the best overall subsets. Thus, a subset of 5 variables might be made up of the best subset of 3 variables from one group plus single variables from two other groups.

To do this we must first find when regression sums of squares are additive. Does this occur only when the variables are orthogonal? Let us consider two variables $X_i$ and $X_j$. Their individual regression sums of squares are the squares of the lengths of the projections of $Y$ on $X_i$ and $X_j$, respectively, i.e.

$$S_i = (\boldsymbol{X_i y})^2/(\boldsymbol{X_i' X_i})$$

$$S_j = (\boldsymbol{X_j y})^2/(\boldsymbol{X_j' X_j}).$$

Let $S_{i.j}$ be the incremental regression sum of squares when variable $X_i$ is added after $X_j$. Then

$$S_{i.j} = (\boldsymbol{X_{i.j} y_{.j}})^2/(\boldsymbol{X_{i.j}' X_{i.j}}),$$

where $\boldsymbol{X_{i.j}}$ is that part of $\boldsymbol{X_i}$ which is orthogonal to $\boldsymbol{X_j}$, i.e. the vector of residuals after $\boldsymbol{X_i}$ is regressed against $\boldsymbol{X_j}$, i.e. $\boldsymbol{X_{i.j}} = \boldsymbol{X_i} - b_{ij}\boldsymbol{X_j}$ where $b_{ij}$ is the least-squares regression coefficient, and $\boldsymbol{y_{.j}}$ is similarly that part of $\boldsymbol{y}$ orthogonal to $\boldsymbol{X_j}$. Then

$$\begin{aligned} S_{i.j} &= \frac{[(\boldsymbol{X_i} - b_{ij}\boldsymbol{X_j})'(\boldsymbol{y} - b_{yj}\boldsymbol{X_j})]^2}{(\boldsymbol{X_i} - b_{ij}\boldsymbol{X_j})'(\boldsymbol{X_i} - b_{ij}\boldsymbol{X_j})} \\ &= \frac{(\boldsymbol{X_i' y} - b_{ij}\boldsymbol{X_j' y})^2}{\boldsymbol{X_i' X_i} - b_{ij}\boldsymbol{X_j' X_i}}. \end{aligned}$$

Now we introduce direction cosines $r_{ij}$, $r_{iy}$, $r_{jy}$, which are the same as correlation coefficients if the variables have zero means, where

$$\begin{aligned} r_{ij} &= \boldsymbol{X_i' X_j}/(\| \boldsymbol{X_i} \| . \| \boldsymbol{X_j} \|) \\ r_{iy} &= \boldsymbol{X_i' y}/(\| \boldsymbol{X_i} \| . \| \boldsymbol{y} \|) \\ r_{jy} &= \boldsymbol{X_j' y}/(\| \boldsymbol{X_j} \| . \| \boldsymbol{y} \|), \end{aligned}$$

where the norms are in the $L_2$-sense. Now we can write

$$S_{i.j} = S_i . \frac{(1 - r_{ij} r_{jy}/r_{iy})^2}{1 - r_{ij}^2}. \tag{3.7}$$

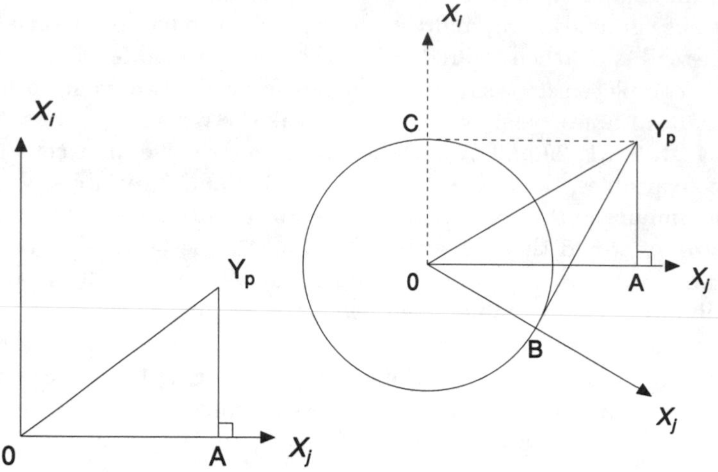


Figure 3.1 *Situations in which regression sums of squares are additive.*

Thus, $S_{i.j} = S_i$, i.e. the regression sums of squares are additive, when

$$1 - r_{ij}^2 = (1 - r_{ij}r_{jy}/r_{iy})^2.$$

A little rearrangement shows that this condition is satisfied either when $r_{ij} = 0$, i.e. $X_i$ and $X_j$ are orthogonal, or when

$$r_{ij} = 2r_{iy}r_{jy}/(r_{iy}^2 + r_{jy}^2). \tag{3.8}$$

Now let us look at what this solution means graphically. In Figure 3.1 (left), $OY_P$ represents the projection of vector $Y$ on the plane defined by variables $X_i$ and $X_j$. $OA$ is the projection of $Y_P$ onto direction $X_j$ and hence its square is $S_j$. Similarly the square of the length of $AY_P$ is the incremental sum of squares, $S_{i.j}$, when variable $X_i$ is added after $X_j$. Now let us work out where $X_i$ must be in this plane so that the projection of $Y_P$ onto $X_i$ is of length equal to $AY_P$. Draw a circle with the centre at $O$ and the radius equal to the length of $AY_P$, then draw tangents to the circle from the point $Y$. There are two such tangents; the one to point $B$ represents the solution (3.8), while the other shown by a broken line in Figure 3.1 (right) is the case in which $X_i$ and $X_j$ are orthogonal. Additional solutions can be obtained by reversing the directions of $X_i$ or $X_j$ or both.

We have only shown so far that if variables $X_i$ and $X_j$ satisfy (3.8), then their regression sums of squares are additive. Does this also apply to three variables $X_i$, $X_j$ and $X_k$, for which (3.8) is satisfied for all three pairs of variables? In other words, is the regression sum of squares for all three variables equal to the sum of their individual sums of squares? The answer is yes. This can be shown by using the methods above to derive the incremental sum of squares for variable $X_i$ after $X_j$ and $X_k$, and then substituting the three conditions for $r_{ij}$, $r_{ik}$ and $r_{jk}$ obtained from (3.8) for pairwise additivity. This method of proof is extremely tedious and it would seem that there should be a simpler geometric argument which will allow the result to be extended recursively to any number of variables that either satisfy (3.8) or are orthogonal for all pairs. We note that in the 2-variable case illustrated in Figure 3.1 (right), $OY_P$ is the diameter of a circle on which A and B both lie. In the 3-variable case, all of the points of projection lie on a sphere. Note also that if either $X_i$ or $X_j$, but not both, is reflected about $OY_P$, then the two $X$-vectors are orthogonal.

This method is still an untried suggestion for reducing the computational load when there are large numbers of variables from which to select. In practice, satisfying (3.8) would have to be replaced with approximate satisfaction of the condition, and limits found for the deviation from perfect additivity of the regression sums of squares. The author is grateful to Ruben Gabriel for stimulating discussion of this idea, but leaves it to others to develop further.

## 3.10 Ridge regression and other alternatives

One technique which has attracted a considerable amount of interest is the ridge regression technique of Hoerl and Kennard (1970a,b). They suggested that, using all the available variables, biased estimators, $b(d)$, of the regression coefficients be obtained using

$$\boldsymbol{b(d)} = (\boldsymbol{X'X} + d\boldsymbol{I})^{-1}\boldsymbol{X'y} \tag{3.9}$$

for a range of positive values of the scalar $d$. They recommended that the predictor variables should first be standardized to have zero mean and so that the sum of squares of elements in any column of $\boldsymbol{X}$ should be one, i.e. that $\boldsymbol{X'X}$ should be replaced with the correlation matrix. $b(d)$ is then plotted against $d$; this plot was termed the 'ridge trace'. Visual examination of the trace usually shows some regression coefficients which are 'stable', that is, they only change slowly, and others which either rapidly decrease or change sign. The latter variables are then deleted.

The basic justification for ridge regression is similar to that for subset selection of trading off bias against variance. For small values of $d$ the amount of bias may be very small while the reduction in variance is very substantial. This applies particularly when the SSP-matrix is very ill-conditioned. To understand what is happening when we use ridge estimation, it is useful to look at the singular-value decomposition (s.v.d.),

$$\underset{n \times k}{\boldsymbol{X}} = \underset{n \times k}{\boldsymbol{U}} \; \underset{k \times k}{\boldsymbol{\Lambda}} \; \underset{k \times k}{\boldsymbol{V}}$$

where the columns of $\boldsymbol{U}$ and $\boldsymbol{V}$ are orthogonal and normalized, that is $\boldsymbol{U'U} = \boldsymbol{I}$ and $\boldsymbol{V'V} = \boldsymbol{I}$, and $\boldsymbol{\Lambda}$ is a diagonal matrix with the singular values on the diagonal. Then

$$X'X = V'\Lambda^2V.$$

In terms of the s.v.d., the biased estimates of the regression coefficients are

$$b(d) = V'(\Lambda^2 + dI)^{-1}\Lambda U'y$$

with covariance matrix

$$\begin{aligned} &= \sigma^2 V'(\Lambda^2 + dI)^{-2}\Lambda^2 V \\ &= \sigma^2 V'.diag\{\lambda_i^2/(\lambda_i^2 + d)^2\}.V, \end{aligned}$$

where $\lambda_i$, $i$=1, 2, ..., $k$ are the singular values. The smallest singular values dominate the variance, but adding $d$ reduces their contribution substantially. For instance, for Jeffers' (1967) example on the compressive strength of pit-props, the singular values range from 2.05 to 0.20 so that for the least-squares estimator ($d = 0$), the covariance matrix of the standardized regression coefficients is

$$\sigma^2 V' \begin{pmatrix} 0.237 & & & & \\ & 0.421 & & & \\ & & \cdots & & \\ & & & 24.4 & \\ & & & & 25.6 \end{pmatrix} V,$$

whereas the covariance matrix for the ridge estimators when $d = 0.05$ is

$$\sigma^2 V' \begin{pmatrix} 0.232 & & & & \\ & 0.403 & & & \\ & & \cdots & & \\ & & & 4.95 & \\ & & & & 4.02 \end{pmatrix} V.$$

There has been a substantial amount of literature on the use of ridge regression (and on other biased estimators) when all $k$ variables are retained. See for instance Lawless (1978), Hocking et al. (1976), Lawless and Wang (1976), Hoerl et al. (1975), McDonald and Galerneau (1975), and the references given in section 1.3. In comparison, little attention has been paid to the idea of using the ridge trace to reduce the number of variables. We shall see later that the use of ridge regression and subset selection can not only reduce variance but can reduce the bias which the selection introduces.

Feiveson (1994) has given two algorithms for performing ridge regression analyses, particularly for the case in which the number of available predictors exceeds the number of cases. He uses the normal equations unfortunately and surprisingly as the paper is published in a numerical analysis journal.

One situation in which ridge regression will not perform well when used for subset selection is when $Y$ is strongly related to some linear combination of the predictor variables, such as $X_1 - X_2$, but has only small correlations with the individual variables. In the example given in section 3.2, ridge regression leads to the deletion of both $X_1$ and $X_2$. For this case, the squares of the singular values are 1.99999895, 1.0 and 0.00000105. The row of $\boldsymbol{V}$ corresponding to the smallest singular value is almost exactly equal to $(\boldsymbol{X_1} - \boldsymbol{X_2})/\sqrt{2}$, and once $d$ exceeds about $10^{-5}$, the regression coefficient for $X_3$ dominates.

If, instead of defining our biased estimators as in (3.9), we had used

$$\tilde{d} = (1 + d)(\boldsymbol{X'X} + d\boldsymbol{I})^{-1}\boldsymbol{X'y} ,$$

where the $X$ and $Y$ variables are normalized as before, then as $d \to \infty$, the individual elements of $\tilde{d}$ tend to the regression coefficients which are obtained if $Y$ is regressed separately against each $X$-variable one at a time. This means that ridge regression used as a selection technique tends to select those variables which both (i) yield regression coefficients with the same signs in single variable regressions and with all variables in the regression, and (ii) which show up early in forward selection.

There are two popular choices for the value of $d$ in (3.9). These are

$$d = k\hat{\sigma^2}/(\hat{\boldsymbol{\beta}}'\hat{\boldsymbol{\beta}}) \tag{3.10}$$

and

$$d = k\hat{\sigma^2}/R^2, \tag{3.11}$$

where $\hat{\beta}$ is the vector of LS regression coefficients, and $\hat{\sigma^2}$ is the estimated residual variance, when the $X$ and $Y$ variables have been scaled to have zero means and unit standard deviations, and $R^2$ is the usual coefficient of determination for LS regression. If the true values for $\sigma$ and $\boldsymbol{\beta}$ are known, then (3.10) minimizes the mean squared error of $\hat{\boldsymbol{\beta}}$ (Hoerl et al. 1975). Lawless and Wang (1976) suggested estimator (3.11). An algorithm for choosing $d$ which can use either (3.10) or (3.11) has been given by Lee (1987).

Suppose we write $\boldsymbol{\alpha} = \boldsymbol{V\beta}$ so that our regression model is

$$Y = \boldsymbol{XV\alpha} + \epsilon,$$

where the columns of $\boldsymbol{XV'}$ are orthogonal. The ridge estimates of the elements of $\boldsymbol{\alpha}$ are

$$\alpha_i(d) = \frac{\lambda_i^2}{\lambda_i^2 + d}.\hat{\alpha}_i,$$

where $\hat{\alpha}_i$ is the LS estimate. This can be generalized by replacing $d$ with $d_i$. Hocking et al. (1976) show that the values of $d_i$ which minimize either the mean squared error

$$E(\boldsymbol{\beta} - \hat{\boldsymbol{\beta}})'(\boldsymbol{\beta} - \hat{\boldsymbol{\beta}})$$

or the mean squared error of prediction

$$E(\boldsymbol{\beta} - \hat{\boldsymbol{\beta}})'\boldsymbol{X'X}(\boldsymbol{\beta} - \hat{\boldsymbol{\beta}})$$

are given by

$$d_i = \sigma^2/\alpha_i^2. \tag{3.12}$$

Of course, both $\sigma^2$ and $\alpha_i^2$ will be unknown in practice. If $\alpha_i$ is replaced first with the LS estimate, $\hat{\alpha}_i$, and then with successive ridge estimates, $\alpha_i(d_i)$,

Hemmerle (1975) has shown that these estimates converge to

$$\alpha_i^* = \begin{cases} 0 & \text{if } t_i^2 \leq 4 \\ \{1/2 + (1/4 - 1/t_i^2)^{1/2}\}\hat{\alpha}_i & \text{otherwise} \end{cases} \tag{3.13}$$

where $t_i = \hat{\alpha}_i$ / s.e.$(\hat{\alpha}_i)$. That is, $t_i$ is the usual $t$-statistic in a multiple regression; in this case it is the $t$-value for the $i^{th}$ principal component of $\boldsymbol{X}$. Thus the application of an iterative ridge regression procedure can be equivalent to a subset selection procedure applied to principal components, followed by some shrinkage of the regression coefficients.

A further ridge estimator of this kind is

$$\alpha_i^* = \begin{cases} 0 & \text{if } t_i^2 \leq 1 \\ (1 - 1/t_i^2)\hat{\alpha}_i & \text{otherwise.} \end{cases} \tag{3.14}$$

This has been proposed by Obenchain (1975), Hemmerle and Brantle (1978), and Lawless (1978).

If $Y$ is predicted well by a linear combination of a small number of eigenvectors, then these eigenvectors can be regarded as new variables. Unfortunately though, these eigenvectors are linear combinations of all the variables, which means that all of the variables will need to be measured to use the predictor. This may be defeating the purpose of selecting a subset.

## 3.11 The nonnegative garrote and the lasso

The nonnegative garrote (or nn-garrote) and the lasso (least absolute shrinkage and selection operator) were introduced about the same time by Breiman (1995) and Tibshirani (1996), respectively. They both impose a limit on the sum of absolute values of the regression coefficients, and lower this limit (tighten the garrote) until some kind of optimum is reached. (N.B. The spelling used here for garrote is that used in the Breiman paper. Tibshirani uses the spelling 'garotte', and some dictionaries also allow 'garrotte'.)

Before any method such as this can be used, there must be some thought about scaling. If for any particular data set, we have some least-squares regression coefficients with values around 100 and others around 0.01, then a change of, say 1, in the coefficients in the first set will have very little effect upon the fit, while a change of the same size for the second set will have a dramatic effect. If we find the least-squares fit subject to a constraint on the sum of the absolute values of the regression coefficients, then the large coefficients will be shrunk substantially while the small ones will be almost unchanged. Breiman and Tibshirani use different approaches to this scaling problem.

If the least-squares regression coefficients for the full model are denoted as $b_j$, $j = 1, \ldots, k$ then the nn-garrote minimizes

$$\sum_i \left( y_i - \sum_j c_j b_j x_{ij} \right)^2$$

subject to the constraints

$$\begin{aligned} c_j &\geq 0 \\ \sum_j c_j &\leq s \end{aligned}$$

for some $s$. Notice that a different shrinkage factor is being applied to each coefficient. If $k$ is the number of predictors, including the intercept term if one is being fitted, then we are only interested in values of $s < k$.

The method of fitting the garrote employed by Breiman was by adapting Lawson and Hanson's (1974) NNLS (nonnegative least squares) code. This well-known code in Fortran, which can be freely downloaded from the *lawson-hanson* directory at *http://www.netlib.org*, handles the constraint that the $c_j$'s most be nonnegative. It does not handle the constraint on the sum of the $c_j$'s. Breiman has made a small package of Fortran subroutines available for the computations. These can be downloaded as file *subgar.f* from *stat-ftp.berkeley.edu* in directory */pub/user/breiman.* In this file it is stated that a barrier method is used to impose the constraint on the sum of the $c_j$'s.

The method used in the lasso is to first normalize each predictor $X_j$ by subtracting the mean and scaling so that $\sum_i^n x_{ij}^2 = n$, where $n$ is the number of cases. Then the sum:

$$\sum_i \left( y_i - \sum_j \beta_j x_{ij} \right)^2$$

is minimized subject to the constraint

$$\sum_j | \beta_j | \leq s$$

for some $s$. Here, we are interested in values of $s$ which are less than the sum of absolute values of the least-squares regression coefficients for the same normalized $X$-variables.

This can be thought of as a quadratic programming problem with linear inequality constraints. It has $2^k$ linear constraints:

$$\begin{aligned} \beta_1 + \beta_2 + \beta_3 + \; ... \; + \beta_k &\leq s \\ -\beta_1 + \beta_2 + \beta_3 + \; ... \; + \beta_k &\leq s \\ \beta_1 - \beta_2 + \beta_3 + \; ... \; + \beta_k &\leq s \\ -\beta_1 - \beta_2 + \beta_3 + \; ... \; + \beta_k &\leq s \\ \beta_1 + \beta_2 - \beta_3 + \; ... \; + \beta_k &\leq s; \end{aligned}$$

that is, with every combination of signs in front of the $\beta$'s. In practice, it is only necessary to use a very small number of these constraints. Starting with the unconstrained least-squares values for the $\beta$'s, the signs in the first constraint are those of the initial $\beta$'s. As the sum, $s$, is reduced, more constraints are added as the $\beta$'s change sign. However, only one of these constraints is active for any value of $s$. A carefully coded algorithm can progress from one value of $s$ to the next lower at which a coefficient $\beta_j$ is reduced to zero and make sure that the correct $\beta_j$'s are being kept at zero and the correct linear constraint is being applied.

Public domain source code for constrained least-squares is available from several sources (mainly in Fortran) for solving such problems. Code available from netlib includes the Lawson and Hanson code used by Tibshirani, contained in their book and also in TOMS algorithm 587, routine DQED by Hanson and Krogh, and routine DGGLSE from Lapack. Software for the complete lasso algorithm is available for the S-plus package from the S-Archive at statlib, which is at:

`http://lib.stat.cmu.edu`

A detailed report on the lasso, Tibshirani (1994), can be downloaded from:

`http://www-stat.stanford.edu/~tibs/research.html`

Alternatively, the problem may be considered as one in linear least squares with just one nonlinear constraint. The ENLSIP package could probably be used for this approach, though the author has not tried this. ENLSIP is for nonlinear least-squares fitting with nonlinear equality or inequality constraints. The Fortran source code can be downloaded from:

`ftp://ftp.cs.umu.se/pub/users/perl/ENLSIP.tar.gz`

A report by Wedin and Lindstroem (1988, revised 1993) can be downloaded from:

`http://www.cs.umu.se/~perl/reports/alg.ps.gz`

A further alternative algorithm is that proposed by Osborne et al. (2000), which appears to be very efficient for handling problems with large number of predictors, particularly for small values of the constrained sum when many of the regression coefficients have been shrunk to zero values.

The author's own algorithm for the lasso is described in the appendix to this chapter.

With both the garrote and the lasso, as $s$ is reduced, most of the regression coefficients shrink with progressively more being shrunk to zero. Thus the methods are effectively selecting subsets of the variables. Quite often a few of the coefficients increase in value initially. The coefficients cannot change sign with the garrote; they can with the lasso. Table 3.9 shows the values of $s$ at which different variables are dropped using the garrote for the POLLUTE data set, and for the OZONE data set used by Breiman. Notice that when the predictors are correlated, it is common for more than one variable to drop out at a time.

As a subset selection procedure, the garrote behaves similarly to backward elimination. For the POLLUTE data, backward elimination drops variable 14 (sulphur dioxide concentration) fairly early, so does the garrote. This variable appears in many of the best-fitting subsets.

Table 3.9 *Values of s at which variables are dropped using the garrote*

| POLLUTE data set (15 vars. + const.) | | OZONE data set (44 variables + constant) | | | |
|---|---|---|---|---|---|
| $s$ | Variable(s) | $s$ | Variable(s) | $s$ | Variable(s) |
| 14.6 | 11 | 43.7 | 12 | 15.8 | 8, 38 |
| 13.5 | 15 | 42.2 | 10 | 14.8 | 41 |
| 12.9 | 10 | 41.8 | 16 | 13.6 | 42 |
| 11.1 | 7 | 40.9 | 19 | 12.7 | 14 |
| 9.4 | 4 | 39.2 | 30 | 11.6 | 27 |
| 9.2 | 8 | 36.6 | 23 | 10.1 | 43 |
| 8.5 | 14 | 36.3 | 31 | 9.5 | 29 |
| 6.2 | 3 | 34.7 | 1, 2 | 8.9 | 11 |
| 5.7 | 5 | 32.9 | 6 | 7.9 | 32 |
| 4.9 | 2, 6 | 29.5 | 20, 28 | 7.5 | 40 |
| 4.1 | 1 | 28.4 | 39 | 6.6 | 22 |
| 3.0 | 9 | 26.5 | 3, 18 | 6.4 | 24 |
| 1.8 | 12 | 25.7 | 13 | 5.3 | 35 |
| | Leaving | 24.4 | 21, 44 | 4.7 | 9 |
| | 13 + const. | 23.1 | 26 | 3.8 | 36 |
| | | 21.4 | 17 | 3.1 | 33 |
| | | 19.9 | 25 | 2.4 | 15 |
| | | 18.4 | 5 | 1.9 | 4 |
| | | 17.6 | 37 | | Leaving |
| | | | | | 7 + const. |

This is the first time that the OZONE data set has been used here. The data consist of 8 basic meteorological variables measured on 330 days in 1976 in Los Angeles, together with one dependent variable, the maximum one-hour average ozone concentration. The first 8 variables are:

$X_1$: 500 mb height
$X_2$: wind speed
$X_3$: humidity
$X_4$: surface temperature
$X_5$: inversion height
$X_6$: pressure gradient
$X_7$: inversion temperature
$X_8$: visibility

These variables have been centered and then the 36 remaining variables have been constructed as squares and interactions in the order: $X_1^2$, $X_1X_2$, $X_2^2$, $X_1X_3$, $X_2X_3$, $X_3^2$, ..., $X_8^2$. Each of these constructed variables was centered before use. (N.B. This data set has been used in a number of publications by

Table 3.10 *Subset selection applied to the OZONE data.*

| No. vars. | Forward selection variables (RSS) | Backward elimin. variables (RSS) | Best found variables (RSS) |
|---|---|---|---|
| 1 | 4 (8246.) | 7 (9378.) | 4 (8246.) |
| 2 | 4, 12 (7165.) | 3, 7 (7429.) | 4, 12 (7165.) |
| 3 | 4, 5, 12 (6387.) | 3, 7, 32 (6140.) | 3, 7, 32 (6140.) |
| 4 | 4, 5, 12, 33 (5855.) | 3, 7, 22, 32 (5664.) | 6, 7, 29, 32 (5565.) |
| 5 | 3, 4, 5, 12, 33 (5466.) | 3, 7, 14, 22, 32 (5308.) | 6, 7, 22, 29, 32 (5186.) |
| 6 | 3, 4, 5, 12, 32, 33 (5221.) | 3, 7, 14, 22, 29, 32 (5174.) | 3, 6, 7, 22, 29, 32 (5039.) |
| 7 | 3, 4, 5, 7, 12, 32, 33 (5118.) | 3, 7, 14, 22, 26, 29, 32 (5001.) | 3, 6, 7, 14, 22, 29, 32 (4984.) |
| 8 | 3, 4, 5, 7, 12, 29, 32, 33 (4986.) | 3, 4, 7, 14, 22, 26, 29, 32 (4896.) | 3, 4, 7, 14, 26, 29, 32, 33 (4883.) |

several authors. The order of the variables is not consistent from one source to another.)

The results shown in Table 3.9 are from using the full data set. Breiman used cross-validation, leaving out 10% of the data at a time, to choose the value of the sum $s$ of the scaling factors for the regression coefficients, and hence his results are not directly comparable to those here. There will be more discussion of cross-validation in later chapters.

Table 3.10 shows the subsets of variables picked from the OZONE data using forward selection and backward elimination, together with the best-fitting subsets so far found by the author using a range of subset selection algorithms. For this data, the Efroymson stepwise procedure using $F$-to-enter and $F$-to-drop both equal to 4.0, selected the subset of 7 variables (plus a constant) numbered 3, 4, 5, 7, 29, 32 and 33 which gives a residual sum of squares of 4986, which is very close to the best that has been found for a subset of 7 variables. If we look at the last 8 variables left in by the garrote we see that they are exactly the same last 8 variables left in by backward elimination. Note also that variable number 6, which features in many of the best-fitting subsets, is eliminated fairly early by the garrote. It is also eliminated early by backward elimination.

Table 3.11 shows the values of the constraint sum, $s$, at which the variables are dropped, and in one case reintroduced, for the POLLUTE data set using the lasso. The value '4.43' shown on the first line of the table is the sum of absolute values of the regression coefficients for the unconstrained least-squares fit. For this data set, only one variable was introduced again after it had been dropped, and none of the regression coefficients changed sign.

Figure 3.2 shows the regression coefficients for 11 out of the 15 predictors, against the value of $s$, for the POLLUTE data set using the lasso. We note that variables numbered 1, 2, 6, 9 and 14 remain in the regression until the

Table 3.11 *Values of s at which variables are dropped (and reintroduced) using the lasso applied to the POLLUTE data set.*

| Value of $s$ | Variable dropped | Variable reintroduced | Value of $s$ | Variable dropped | Variable reintroduced |
|---|---|---|---|---|---|
| 4.43 | (none) | | 1.24 | 7 | |
| 3.87 | 11 | | 0.95 | 8 | |
| 2.71 | 15 | | 0.83 | 2 | |
| 2.27 | 13 | | 0.39 | 1 | |
| 1.97 | 4 | | 0.36 | 14 | |
| 1.81 | 12 | 15 | 0.17 | 6 | |
| 1.68 | 5 | | 0.00 | 9 | |
| 1.63 | 15 | | | | |
| 1.60 | 10 | | | | |
| 1.29 | 3 | | | | |

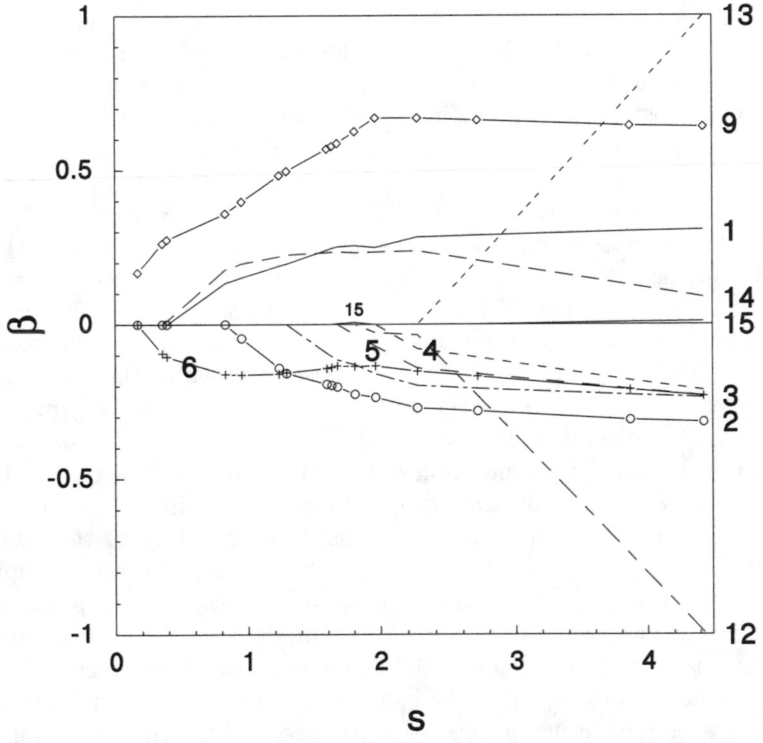


Figure 3.2 *Regression coefficients for most of the standardized variables against the value, s, of the constraint for the POLLUTE data set using the lasso*

Table 3.12 *Tibshirani examples*

| Example | Predictors | nonzero $\beta$'s |
|---|---|---|
| 1 | 8 | 3 |
| 2 | 8 | 8 |
| 3 | 8 | 1 |
| 4 | 40 | 20 |

value of $s$ is very small. These variables appear in nearly all of the best-fitting subsets containing small numbers of predictors.

The lasso applies its scaling to the $X$-variables whereas the garrote scales the regression coefficients. In the physical sciences, it is fairly common to find that the dependent variable is strongly related to differences between variables, say to the difference $(X_1 - X_2)$. In such cases, in using the lasso it will be sensible to centre and scale $(X_1 - X_2)$ rather than the individual variables. As the differences between the variables will often be an order of magnitude smaller than the individual centered variables, the regression coefficient of $(X_1 - X_2)$ is likely to be very large. As the objective is to shrink the sum of absolute values of the regression coefficients, the coefficients of such differences will be shrunk substantially and such variables may be eliminated very early.

A very important aspect of both the garrote and the lasso, which has not been mentioned so far, is the 'stopping rule' or the method of deciding how much shrinkage to apply. Breiman and Tibshirani consider several methods. These include cross-validation; that is, the garrote or lasso is applied to most of the data set, say 80% to 90%, and the remaining 20% to 10% of the data are predicted. This is repeated many times. The shrinkage factor that yields the best predictions of the omitted data is then chosen. Another method is that of the 'little bootstrap' in which extra noise is added to the dependant variable to see how well the predictions fit the noisy values. These topics will be discussed in more detail in later chapters.

Tibshirani (1996) presents simulation results which compare the garrote, the lasso, ridge regression, and all-subsets regression with 5-fold cross-validation used as the stopping rule, together with least-squares fitting of the model containing all of the predictors. The numbers of predictors in the four examples were as shown in Table 3.12. In all cases, the predictor variables were positively correlated. Except for example 2, in which all of the true $\beta$'s were nonzero, the lasso usually picked too many variables, but nearly always included the true model. The garrote usually picked smaller models than the lasso. Ridge regression was best in terms of mean-squared prediction error for examples 2 and 4 but worst for examples 1 and 3. All-subsets regression was worst for example 2. In terms of mean-squared prediction error, there was not much difference between the lasso and the garrote.

There have been several extensions and applications in the literature. Fu

Table 3.13 *Florida cloud-seeding data*

| Date | $X_1$ | $X_2$ | $X_3$ | $X_4$ | $X_5$ | $Y$ |
|---|---|---|---|---|---|---|
| 1 July 1971 | 2.0 | 0.041 | 2.70 | 2 | 12 | 0.32 |
| 15 July 1971 | 3.0 | 0.100 | 3.40 | 1 | 8 | 1.18 |
| 17 July 1973 | 3.0 | 0.607 | 3.60 | 1 | 12 | 1.93 |
| 9 Aug. 1973 | 23.0 | 0.058 | 3.60 | 2 | 8 | 2.67 |
| 9 Sept. 1973 | 1.0 | 0.026 | 3.55 | 0 | 10 | 0.16 |
| 25 June 1975 | 5.3 | 0.526 | 4.35 | 2 | 6 | 6.11 |
| 9 July 1975 | 4.6 | 0.307 | 2.30 | 1 | 8 | 0.47 |
| 16 July 1975 | 4.9 | 0.194 | 3.35 | 0 | 12 | 4.56 |
| 18 July 1975 | 12.1 | 0.751 | 4.85 | 2 | 8 | 6.35 |
| 24 July 1975 | 6.8 | 0.796 | 3.95 | 0 | 10 | 5.74 |
| 30 July 1975 | 11.3 | 0.398 | 4.00 | 0 | 12 | 4.45 |
| 16 Aug. 1975 | 2.2 | 0.230 | 3.80 | 0 | 8 | 1.16 |
| 28 Aug. 1975 | 2.6 | 0.136 | 3.15 | 0 | 12 | 0.82 |
| 12 Sept. 1975 | 7.4 | 0.168 | 4.65 | 0 | 10 | 0.28 |

(1998) studied the bridge in which $\sum |\beta_j|^\gamma$ is constrained, for $\gamma \geq 1$. $\gamma = 1$ corresponds to the lasso. The bridge had previously been proposed by Frank and Friedman (1993). Osborne et al. (1998) have used the lasso for knot selection in fitting splines. The S-Plus code for the lasso at statlib includes code for fitting generalized linear models and proportional hazards models.

## 3.12 Some examples

### *3.12.1 Example 1*

The data for this example are from a paper by Biondini et al. (1977). In a cloud-seeding experiment, daily target rainfalls were measured as the dependent variable. An equation was required for predicting the target rainfall on the days on which clouds were seeded, by using equations developed from the data for days on which clouds were not seeded. If seeding affected rainfall then the actual rainfalls on the seeded days would be very different from those predicted. Five variables were available for use, but by including linear, quadratic and interaction terms, the total number of variables was increased to 20. There were 58 observations in all, but these included those on seeded days. Also, the authors decided to subclassify the days according to radar echoes. Table 3.13 contains the data on the 5 variables and the rainfall for the largest subclassification of the data which contained 14 observations.

Table 3.14 shows the results obtained by using forward selection, sequential replacement, and exhaustive search algorithms on this data. The variables $X_6$ to $X_{20}$ were, in order, $X_1^2$ to $X_5^2$, $X_1X_2$, $X_1X_3$, $X_1X_4$, $X_1X_5$, $X_2X_3$, $X_2X_4$, $X_2X_5$, $X_3X_4$, $X_3X_5$, $X_4X_5$.

In this case, sequential replacement has only found one subset which fits

Table 3.14 *RSS's for subsets of variables for the Florida cloud-seeding data. (Numbers in brackets are the numbers of the selected variables)*

| No. of variables | Forward selection | Sequential replacement | Exhaustive search |
|---|---|---|---|
| Const. | 72.29 | 72.29 | 72.29 |
| 1 | 26.87<br>(1,5) | 26.87<br>(1,5) | 26.87<br>(1,5) |
| 2 | 21.56<br>(14,15) | 21.56<br>(14,15) | 21.56<br>(14,15) |
| 3 | 19.49<br>(14,1,5,17) | 19.49<br>(14,1,5,17) | 12.61<br>(9,1,7,20) |
| 4 | 11.98<br>(12,14,15,17) | 11.98<br>(12,14,15,17) | 11.49<br>(9,10,17,20) |
| 5 | 9.05<br>(6,12,1,4,15,17) | 8.70<br>(1,12,1,5,17,19) | 6.61<br>(1,2,6,12,15) |

better than those found using forward selection. We see that neither forward selection nor sequential replacement found the subset of variables 9, 17 and 20, which gives a much better fit than variables 14, 15 and 17. In fact, the five best-fitting subsets of 3 variables (see Table 3.16) all fit much better than those found by forward selection and sequential replacement.

We note also that the variables numbered from 11 upward involve products of two different original variables, i.e. they are interactions. The first single variables selected by forward selection and sequential replacement do not appear until we reach subsets of 5 variables, and then only variable $X_1$ or its square; single variables start appearing from subsets of 3 variables in the best subsets found from the exhaustive search.

How well does the Efroymson algorithm perform? That depends upon the values used for $F_d$ and $F_e$. If $F_e$ is greater than 2.71, it stops with only one variable selected; that is, variable 15. If $F_e$ is between 1.07 and 2.71, then the 2-variable subset of variables 14 and 15 is selected. If $F_e$ is less than 1.07, the algorithm proceeds as for forward selection. To find any of the good subsets of 3 variables, it would need to do some deletion. For instance, after the subset of variables 14, 15 and 17 has been found, a value of $F_d$ greater than 2.74 is needed to delete variable 14. This would mean that $F_d > F_e$, and this results in indefinite cycling.

Now let us look at the Hoerl and Kennard method. The ridge trace for these data is shown in Figure 3.3. For this figure, each variable was scaled and shifted

Table 3.15 *Replacement algorithm using random starts, applied to the Florida data*

| Subset of 3 variables | RSS | Frequency | Subset of 4 variables | RSS | Frequency |
|---|---|---|---|---|---|
| 9, 17, 20 | 12.61 | 41 | 9, 10, 17, 20 | 11.49 | 41 |
| 5, 10, 11 | 16.29 | 3 | 5, 10, 11, 16 | 11.85 | 3 |
| 14, 15, 17 | 19.49 | 52 | 12, 14, 15, 17 | 11.98 | 52 |
| 3, 8, 11 | 20.34 | 4 | 3, 8, 11, 19 | 17.70 | 4 |

so that it had zero mean and unit standard deviation. We will try to apply the criteria of Hoerl and Kennard, though these were not precisely spelt out in their paper. They would probably retain variables 3, 8, 12 and 14, which have consistently large regression coefficients for all values of $d$. We would certainly not consider variables 2, 11 or 19, as their regression coefficients change sign. There are many subsets of 4 variables which fit much better than variables 3, 8, 12 and 14.

A method which is usually less expensive than the exhaustive search is that of using a replacement algorithm starting from randomly chosen subsets of variables. Using 100 random starting subsets of three variables gave the following frequencies of subsets of three and four variables.

We note that the most frequently occurring stationary subset was not that which gave the best fit, though in this case the use of random starts has found the best-fitting subsets of both three and four variables.

So far, we have only looked at the best subsets of each size found by these various methods. Table 3.16 lists the five best-fitting subsets of each size from one to five variables. We notice that in most cases there is very little separating the best from the second or even the fifth-best. In cases such as this, we can expect very substantial selection bias if we were to calculate the least-squares regression coefficients. This will be examined in more detail in Chapter 5.

At this stage, we will make only a few minor comments on these results as the topic of inference is treated in the next chapter. We notice that there is a very big drop in $RSS$ from fitting a constant to fitting one variable, either $X_{15} = X_2X_3$ or $X_{11} = X_1X_2$, and only a gradual decline after that. This suggests that we may not be able to do better than use one of these two variables as the predictor of $Y$. A further practical comment is that rainfalls (variable $Y$ is rainfall) have very skew distributions, and it is common practice to use a transformation such as a cube-root or logarithm. It looks as if such a transformation would give improved fits in this case.

### *3.12.2 Example 2*

This is the STEAM data set of Chapter 2 (see Table 2.2 for the source). Table 3.17 shows how well forward selection, backward elimination, and sequential

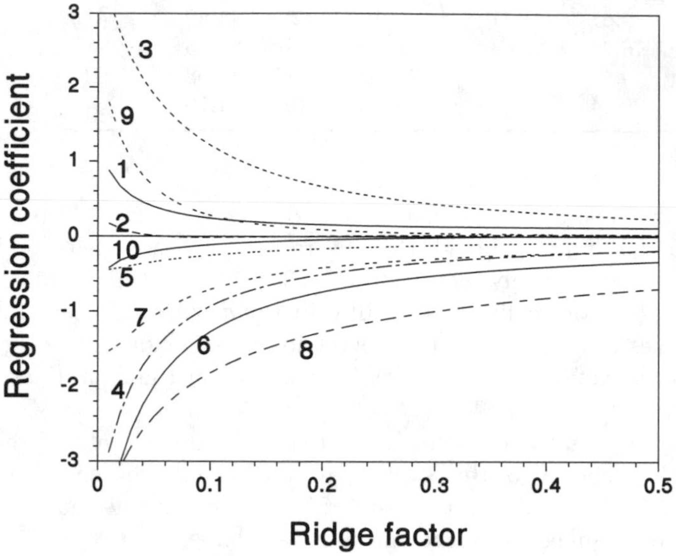


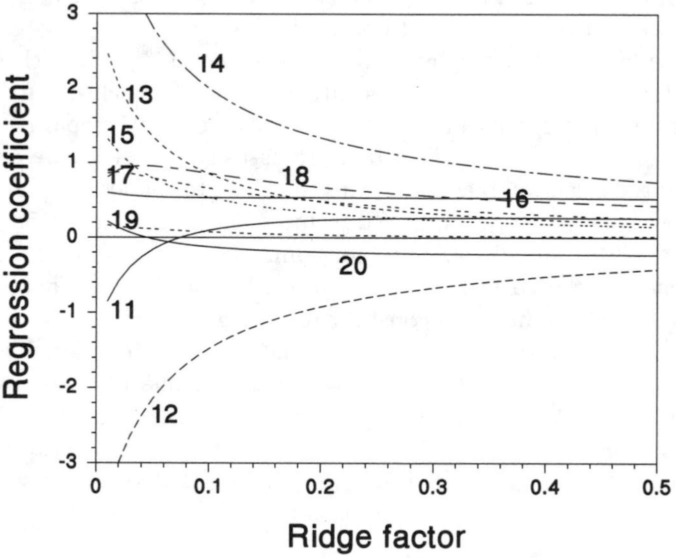


Figure 3.3 *Ridge trace for the CLOUDS data set. Upper figure is for variables* $X_1$ *to* $X_{10}$, *lower figure for variables* $X_{11}$ *to* $X_{20}$.

Table 3.16 *Five best-fitting subsets of sizes from one to five variables for the cloud-seeding data*

| No. of variables | RSS | Variables |
|---|---|---|
| 1 | 26.87 | 15 |
| | 27.20 | 11 |
| | 32.18 | 2 |
| | 34.01 | 7 |
| | 42.99 | 17 |
| 2 | 21.56 | 15, 14 |
| | 21.81 | 15, 1 |
| | 22.29 | 15, 12 |
| | 22.73 | 15, 6 |
| | 23.98 | 15, 13 |
| 3 | 12.61 | 9, 17, 20 |
| | 15.56 | 2, 9, 20 |
| | 16.12 | 9, 15, 20 |
| | 16.29 | 5, 10, 11 |
| | 17.24 | 7, 9, 20 |
| 4 | 11.49 | 9, 10, 17, 20 |
| | 11.63 | 5, 9, 17, 20 |
| | 11.77 | 8, 9, 17, 20 |
| | 11.85 | 5, 10, 11, 16 |
| | 11.97 | 2, 9, 10, 20 |
| 5 | 6.61 | 1, 2, 6, 12, 15 |
| | 8.12 | 9, 12, 14, 15, 20 |
| | 8.44 | 1, 2, 12, 13, 15 |
| | 8.70 | 1, 12, 15, 17, 19 |
| | 8.82 | 1, 3, 6, 8, 13 |

replacement performed in relation to the best-fitting subsets of each size obtained by exhaustive search. We note that for this data set there are more observations than variables and backward elimination is feasible.

For this set of data, forward selection, backward elimination and sequential replacement have all performed fairly well, with sequential replacement having found the best-fitting subsets of all sizes. We notice that the $RSS$ is almost constant for three or more variables for the best-fitting subsets of each size. This suggests that the best predictor is one containing not more than three variables. The $RSS$ with all 9 variables in the model is 4.87, and this has 15 degrees of freedom. Dividing the $RSS$ by its number of degrees of freedom

Table 3.17 *RSS's for subsets of variables selected using various procedures for the STEAM data. (Numbers in brackets are the numbers of the selected variables)*

| No. of variables | Forward selection | Backward elimination | Sequential replacement | Exhaustive search |
|---|---|---|---|---|
| Const. | 63.82 | 63.82 | 63.82 | 63.82 |
| 1 | 18.22<br>(7) | 18.22<br>(7) | 18.22<br>(7) | 18.22<br>(7) |
| 2 | 8.93<br>(1,7) | 8.93<br>(1,7) | 8.93<br>(1,7) | 8.93<br>(1,7) |
| 3 | 7.68<br>(1,5,7) | 7.68<br>(1,5,7) | 7.34<br>(4,5,7) | 7.34<br>(4,5,7) |
| 4 | 6.80<br>(1,4,5,7) | 6.93<br>(1,5,7,9) | 6.80<br>(1,4,5,7) | 6.80<br>(1,4,5,7) |
| 5 | 6.46<br>(1,4,5,7,9) | 6.54<br>(1,5,7,8,9) | 6.41<br>(1,2,5,7,9) | 6.41<br>(1,2,5,7,9) |

gives a residual variance estimate of 0.32, and this is of the same order of magnitude as the drops in $RSS$ from three to four variables, and from four to five variables. If we regressed the least-squares residuals from fitting the first three variables against a variable which consisted of a column of random numbers, we would expect a drop in $RSS$ of about 0.32.

The five best-fitting subsets of one, two, and three variables are shown in Table 3.18. We see that there is no competition in the case of a single variable, that there are three close subsets of two variables, and at least five close subsets of three variables. If least-squares estimates of regression coefficients were used in a predictor, very little bias would result if the best subset of either one or two variables were used. There would be bias if the best subset of three variables were used.

### *3.12.3 Example 3*

The data for this example are for the DETROIT data set (see Table 2.2 for the source). The data are annual numbers of homicides in Detroit for the years 1961 to 1973, and hence contain 13 observations. As there are 11 predictor variables available, there is only one degree of freedom left for estimating the residual variance if a constant is included in the model. Table 3.19 shows the performance of forward selection, sequential replacement, backward elimination, and exhaustive search on this data.

Table 3.18 *Five best-fitting subsets of one, two, and three variables for the STEAM data*

| No. of variables | RSS | Variables |
|---|---|---|
| 1 | 18.22 | 7 |
| | 37.62 | 6 |
| | 45.47 | 5 |
| | 49.46 | 3 |
| | 53.88 | 8 |
| 2 | 8.93 | 1, 7 |
| | 9.63 | 5, 7 |
| | 9.78 | 2, 7 |
| | 15.60 | 4, 7 |
| | 15.99 | 7, 9 |
| 3 | 7.34 | 4, 5, 7 |
| | 7.68 | 1, 5, 7 |
| | 8.61 | 1, 7, 9 |
| | 8.69 | 1, 4, 7 |
| | 8.71 | 5, 7, 8 |

In this example, none of the 'cheap' methods has performed well, particularly in finding the best-fitting subsets of three and four variables. For larger subsets, sequential replacement has been successful. Backward elimination dropped variable number 6 first, and this appears in many of the better-fitting subsets, while it left in variables 3, 8 and 10, which appear in few of the best subsets, until a late stage. In the experience of the author, the so-called 'cheap' methods of variable selection usually perform badly when the ratio of the number of observations to the number of variables is less than or close to one. In such cases, the best-fitting subset of $p$ variables often does not contain the best-fitting subset of $p-1$ variables; sometimes they have no variables in common, and methods which add or drop one variable at a time either cannot find the best-fitting subsets or have difficulty in finding them. One remarkable feature of this data set is that the first variable selected by forward selection, variable 6, was the first variable deleted in the backward elimination.

Table 3.20 shows the five best-fitting subsets of each size up to five variables. The subsets of three variables look extraordinary. Not merely is the subset of variables 2, 4, and 11 far superior to any other subset of three variables, but no subsets of one or two of these three variables appear in the best-fitting subsets of one or two variables. Variable 2 has the lowest correlation in absolute value with the dependent variable; this correlation is 0.21, the next smallest is 0.55,

Table 3.19 *RSS's for subsets of variables for the DETROIT data set. (Numbers in brackets are the numbers of the selected variables)*

| No. of vars. | Forward selection | Backward elimination | Sequential replacement | Exhaustive search |
|---|---|---|---|---|
| Const. | 3221.8 | 3221.8 | 3221.8 | 3221.8 |
| 1 | 200.0<br>(6) | 680.4<br>(11) | 200.0<br>(6) | 200.0<br>(6) |
| 2 | 33.83<br>(4,6) | 134.0<br>(4,11) | 33.83<br>(4,6) | 33.83<br>(4,6) |
| 3 | 21.19<br>(4,6,10) | 23.51<br>(3,4,11) | 21.19<br>(4,6,10) | 6.77<br>(2,4,11) |
| 4 | 13.32<br>(1,4,6,10) | 10.67<br>(3,4,8,11) | 13.32<br>(1,4,6,10) | 3.79<br>(2,4,6,11) |
| 5 | 8.20<br>(1,2,4,6,10) | 8.89<br>(3,4,7,8,11) | 2.62<br>(1,2,4,9,11) | 2.62<br>(1,2,4,9,11) |
| 6 | 2.38<br>(1,2,4,6,10,11) | 6.91<br>(3,4,7,8,9,11) | 1.37<br>(1,2,4,6,7,11) | 1.37<br>(1,2,4,6,7,11) |

and most of the others exceed 0.9 in absolute value. The $RSS$'s for subsets from these three variables are shown in Table 3.21.

These all compare very unfavourably with the better-fitting subsets of one and two variables. Is this just a case of remarkable overfitting?

In view of the lack of competition from other subsets, the bias in the regression coefficients is likely to be very small if this subset of three variables is used for prediction with least-squares estimates for the parameters. The three variables in this case are % unemployed (variable 2), number of handgun licences per 100,000 population (variable 4), and average weekly earnings (variable 11). All three regression coefficients are positive.

### *3.12.4 Example 4*

The data for this example is the POLLUTE data set (see Table 2.2 for the source). The dependent variable is an age-adjusted mortality rate per 100,000 population. The data are for 60 Metropolitan Statistical Areas in the United States. The predictor variables include socioeconomic, meteorological, and pollution variables.

Table 3.22 shows the subsets of variables selected by forward selection, backward elimination, sequential replacement, and exhaustive search. In this case,

Table 3.20 *Five best-fitting subsets of each size from one to five variables for the DETROIT data*

| No. of variables | RSS | Variables |
|---|---|---|
| 1 | 200.0 | 6 |
| | 227.4 | 1 |
| | 264.6 | 9 |
| | 277.7 | 8 |
| | 298.3 | 7 |
| 2 | 33.83 | 4, 6 |
| | 44.77 | 2, 7 |
| | 54.45 | 1, 9 |
| | 55.49 | 5, 6 |
| | 62.46 | 3, 8 |
| 3 | 6.77 | 2, 4, 11 |
| | 21.19 | 4, 6, 10 |
| | 23.05 | 1, 4, 6 |
| | 23.51 | 3, 4, 11 |
| | 25.04 | 4, 6, 11 |
| 4 | 3.79 | 2, 4, 6, 11 |
| | 4.58 | 1, 2, 4, 11 |
| | 5.24 | 2, 4, 7, 11 |
| | 5.41 | 2, 4, 9, 11 |
| | 6.38 | 2, 4, 8, 11 |
| 5 | 2.62 | 1, 2, 4, 9, 11 |
| | 2.64 | 1, 2, 4, 6, 11 |
| | 2.75 | 1, 2, 4, 7, 11 |
| | 2.80 | 2, 4, 6, 7, 11 |
| | 3.12 | 2, 4, 6, 9, 11 |

Table 3.21 *RSS's for combinations of variables numbered 2, 4, and 11 from the DETROIT data set*

| Variable | 2 | 4 | 11 | 2,4 | 2,11 | 4,11 |
|---|---|---|---|---|---|---|
| RSS | 3080 | 1522 | 680 | 1158 | 652 | 134 |

Table 3.22 *RSS's for subsets of variables for the POLLUTE data. (Numbers in brackets are the numbers of the selected variables)*

| No. of vars. | Forward selection | Backward elimination | Sequential replacement | Exhaustive search |
|---|---|---|---|---|
| Const. | 228308 | 228308 | 228308 | 228308 |
| 1 | 133695<br>(9) | 133695<br>(9) | 133695<br>(9) | 133695<br>(9) |
| 2 | 99841<br>(6,9) | 127803<br>(9,12) | 99841<br>(6,9) | 99841<br>(6,9) |
| 3 | 82389<br>(2,6,9) | 91777<br>(9,12,1,3) | 82389<br>(2,6,9) | 82389<br>(2,6,9) |
| 4 | 72250<br>(2,6,9,14) | 78009<br>(6,9,12,13) | 69154<br>(1,2,9,14) | 69154<br>(1,2,9,14) |
| 5 | 64634<br>(1,2,6,9,14) | 69136<br>(2,6,9,12,13) | 64634<br>(1,2,6,9,14) | 64634<br>(1,2,6,9,14) |
| 6 | 60539<br>(1,2,3,6,9,14) | 64712<br>(2,5,6,9,12,13) | 60539<br>(1,2,3,6,9,14) | 60539<br>(1,2,3,6,9,14) |

which is the only's one with substantially more observations than variables, sequential replacement has found the best-fitting subsets of all sizes, while forward selection has only failed in the case of the subset of 4 variables where the subset selected is the second-best of that size. Backward elimination has performed poorly, having dropped variable 14 from its subset of 10 variables and of course from all smaller subsets.

In this case, we have a good estimate of the residual variance. The $RSS$ with all 15 variables and a constant in the model is 53680 which has 44 degrees of freedom, giving a residual variance estimate of 1220. The drop in $RSS$ from the best subset of six variables to the best of seven (not shown) is less than twice the residual variance, suggesting that the best subset for prediction is of not more than six variables. The drop in $RSS$ from the best subset of five variables to the best of six is also not very impressive, being less than four times the residual variance in each case.

From Table 3.23 showing the five best-fitting subsets of each size, we see that in this case there is one dominant variable, variable number 9. This variable is the percentage of the population which is nonwhite. Without this variable, the best fitting subsets are shown in Table 3.24.

These are all inferior to the best subsets including variable 9. Some of

Table 3.23 *Five best-fitting subsets of each size from one to five variables for the POLLUTE data*

| No. of variables | RSS | Variables |
|---|---|---|
| 1 | 133695 | 9 |
| | 168696 | 6 |
| | 169041 | 1 |
| | 186716 | 7 |
| | 186896 | 14 |
| 2 | 99841 | 6, 9 |
| | 103859 | 2, 9 |
| | 109203 | 9, 14 |
| | 112259 | 4, 9 |
| | 115541 | 9, 10 |
| 3 | 82389 | 2, 6, 9 |
| | 83335 | 1, 9, 14 |
| | 85242 | 6, 9, 14 |
| | 88543 | 2, 9, 14 |
| | 88920 | 6, 9, 11 |
| 4 | 69154 | 1, 2, 9, 14 |
| | 72250 | 2, 6, 9, 14 |
| | 74666 | 2, 5, 6, 9 |
| | 76230 | 2, 6, 8, 9 |
| | 76276 | 1, 6, 9, 14 |
| 5 | 64634 | 1, 2, 6, 9, 14 |
| | 65660 | 1, 2, 3, 9, 14 |
| | 66555 | 1, 2, 8, 9, 14 |
| | 66837 | 1, 2, 9, 10, 14 |
| | 67622 | 2, 4, 6, 9, 14 |

Table 3.24 *Best-fitting subsets omitting variable 9 (% nonwhite)*

| Vars. | 6 | 1,14 | 1,4,14 | 1,4,7,14 | 1,2,4,11,14 | 1,2,3,4,11,14 |
|---|---|---|---|---|---|---|
| RSS | 168696 | 115749 | 102479 | 92370 | 87440 | 81846 |

the other variables most frequently selected are annual rainfall (variable 1), January (2) and July temperatures (3), median number of years of education (6), and sulphur dioxide concentration (14).

As the values of the $RSS$'s are relatively close together for subsets of the same size, it could appear that there is close competition for selection. However, if we compare the differences between $RSS$'s with our residual variance estimate of 1220, we see that only a small number are close.

### *3.12.5 Example 5*

An important example of subset selection is in near-infrared spectroscopy (NIRS). In this field, absorbances are measured at a large number of evenly-spaced wavelengths. These absorbances, or usually their logarithms, are used as the predictor variables, one predictor variable corresponding to each wavelength.

Hans Thodberg from the Danish Meat Research Institute has made a set of data publicly available. The Tecator data set, recorded using a Tecator Infratec Food and Feed Analyser, contains the logarithms to base 10 of absorbances at 100 wavelengths in the range 850 to 1050nm, of 240 samples of finely chopped meat. The data may be downloaded from the 'datasets' directory at statlib, either by ftp or WWW. The URL is:

`http://lib.stat.cmu.edu/datasets/tecator`

The samples are divided into 5 groups as follows.

| Group | Name | Samples |
|---|---|---|
| C | Training | 129 |
| M | Monitoring | 43 |
| T | Testing | 43 |
| E1 | Extrapolation, fat | 8 |
| E2 | Extrapolation, protein | 17 |

The objective is to predict moisture, fat and protein content of the meat from the absorbances. These three quantities were obtained using analytic chemical techniques for each of the samples, and their values are recorded in the data set. We will only be looking at the prediction of the percentage of fat.

Figure 3.4 shows the spectra of $-\log_{10}$ of the absorbances against the wavelength number. The plot is for data set M as it is clearer with only 43 samples, compared with the 129 samples of data set C. The lines of dots are those of samples with low fat % (<8%), the long and short dashes are for middle percentages of fat (8 to 25%), and the solid lines are for samples with more than 25% fat.

There is no obvious connection between the vertical position of a line and

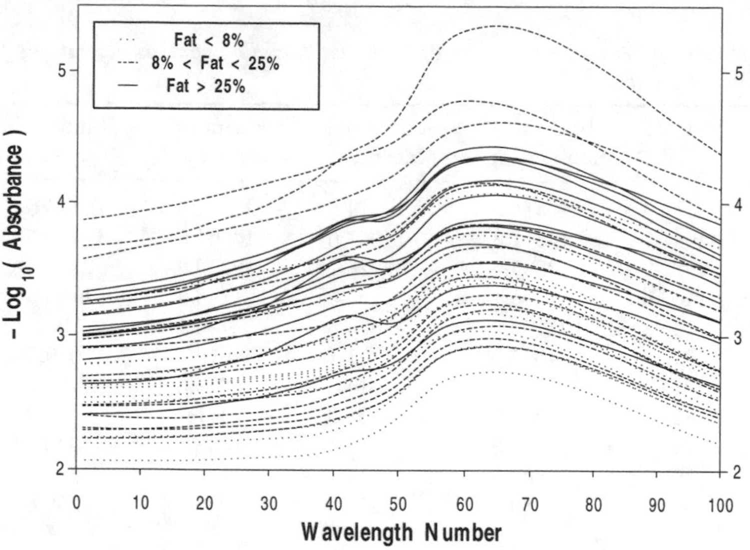


Figure 3.4 *Spectrum of log-absorbances for Tecator data set C.*

the percentage of fat for that sample. Visual inspection shows 'humps' between wavelengths 30 and 50 for most of the samples with high fat.

Various subset selection algorithms were applied to set C of this data. The residual sums of squares for some of the best-fitting subsets are shown in Table 3.25. The exhaustive search algorithm has not been used as the number of subsets to be searched is far too large. For instance, the number of subsets of 6 variables out of 100 is approximately $1.2 \times 10^9$. Though only a small fraction of these are actually searched using a branch-and-bound algorithm, it still proved to be too large a number.

The last column in this table, headed 'Random, 2-at-a-time', was obtained by randomly picking a subset of each size, and then replacing two variables at a time until no reduction in the residual sum of squares could be obtained. For each size of subset, 10 replications were done. It is likely that most of the entries in this column are the minimum RSS's that would be obtained using exhaustive searches.

Table 3.25 shows that the replacement algorithms can find very much better subsets than algorithms that only add or delete a single variable at a time. Looking at the least-squares regression coefficients helps us understand why this happens. Let us look at the best-fitting subset of 6 wavelengths. '6' is selected as there are certain commercial instruments used in near-infrared spectroscopy (NIRS) which use 6 wavelengths. The wavelengths which give the minimum RSS of 475.3 and their least-squares regression coefficients are:

Table 3.25 *RSS's for best-fitting subsets of each size found by various algorithms for data set C of the Tecator data*

| No. of vars. | Forward selection | Backward elimination | Sequential replacement | Sequential 2-at-a-time | Random 2-at-a-time |
|---|---|---|---|---|---|
| Const. | 20514.6 | 20514.6 | 20514.6 | 20514.6 | 20514.6 |
| 1 | 14067.6 | 14311.3 | 14067.6 | 14067.6 | 14067.6 |
| 2 | 2982.9 | 3835.5 | 2228.2 | 2228.2 | 2228.2 |
| 3 | 1402.6 | 1195.1 | 1191.0 | 1156.3 | 1156.3 |
| 4 | 1145.8 | 1156.9 | 833.6 | 799.7 | 799.7 |
| 5 | 1022.3 | 1047.5 | 711.1 | 610.5 | 610.5 |
| 6 | 913.1 | 910.3 | 614.3 | 475.3 | 475.3 |
| 8 | 852.9 | 742.7 | 436.3 | 417.3 | 406.2 |
| 10 | 746.4 | 553.8 | 348.1 | 348.1 | 340.1 |
| 12 | 595.1 | 462.6 | 314.2 | 314.2 | 295.3 |
| 15 | 531.6 | 389.3 | 272.6 | 253.3 | 252.8 |
| ... | | | | | |
| 100 | 27.6 | | | | |

| Wavelength number | Regression coefficient |
|---|---|
| 1 | –65.2 |
| 31 | +7558 |
| 33 | –12745 |
| 36 | +5524 |
| 53 | –1900 |
| 54 | +1628 |

Thus we have three wavelengths (31, 33 and 36) which are essentially measuring the curvature in part of the spectrum, and two (53 and 54), which are measuring the slope in another part. Quite often in NIRS, if six wavelengths are chosen, the best ones are in two groups of three measuring the curvature in two different parts of the spectrum. Adding, deleting, or replacing one variable at a time rarely finds such subsets.

Principal components have often been used as predictors in NIRS, and other fields, for instance, in meteorology where they are known as empirical orthogonal fields. The Tecator data set includes the first 22 principal components. Table 3.26 lists the best-fitting subsets of principal components. We note that the principal components do not appear to be good predictors. It takes 14 principal components to reduce the RSS to a value that can be obtained using only 6 wavelengths.

Another method of fitting models that has been popular in NIRS is Partial Least Squares (PLS). See, for instance, Martens and Naes (1989), Brown

Table 3.26 *RSS's for best-fitting subsets of principal components for data set C of the Tecator data*

| Size | RSS | Principal component numbers |
|---|---|---|
| 1 | 13119.3 | 3 |
| 2 | 7715.3 | 3 4 |
| 3 | 2720.8 | 1 3 4 |
| 4 | 1915.3 | 1 3 4 5 |
| 5 | 1449.7 | 1 2 3 4 5 |
| 6 | 1161.8 | 1 2 3 4 5 6 |
| 8 | 868.4 | 1 2 3 4 5 6 11 18 |
| 10 | 686.9 | 1 2 3 4 5 6 9 11 14 18 |
| 12 | 564.7 | 1 2 3 4 5 6 9 11 14 15 18 19 |
| 15 | 445.9 | 1 2 3 4 5 6 8 9 10 11 14 15 16 18 |
| ... | | |
| 22 | 378.6 | |

Table 3.27 *RSS's for partial least squares (PLS) for data set C of the Tecator data*

| Size | RSS | Size | RSS |
|---|---|---|---|
| 1 | 15561.0 | 6 | 1141.9 |
| 2 | 7673.9 | 8 | 940.6 |
| 3 | 3431.2 | 10 | 746.6 |
| 4 | 2226.4 | 12 | 557.5 |
| 5 | 1220.9 | 15 | 361.4 |

(1993), or Helland (1990). At each stage, the direction of steepest descent is found. This direction can be considered as a variable which is a linear combination of all of the log-absorbances. The variable to be fitted, fat percentage in our case, is regressed linearly against this variable. The residuals are calculated and then another direction of steepest descent is found. Table 3.27 shows the progressive residual sums of squares (RSS). The performance of PLS in this example is similar to that of principal components and inferior to that of using individual wavelengths. In the region of interest, to obtain a similar RSS to that obtained using the best $k$ individual wavelengths, we need about double the number of terms in PLS.

In any real life modelling exercise, the search for models which fit well must always be accompanied by the use of established regression diagnostics. Figure 3.5 illustrates how closely the least-squares model using 6 wavelengths fits the C data set, while Figure 3.6 shows the residuals from the fitted line. The residual for each point is that obtained using the least-squares fit with that point excluded from the least-squares calculations. Visual examination of these figures shows evidence of curvature. At the left-hand end, the actual

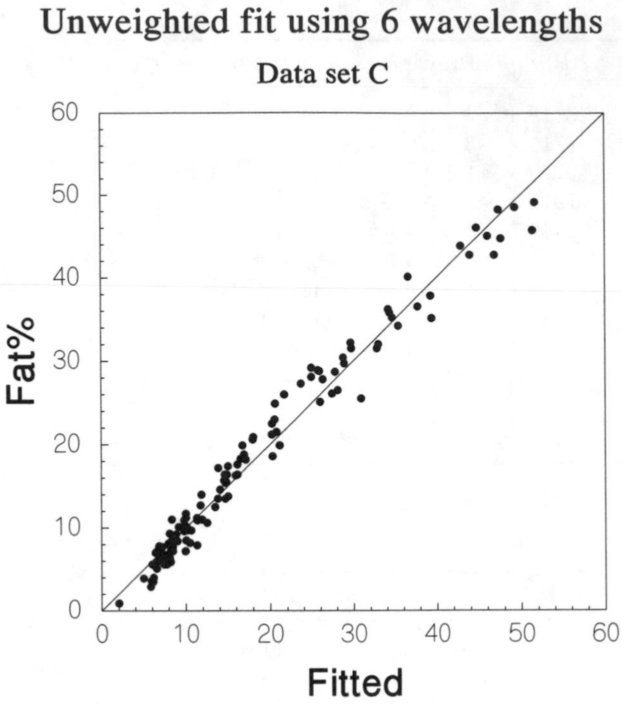


Figure 3.5 *Fitted values for the best-fitting 6 wavelengths for Tecator data set C.*

fat percentage is usually lower than the fitted value, and the same can be seen at the right-hand end. This curvature is highly significant using conventional significance testing. It is not surprising that the relationship is nonlinear over such a wide range of fat concentrations; there is no theory to say that the relationship should be linear.

Let us apply a transformation to adjust for this curvature. We could do this either by transforming the log-absorbances or the fat concentrations. The latter is much easier. Let $Y$ be the transformed fat percentage. We need only a small amount of curvature. A very simple transformation is

$$Y = \text{Fat} \ + \ a(\text{Fat} - f_0)^2.$$

Let us choose $a$ and $f_0$ to lift the fat percentage by 2% when Fat = 5%, and to lift it by 6% when Fat = 45%. That is, we want to solve the pair of equations

$$\begin{aligned} a(5 - f_0)^2 &= 2 \\ a(45 - f_0)^2 &= 6 \end{aligned}$$

By dividing one equation by the other, $a$ is eliminated and the resulting linear

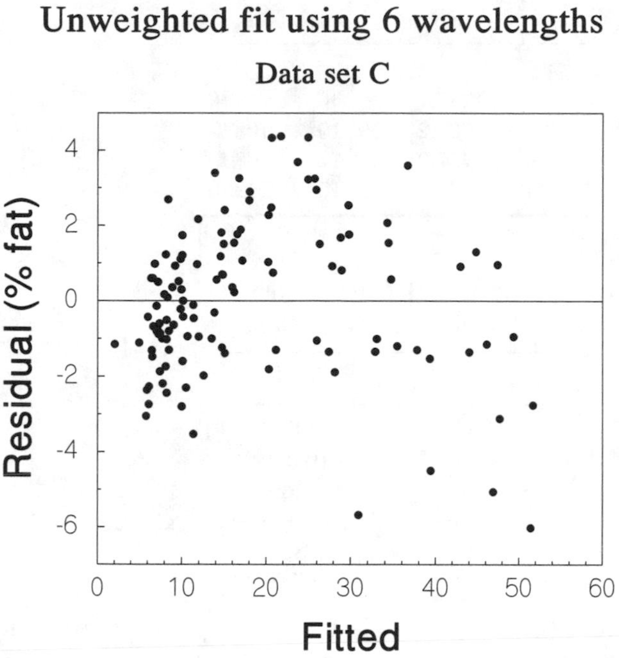


Figure 3.6 *Residuals from the best-fitting 6 wavelengths for Tecator data set C.*

equation gives that $f_0 = 19.6$. Substitution in either original equation then yields the value of $a \approx 0.01$. For simplicity, we will use the transformation

$$Y = \text{Fat} + \left(\frac{\text{Fat} - 20}{10}\right)^2.$$

The inverse transformation is simply obtained by solving a quadratic equation and taking the right root. This turns out to be

$$\text{Fat} = -30 + \sqrt{500 + 100Y}.$$

This will be needed to model the adjusted fat percentage, $Y$, and calculate fitted values for fat percentage from it.

Table 3.28 shows the residual sums of squares obtained for the best-fitting subsets for the transformed fat percentage. Even though the transformation has only changed most of the values of fat percentage by one or two percent, the reduction in the RSS has been appreciable.

If we look again at the least-squares regression coefficients for the best-fitting subset of 6 wavelengths, we find

Table 3.28 *RSS's for best-fitting subsets of wavelengths for data set C of the Tecator data after transforming fat percentage*

| Size | RSS | Size | RSS |
|---|---|---|---|
| 1 | 16137.4 | 6 | 284.5 |
| 2 | 1785.1 | 8 | 218.6 |
| 3 | 729.2 | 10 | 198.7 |
| 4 | 437.3 | 12 | 168.5 |
| 5 | 345.9 | 15 | 142.6 |

| Wavelength number | Regression Coefficient |
|---|---|
| 1 | –45.4 |
| 31 | +6076 |
| 33 | –10461 |
| 36 | +4663 |
| 52 | –583.1 |
| 55 | +348.8 |

This subset is almost the same as the last subset, while the regression coefficients have reduced slightly. At first, it appears that the last two regression coefficients have shrunk dramatically, but previously we had two consecutive wavelengths (numbers 53 and 54), whereas now we are estimating the slope from wavelengths which are 3 apart. The regression coefficients have shrunk overall and the transformation has had its greatest effect upon the largest and smallest fat percentages, thus shrinking its range.

The sum of the regression coefficients above (calculated to greater accuracy) is –1.6. As several of the individual coefficients are in thousands, this is almost complete cancellation.

We will return to this data set in later chapters. For the moment we will take a look at how well the model chosen using data set C fits data sets M and T. Figure 3.7 shows the fit of the 6 and 12 wavelength models to the M and T data sets, after using a quadratic adjustment to the fat percentages, and then adjusting back. The regression coefficients were calculated using only the M data set. The fit using 12 wavelengths is a little better.

## 3.13 Conclusions and recommendations

It could be argued that the above examples, except for the POLLUTE data set, are not very typical, having low ratios of numbers of observations to variables; however, these ratios are fairly representative of the author's experience as a consulting statistician. What is perhaps unrepresentative is the low num-

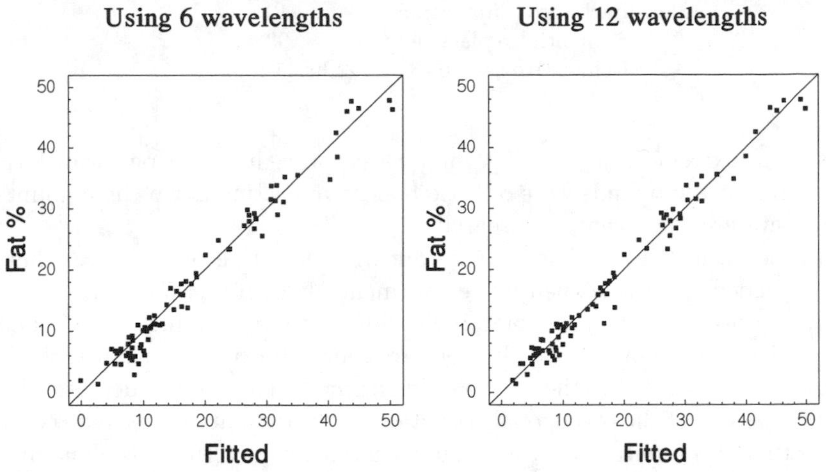


Figure 3.7 *The best-fitting models of 6 and 12 wavelengths to data set C fitted to Tecator data sets M and T, adjusting for nonlinearity*

ber of available predictor variables in these examples. In meteorology, users of variable-selection procedures often have a choice from more than 100 variables, while in using infrared spectroscopy as a substitute for chemical analysis, the author has occasionally encountered examples with over a thousand variables.

In our examples, the 'cheap' methods considered, forward selection, backward elimination, and sequential replacement have usually not found all of the best-fitting subsets and have sometimes fared poorly. Of these three methods, sequential replacement has been the most successful. A more extensive comparison has been made by Berk (1978b) who examined nine published cases, though all had more observations than variables. In three of the nine cases, forward selection and backward elimination found the best-fitting subsets of all sizes. In most other cases, the differences between the $RSS$'s for the subsets selected by forward selection and backward elimination and the best subsets were very small.

An exhaustive search is very time consuming. The following times were recorded for the four principal procedures considered here on a set of data from the near-infrared analysis of wool samples. There were 25 variables, which were reflectances at different wavelengths, and 72 observations. The best-fitting subsets of 6 variables, plus a constant, were sought and the best 5 subsets of each size were found. The times were on a Cromemco Z-2D microcomputer based on a Z80A microprocessor. The procedures were used in the order listed which meant that the subsets found by the first three procedures were available as initial bounds for the exhaustive search.

When the procedures were repeated with only the single best-fitting subset of each size being sought, the exhaustive search took 2 hours 28 minutes.

| | |
|---|---|
| Forward selection | 20 sec. |
| Backward elimination | 113 sec. |
| Sequential replacement | 93 sec. |
| Exhaustive search | 2 hr.45 min. |

This is an extreme example. The first three procedures all performed very badly so that the bounds were of little benefit in cutting down the amount of computation in the exhaustive search.

In general, if it is feasible to carry out an exhaustive search, then that is recommended. As the sequential replacement algorithm is fairly fast, it can always be used first to give some indication of the maximum size of subset that is likely to be of interest. If there are more observations than variables, then the $RSS$ with all of the variables in the model should be calculated first. This comes out of the orthogonal reduction as the data are input and requires no additional computation. It is useful as a guide to determine the maximum size of subset to examine.

It is recommended that several subsets of each size be recorded. The alternative subsets are useful both for testing whether the best-fitting subset is significantly better than others (see Spj$\phi$tvoll's test in Chapter 4) or for finding the subset which may give the best predictions for a specific set of future values of the predictor variables (see Chapter 6). The importance of this and the use of the extra information contained in second-best and subsequent subsets will become clear in Chapter 5.

When it is not feasible to carry out the exhaustive search, the use of random starts followed by sequential replacement, or of two-at-a-time replacement, can be used, though there can be no guarantee of finding the best-fitting subsets. A further alternative is that of grouping variables which was discussed in section 3.8 but this requires further research.

In all cases, graphical or other methods should be used to assess the adequacy of the fit obtained. Some suitable methods will be described in Chapter 6. These examinations often uncover patterns in the residuals that may indicate the suitability of using a transformation, or of using some kind of weighting, or of adding extra variables, such as quadratic or interaction terms. Unfortunately, inference becomes almost impossible if the total subset of available predictors is augmented subjectively in this way.

# Appendix A
# An algorithm for the Lasso

For the lasso, we want to find values of the regression coefficients, $\beta_j(s)$, to minimize

$$S = \sum_{i=1}^{n}(y_i - \sum_{j=1}^{k}\beta_j x_{ij})^2 \tag{A.1}$$

subject to the constraint

$$\sum_{j=1}^{k} |\beta_j| \le s . \tag{A.2}$$

The $X$ and $Y$ variables have already been standardized so that each column of $\boldsymbol{X}$ and the vector $\boldsymbol{y}$ have zero mean and standard deviation equal to one.

Basically, the algorithm is as follows:

Start by calculating the unconstrained (OLS) regression coefficients, and the sum, $s_0$, of their absolute values. For $s < s_0$, the constraint becomes an equality constraint. $s$ will be reduced from $s_0$ to zero in steps. At each step, a regression coefficient is reduced to zero. The constraint is enforced between steps by choosing one of the predictor variables, say number $j_0$, that has a nonzero value at the end of the previous step, and 'eliminating' it. Let $\delta_j$ be an indicator variable which can take values +1, 0 or -1 according to whether the value of $\beta_j$ is positive, zero or negative. Variable number $j_0$ is 'eliminated' by substituting

$$\delta_{j_0}\beta_{j_0} = s - \sum_{j \ne j_0}^{k} \delta_j \beta_j .$$

Then instead of solving a constrained linear least squares problem, we minimize

$$S = \sum_{i=1}^{n} (y_i - \sum_{j=1}^{k} \beta_j x_{ij})^2$$

$$= \sum_{i=1}^{n} \left( y_i - \sum_{j \ne j_0} \beta_j x_{ij} - \delta_{j_0} (s - \sum_{j \ne j_0} \delta_j \beta_j) x_{i,j_0} \right)^2$$

This is then an unconstrained linear least-squares problem in one less variable, with the values $y_i$ of the $Y$-variable replaced with

$y_i - \delta_{j_0} s x_{i,j_0}$

and the values $x_{ij}$ replaced with

$x_{ij} - \delta_{j_0} \delta_j x_{i,j_0}$.

Notice that the new values of $Y$ are related to $s$, whereas the new $X$-variables are not. Hence the least-squares solution is linear in $s$, until one of the regression coefficients is reduced to zero and one or more of the $\delta_j$'s changes.

In most cases, when one of the $\beta_j$'s is reduced to zero, we just set the corresponding $\delta_j$ to zero and just drop that variable from the regression using the methods described in chapter two. There are two cases in which we do something different. Firstly, the regression coefficient reduced to zero may be number $j_0$. In this case, the values of the replaced $X$ and $Y$ variables must be recalculated. To try to ensure that this does not happen very often, choose $j_0$ to be the number of the largest regression coefficient in absolute value.

The other case is more interesting, and that is that in which a variable that had been dropped at a higher value of $s$ can be reintroduced. When does this happen? Suppose we have the solution $\boldsymbol{\beta}(s)$ at one value of $s$ (less than $s_0$). Now let us look at what happens if we change any value of $\beta_j$ by a small amount, leaving all of the other $\beta_j$'s unchanged. This will of course violate the equality constraint. The rate of change of the sum of squares, $S$, with respect to this regression coefficient is

$$\frac{dS}{d\beta_j} = -2\sum_{i=1}^{n} e_i x_{ij}$$

where $e_i$ is the $i^{th}$ residual. Suppose that the numerical value of this is 10 for one variable, and less than 5 in absolute value for all the others. Then a small change in this regression coefficient is highly desirable. It will reduce the sum of squares substantially and we can make changes in one or more of the other regression coefficients to satisfy the equality constraint and still reduce the sum of squares. This means that between steps at which coefficients are reduced to zero, the derivatives with respect to all of the nonzero regression coefficients are exactly equal in absolute value (usually some will be positive, and some negative). At the values of $s$ at which a regression coefficient is reduced to zero, we reintroduce a variable if its derivative is greater in absolute value than the absolute value for the variables still in the model.

The steps can easily be determined. The algorithm starts at $s = s_0$. We try a moderately large step to say $s = s_0/2$. Several of the regression coefficients may change sign. As the solutions for the $\beta_j$'s change linearly with $s$, linear interpolation tells us when the first reaches zero. It may be at say $0.93s$, or perhaps at a value less than $s = s_0/2$. We move to this value of $s$, drop and perhaps reintroduce a variable, then proceed to half of this new value of $s$.

A simplified version of this algorithm could be used for the garrote. Instead of centering and scaling the $X$-variables, each one is multiplied by its least-squares coefficient when there are no constraints. The $\beta_j$'s are then replaced with the coefficients $c_j$. The $\delta_j$'s are unnecessary as the $c_j$'s must be positive or zero.

CHAPTER 4

# Hypothesis testing

## 4.1 Is there any information in the remaining variables?

Suppose that by some method we have already picked $p$ variables, where $p$ may be zero, out of $k$ variables available to include in our predictor subset. If the remaining variables contain no further information which is useful for predicting the response variable, then we should certainly not include any more. But how do we know when the remaining variables contain no further information? In general, we do not; we can only apply tests and take gambles based upon the outcome of those tests.

The simplest such test known to the author is that of augmenting the set of predictor variables with one or more artificial variables whose values are produced using a random number generator. When the selection procedure first picks one of these artificial variables, the procedure is stopped and we go back to the last subset containing none of the artificial variables. Let us suppose that we have reached the stage in a selection procedure when there is no useful information remaining (though we would not know this in a real case), and that there are 10 remaining variables plus one artificial variable. *A priori* the chance that the artificial variable will be selected next is then 1 in 11. Hence it is likely that several useless variables will be added before the artificial variable is chosen. For this method to be useful and cause the procedure to stop at about the right place, we need a large number of artificial variables, say of the same order as the number of real variables. This immediately makes the idea much less attractive; doubling the number of variables increases the amount of computation required by a much larger factor.

In Table 4.1, the $RSS$'s are shown for the five best-fitting subsets of 2, 3, 4 and 5 variables for the four data sets used in examples in section 3.10. The name 'CLOUDS' indicates the cloud-seeding data. The numbers of added variables were:

CLOUDS 5, STEAM 9, DETROIT 11 and POLLUTE 10.

The values of the artificial variables were obtained using a uniform random number generator. Except for the DETROIT data set, the artificial variables indicate that we should stop at three variables. In the case of the DETROIT data, the separation between the best-fitting and the second-best of four variables is not large, and it looks doubtful whether the subset of four variables would have been the best if we had generated more artificial variables. It is of interest to notice that there is no challenge to the best subset of three variables for the DETROIT data; a subset on which we commented in sec-

Table 4.1 *RSS's for the five best-fitting subsets of 2, 3, 4 and 5 variables with artificial variables added; the number of asterisks equals the number of artificial variables in the subset*

| No. of variables | CLOUDS | STEAM | DETROIT | POLLUTE |
|---|---|---|---|---|
| 2 | 21.56 | 8.93 | 33.83 | 99841 |
| | 21.81 | 9.63 | 44.77 | 103859 |
| | 22.29 | 9.78 | 54.46 | 109203 |
| | 22.41 * | 15.39 * | 55.49 | 112259 |
| | 22.42 * | 15.60 | 62.46 | 115541 |
| 3 | 12.61 | 7.34 | 6.77 | 82389 |
| | 15.56 | 7.68 | 21.19 | 83335 |
| | 16.12 | 7.81 * | 23.05 | 85242 |
| | 16.29 | 8.29 * | 23.51 | 87365 * |
| | 16.96 * | 8.29 * | 25.01 * | 88543 |
| 4 | 5.80 ** | 6.41 ** | 3.79 | 69137 * |
| | 6.52 ** | 6.72 * | 4.08 * | 69154 |
| | 8.25 ** | 6.80 | 4.58 | 72250 |
| | 8.25 ** | 6.93 | 5.24 | 74666 |
| | 8.32 * | 7.02 | 5.38 * | 75607 * |
| 5 | 3.69 ** | 5.91 ** | 1.33 * | 61545 * |
| | 4.08 ** | 5.93 * | 1.71 * | 62494 * |
| | 4.09 ** | 6.01 * | 2.15 * | 63285 ** |
| | 4.28 ** | 6.05 *** | 2.62 | 64505 * |
| | 4.40 ** | 6.11 * | 2.64 | 64549 * |

tion 3.10 stood out remarkably. In the case of the STEAM data we must have some doubts about whether the first two subsets of three variables are an improvement over the best-fitting subsets of two variables; the next three subsets containing artificial variables are very close.

We noted in section 3.3 that the widely used '$F$-to-enter' statistic does not have an $F$-distribution or anything remotely like one. However the quantity is a useful heuristic indicator, and we propose to use it and even to abuse it by using it to compare $RSS$'s for different sizes of subsets when the larger subset does not necessarily contain the smaller one. Let us define

$$F\text{-to-enter} = \frac{RSS_p - RSS_{p+1}}{RSS_{p+1}/(n-p-1)}$$

where $RSS_p$ and $RSS_{p+1}$ are the $RSS$'s for subsets of $p$ and $p+1$ variables, and $n$ is the number of observations. If the model includes a constant which has not been counted in the $p$ variables, as is usually the case, then another 1 must be subtracted from the number of degrees of freedom. Comparing the best subsets of three and four variables in Table 4.1, the values of the

$F$-to-enter are 10.6 (CLOUDS), 2.9 (STEAM), 6.3 (DETROIT), and 10.5 (POLLUTE) with 9, 20, 8, and 25 degrees of freedom respectively for the denominators. It is interesting to note that only the third largest of these may be significant at say the 5% level, though we can only guess at this on the basis of the previous test using the addition of artificial variables.

An alternative test to the above is that of Forsythe et al. (1973). If the true model is linear in just $p$ out of the $k$ available predictor variables, these $p$ variables have been correctly selected and are the first $p$ variables in the triangular factorization, then the $(k-p)$ projections of the dependent variable, $Y$, on the remainder of the space spanned by the $X$-variables, are uncorrelated and have variance equal to the residual variance if the true residuals are uncorrelated and homoscedastic (i.e. they all have the same variance). This is easily shown as follows. We suppose that

$$\boldsymbol{y} = \boldsymbol{X_A}\boldsymbol{\beta} + \boldsymbol{\epsilon},$$

where $\boldsymbol{X_A}$ consists of the first $p$ columns of $\boldsymbol{X}$, the true residuals in vector $\boldsymbol{\epsilon}$ have zero mean, the same variance, $\sigma^2$, and are uncorrelated, i.e. $E(\boldsymbol{\epsilon\epsilon'}) = \sigma^2\boldsymbol{I}$ where $\boldsymbol{I}$ is an $n \times n$ identity matrix with $n$ equal to the number of observations. Let $\boldsymbol{X_B}$ consist of the remaining $(k-p)$ columns of $\boldsymbol{X}$, and let the orthogonal reduction of $\boldsymbol{X}$ be

$$(\boldsymbol{X_A},\ \boldsymbol{X_B}) = (\boldsymbol{Q_A},\ \boldsymbol{Q_B})\boldsymbol{R},$$

where the columns of $\boldsymbol{Q_A}$ and $\boldsymbol{Q_B}$ are mutually orthogonal and normalized. Then we have

$$\begin{aligned}\begin{pmatrix}\boldsymbol{Q'_A}\\ \boldsymbol{Q'_B}\end{pmatrix}\boldsymbol{y} &= \begin{pmatrix}\boldsymbol{Q'_A}\\ \boldsymbol{Q'_B}\end{pmatrix}\boldsymbol{X_A\beta} + \begin{pmatrix}\boldsymbol{Q'_A}\\ \boldsymbol{Q'_B}\end{pmatrix}\boldsymbol{\epsilon}\\ &= \begin{pmatrix}\boldsymbol{I}\\ \boldsymbol{0}\end{pmatrix}\boldsymbol{R\beta} + \begin{pmatrix}\boldsymbol{Q'_A}\\ \boldsymbol{Q'_B}\end{pmatrix}\boldsymbol{\epsilon},\end{aligned}$$

where $\boldsymbol{I}$ is a $p \times p$ identity matrix.

The last $(k-p)$ projections are given by

$$\boldsymbol{Q'_B y} = \boldsymbol{Q'_B \epsilon}.$$

Then

$$E(\boldsymbol{Q'_B y}) = \boldsymbol{0}$$

and

$$\begin{aligned}E(\boldsymbol{Q'_B y y' Q_B}) &= E(\boldsymbol{Q'_B \epsilon\epsilon' Q_B})\\ &= \sigma^2\boldsymbol{Q'_B Q_B}\\ &= \sigma^2\boldsymbol{I},\end{aligned}$$

where $\boldsymbol{I}$ is a $(k-p) \times (k-p)$ identity matrix. Notice that there is no need for the matrix $\boldsymbol{R}$ to be triangular; any orthogonal reduction will suffice. Also the only distributional assumptions imposed on the true residuals are that they have finite variance, the same variance for each residual, and that the residuals are uncorrelated.

Table 4.2 *Numbers of times that the maximum reduction in RSS from adding a fourth variable was exceeded when the projections for the remaining variables were permuted*

| Data set | CLOUDS | STEAM | DETROIT | POLLUTE |
|---|---|---|---|---|
| No. of exceedances (out of 1000) | 974 | 847 | 308 | 348 |

If we have selected too many variables, but they include all of those in the true model for $Y$, then the above results still apply as the model then has zeroes for some of the elements of $\boldsymbol{\beta}$.

The projections in the last $(k - p)$ positions of $\boldsymbol{Q}'\boldsymbol{y}$ can be used to test the hypothesis that there is no more useful information for predicting $Y$ if we are prepared to accept that the residuals from the true model are uncorrelated and homoscedastic. Alternatively, they can be used for testing the properties of the true residuals if we are satisfied with the selected variables in the model.

Forsythe et al. (1973) suggested a test for the former as follows. First find the maximum reduction in $RSS$, which can be obtained by adding one further variable; call this $S_{max}$. Then permute the last $(k - p)$ elements of $\boldsymbol{Q}'\boldsymbol{y}$, each time calculating the maximum reduction in $RSS$, which can be achieved by adding one variable to the $p$ already selected, and counting the number of times that $S_{max}$ is exceeded. This can be done quickly and can easily be incorporated into a forward selection routine. If the remaining $(k-p)$ variables contain no further useful information, then $S_{max}$ can be regarded as a random sample from the distribution of maximum reductions in $RSS$. Hence, if we carry out say 1000 permutations, the number of times that $S_{max}$ is exceeded is equally likely to take any value from 0 to 1000.

This test assumes that the last elements of $\boldsymbol{Q}'\boldsymbol{y}$ are interchangeable, which requires that they are identically distributed. In general, this will not be true unless the true residuals have a normal distribution. We have only shown that the elements have the same first two moments. However, as these elements are weighted linear combinations of the true but unknown residuals, by the central limit theorem they will usually have distributions which are close to normal, particularly if the number of observations is more than say double the number of variables. Using the elements of $\boldsymbol{Q_B}$, it is possible, though tedious, to calculate higher moments for each of the last elements of $\boldsymbol{Q}'\boldsymbol{y}$, in terms of the moments of the true residuals. This exercise (or a similar one for least-squares residuals) shows that orthogonal reduction (or least-squares fitting) is very effective in making residuals look normal even when the true residuals are very nonnormal.

The above test has a disadvantage in that the significance level depends upon the order of occurrence of the last $(k - p)$ variables. It will not usually be practical to use all $(k - p)!$ permutations of these variables.

Table 4.2 shows the results obtained using the four sets of data used in the examples in section 3.10. In each case, the best-fitting subset of three variables was used for the subset of selected variables. These results support the belief that there is no further useful information in the remaining variables in these data sets. The high number of exceedances for the cloud-seeding data suggests that there may have already been some overfitting in selecting the subset of three variables.

In the case of the POLLUTE data set, the best single variable to add to the first three does not give the best-fitting subset of four variables. The best-fitting subset of three variables consists of variables 2, 6 and 9, which give an $RSS = 82389$. Adding variable 14 reduces the $RSS$ to 72250, but the subset of variables 1, 2, 9 and 14 gives an $RSS = 69154$. As the residual variance estimate with all 15 variables included is 1220, the difference of 3096 between these two $RSS$'s is relatively large for two subsets of the same size.

It is possible for the remaining variables to contain a pair of variables which together substantially improve the prediction of $Y$, but which are of very little value separately. This would not be detected by the Forsythe permutation test, though a 'two-variables-at-a-time' version of the test could be carried out if this were suspected. The test as presented by Forsythe et al. is probably better to use as a 'carry on' rule rather than as a 'stopping' rule as in the case when $S_{max}$ is only exceeded a small number of times, the remaining variables most probably contain additional information, whereas when the number of exceedances is larger, the situation is uncertain.

One important feature of this permutation test is that it can be used when the number of variables exceeds the number of observations, which is precisely the situation for which it was devised.

When there are more observations than variables, there is information in the residual variation that is not used by the Forsythe permutation test. A similar test is to use the maximum $F$-to-enter. Thus, after $p$ variables have been selected, the maximum $F$-to-enter is found for the remaining $(k - p)$ variables. It is not difficult to write down the form of the distribution of this maximum $F$-to-enter, making the usual assumptions that the true residuals are independently and identically normally distributed, plus the added assumption that the selected $p$ variables represent the true relationship; i.e. we have not already overfitted by adding one or more variables, which by chance had moderately high correlations with the variable to be predicted. This distribution depends upon the values of the $X$-variables so that it is not feasible to tabulate it except perhaps for the case in which the $X$-variables are uncorrelated. However, the distribution can easily be simulated for the particular data set at hand by using a random normal generator for the last $(k - p)$ projections and a random gamma generator for the residual sum of squares with all the variables in the model. The $F$-to-enter statistic is dimensionless, so it is not necessary to estimate the residual variance, the random projections can have unit variance and the sum of squares of residuals can be sampled from a chi-squared distribution without applying any scaling. However, such a test proves only whether a single variable can significantly reduce the residual

Table 4.3 *Lack-of-fit statistic after fitting the best-fitting subsets of three variables (Degs. = degrees, Num. = Numerator, Denom. = Denominator)*

| Data set | No. of variables in subset | Degs. of freedom Num. | Degs. of freedom Denom. | Lack-of-fit F |
|---|---|---|---|---|
| CLOUDS | 3 | 10 | 0 | n.a. |
| STEAM | 3 | 6 | 15 | 1.27 |
| DETROIT | 3 | 8 | 1 | 9.30 |
| POLLUTE | 3 | 12 | 44 | 1.96 |

sum of squares, and will again fail to detect cases where two or more variables collectively can produce a significant reduction. The author can supply a Fortran routine for such a test, but it was not thought to have sufficient value to include here.

A simple alternative to the $F$-to-enter test is one often called the lack-of-fit test. If we have $n$ observations and have fitted a linear model containing $p$ out of $k$ variables, plus a constant, then the difference in $RSS$ between fitting the $p$ variables and fitting all $k$ variables, $RSS_p - RSS_k$, can be compared with $RSS_k$ giving the lack-of-fit statistic

$$\text{lack} - \text{of} - \text{fit } F \;=\; \frac{(RSS_p \;-\; RSS_k)/(k-p)}{RSS_k/(n-k-1)}.$$

Table 4.3 shows values of the lack-of-fit statistic after fitting three variables for each of our four examples. If the subset of three variables had been chosen *a priori*, the usual conditions of independence, constant variance, and normality are satisfied, and there is no useful information in the remaining variables, then the lack-of-fit statistic is sampled from an $F$-distribution with the numbers of degrees of freedom shown in the table. None of these values is significant at the 5% level, though for the POLLUTE data set is very close, it is at the 5.3% point in the tail of its $F$-distribution. The value of the lack-of-fit statistic for the DETROIT data set is large, but it needs to be much larger to become significant as there is only one degree of freedom for the residual sum of squares with all 11 variables in the model.

A further statistic which can be used to test for no further information is the coefficient of determination, $R^2$. This has usually been employed to test whether there is any predictive power in any of the variables. $R^2$ can be defined as

$$R^2 \;=\; 1 \;-\; RSS_p/RSS_1,$$

provided that a constant has been included in the model as the first variable, so that $RSS_1$ is the sum of squares of $Y$ about its mean. There is some ambiguity in the definition of $R^2$ when a constant is not fitted. Sometimes the definition above is used, in which case negative values are possible, while the total sum of squares of values of the variable is sometimes used instead of $RSS_1$.

The distribution of $R^2$ has been tabulated for a number of cases in which the response variable, $Y$, is normally distributed and independent of the $X$-variables. Diehr and Hoflin (1974), Lawrence et al. (1975), and McIntyre et al. (1983) have generated the distribution of $R^2$ using Monte Carlo methods for subset selection using respectively exhaustive search and forward selection, for uncorrelated normally distributed $X$-variables. Zurndorfer and Glahn (1977), and Rencher and Pun (1980) have also looked at the case of correlated $X$-variables using forward selection and the Efroymson algorithm, respectively. It is clear from both of these last two studies that the value of $R^2$ tends to be higher when the $X$-variables are uncorrelated. Table 4.4 shows the values of the average and upper 95% points of the distribution of $R^2$, denoted by $R_s^2$ and $R_{95}^2$, obtained by Rencher and Pun for the case of uncorrelated $X$-variables. The number of $X$-variables, $p$, does not include the constant term which was fitted. Rencher and Pun do not state the values used for the $F$-to-enter and $F$-to-delete parameters in deriving their tables, but do say that they were reduced when necessary to force the required number of variables into subsets. The 95%-points of $R^2$ found by Diehr and Hoflin tend to be somewhat higher than those found by Rencher and Pun, but this is to be expected as they used exhaustive search as their selection procedure.

Tables similar to Table 4.4 have also been produced by Wilkinson and Dallal (1981) for $R^2$ for the case of forward selection, except that their tables are in terms of the number of available variables and the value used for $F$-to-enter.

The values of $R^2$ for the best-fitting subsets of three variables for our four examples are given in Table 4.5. From Table 4.4, we see that the values of $R^2$ for the STEAM, DETROIT, and POLLUTE data sets are all significant at the 5% level. For the CLOUDS data, interpolation between four entries is necessary. It appears that the value of $R^2$ may be just short of the 5% point, though the tables are for the Efroymson algorithm, not for an exhaustive search.

Zirphile (1975) attempted to use extreme-value theory to derive the distribution of $R^2$. He makes the false assumption that the values of $R^2$ for the ${}^kC_p$ different subsets of $p$ variables out of $k$ are uncorrelated, and uses a normal distribution to approximate the distribution of $R^2$ for a randomly chosen subset of variables. The distribution of $R^2$ for a random subset when the $Y$-variable is uncorrelated with the $X$-variables is a beta distribution with

$$\text{prob}(R^2 < z) = \frac{1}{B(a,b)} \int_0^z t^{a-1}(1-t)^{b-1}dt,$$

where $a = p/2$, $b = (n-p-1)/2$ if a constant has been included in the model but not counted in the $p$ variables. Using the beta distribution and fitting constants to their tables, Rencher and Pun obtained the following formula for the upper $100(1-\gamma)\%$ point of the distribution of the maximum $R^2$ using the

Table 4.4 *Values of the average and upper 95% points of the distribution of $R^2$ for subsets selected using Efroymson's algorithm (n = sample size, k = number of available uncorrelated predictor variables)*

| | | Number of variables in selected subset | | | | | | | | | |
|---|---|---|---|---|---|---|---|---|---|---|---|
| | | p = 2 | | p = 4 | | p = 6 | | p = 8 | | p = 10 | |
| n | k | $R_s^2$ | $R_{95}^2$ | $R_s^2$ | $R_{95}^2$ | $R_s^2$ | $R_{95}^2$ | $R_s^2$ | $R_{95}^2$ | $R_s^2$ | $R_{95}^2$ |
| 5 | 5 | .784 | .981 | | | | | | | | |
| | 10 | .900 | .991 | | | | | | | | |
| | 20 | .952 | .995 | | | | | | | | |
| 10 | 5 | .421 | .665 | .540 | .851 | | | | | | |
| | 10 | .567 | .822 | .778 | .955 | .894 | .991 | | | | |
| | 20 | .691 | .877 | .912 | .983 | .984 | .999 | .997 | 1.000 | | |
| | 40 | .791 | .907 | .965 | .991 | .996 | 1.000 | .999 | 1.000 | | |
| 20 | 10 | .299 | .510 | .423 | .658 | .488 | .726 | | | | |
| | 20 | .391 | .562 | .585 | .771 | .701 | .858 | .786 | .920 | | |
| | 40 | .469 | .618 | .700 | .819 | .835 | .920 | .916 | .969 | .963 | .989 |
| 30 | 10 | .202 | .337 | .285 | .451 | .326 | .511 | | | | |
| | 20 | .264 | .388 | .405 | .546 | .496 | .648 | .561 | .716 | | |
| | 40 | .331 | .456 | .514 | .640 | .639 | .759 | .733 | .846 | .803 | .901 |
| 40 | 10 | .147 | .260 | .206 | .339 | .233 | .374 | | | | |
| | 20 | .203 | .319 | .309 | .445 | .376 | .527 | .424 | .583 | | |
| | 40 | .251 | .349 | .397 | .507 | .502 | .626 | .584 | .704 | .651 | .775 |
| 50 | 30 | .184 | .267 | .291 | .397 | .367 | .484 | .424 | .559 | .469 | .612 |
| | 40 | .201 | .288 | .324 | .432 | .413 | .526 | .481 | .598 | .537 | .657 |
| 60 | 30 | .157 | .229 | .247 | .341 | .312 | .425 | .360 | .479 | .398 | .523 |
| | 40 | .169 | .249 | .272 | .372 | .348 | .457 | .406 | .515 | .454 | .569 |

Table 4.5 *Values of $R^2$ for the best-fitting subsets of three variables*

| Data set | n | k | p | $R^2$ |
|---|---|---|---|---|
| CLOUDS | 14 | 20 | 3 | .826 |
| STEAM | 25 | 9 | 3 | .885 |
| DETROIT | 13 | 11 | 3 | .998 |
| POLLUTE | 60 | 15 | 3 | .639 |

Efroymson algorithm as

$$R^2_\gamma = [1 + \log_e \gamma (\log_e N)^{1.8N^{.04}}] \, F^{-1}(\gamma), \tag{4.1}$$

where $N = {}^k C_p$, and $F^{-1}(\gamma)$ is the value of $z$ such that prob$(R^2 < z) = \gamma$.

Values of $F^{-1}(\gamma)$ can be obtained from tables of the incomplete beta function, or from tables of the $F$-distribution as follows. Writing $Reg_p$ to denote the regression sum of squares on $p$ variables, we have

$$R^2 = Reg_p/(Reg_p + RSS_p).$$

Write

$$F = \frac{Reg_p/p}{RSS_p/(n-p-1)}$$

as the usual variance ratio for testing the significance of the subset of $p$ variables if it had been chosen *a priori*. Then

$$R^2 = p/[p + (n-p-1)F]. \tag{4.2}$$

Thus, the value of $R^2$ such that prob$(R^2 < z) = \gamma$ can be calculated from the value of $F$, with $p$ and $(n-p-1)$ degrees of freedom for the numerator and denominator, respectively, such that the upper tail area is $\gamma$. As the reciprocal of a variance ratio also has an $F$-distribution but with the degrees of freedom interchanged, because of the way in which the $F$-distribution is usually tabulated, we use the tables with $(n-p-1)$ and $p$ degrees of freedom for numerator and denominator, respectively, and then take the reciprocal of the $F$-value read from the tables. The upper limit on $R^2$ is then obtained by substitution in (4.2), and then into (4.1).

In the case of the CLOUDS data set the above method gives $R^2_{95} = 0.878$ which means that our value is not significant at the 5% level. Bearing in mind that the Rencher and Pun formula is for the Efroymson algorithm whereas we found the subset using an exhaustive search, our value for $R^2$ is even less significant.

## 4.2 Is one subset better than another?

In the first section of this chapter, we looked at ways of testing the hypothesis that $\beta_{p+1}, ..., \beta_k = 0$, where these $\beta$'s are the regression coefficients of the variables which have not been selected. In some cases, these tests were of dubious value as they required that a subset had been selected *a priori* even though the test is almost always applied using the same data as was used to select the subset. Those tests only tested whether one subset was better than another, in some sense, when one subset was completely contained in the other.

What is meant by saying that one subset is better than another? There are many possible answers. One is that used by Spjøtvoll (1972a), and this is equivalent to that used by Borowiak (1981). One subset is considered better than another if the regression sum of squares of the expected values of $Y$

upon the subset of $X$-variables is larger for that subset; Borowiak used the complementary residual sum of squares. However, unlike Borowiak and most other workers, Spjøtvoll did not make the assumption that the linear model in one of the two subsets was the true model.

Let us suppose that $\boldsymbol{y} = \boldsymbol{X\beta} + \boldsymbol{\epsilon}$, where the residuals, $\epsilon$, are independently and identically normally distributed with zero mean and unknown variance $\sigma^2$. That is, if we use all of the available predictor variables, then we have the true model, though an unknown number of the $\beta$'s may be zero. Then if we denote the least-squares estimate of vector $\boldsymbol{\beta}$ by $\hat{\boldsymbol{\beta}}$, we have from standard theory

$$P\{(\boldsymbol{\beta} - \hat{\boldsymbol{\beta}})'\boldsymbol{X}'\boldsymbol{X}(\boldsymbol{\beta} - \hat{\boldsymbol{\beta}}) \leq ks^2 F_{\alpha,k,n-k}\} = 1 - \alpha, \tag{4.3}$$

where $s^2$ is the sample estimate of the residual variance with all $k$ variables in the model; that is, $s^2 = RSS_k/(n-k)$, and $F_{\alpha,k,n-k}$ is the upper $\alpha$-point (i.e. the tail area $= \alpha$) of the $F$-distribution with $k$ and $(n-k)$ degrees of freedom for the numerator and denominator respectively. If the linear model including all of our predictor variables is not the true model, then the equality in (4.3) should be replaced with '$\geq$'. Spjøtvoll refers readers to pages 136-137 of Scheffe (1959) for a method of proof of this statement.

Let $X_1$, $X_2$ be two subsets of variables we want to compare, with $p_1$ and $p_2$ variables, respectively. For subset $X_i$ the fitted values of $Y$ for given values of the $X$-variables using the least-squares fitted relationship are given by

$$\begin{aligned}\hat{\boldsymbol{y}} &= \boldsymbol{X}_i\hat{\boldsymbol{\beta}}_i \\ &= \boldsymbol{X}_i(\boldsymbol{X}_i'\boldsymbol{X}_i)^{-1}\boldsymbol{X}_i'\boldsymbol{y}.\end{aligned}$$

If the expected values of $Y$ are denoted by $\boldsymbol{\eta}$ then the sum of squares of deviations of the fitted values from the expected $Y$-values for a future set of data with the same values given for the $X$-variables is

$$\begin{aligned}&[\boldsymbol{\eta} - \boldsymbol{X}_i(\boldsymbol{X}_i'\boldsymbol{X}_i)^{-1}\boldsymbol{X}_i'(\boldsymbol{\eta}+\boldsymbol{\epsilon})]'[\boldsymbol{\eta} - \boldsymbol{X}_i(\boldsymbol{X}_i'\boldsymbol{X}_i)^{-1}\boldsymbol{X}_i'(\boldsymbol{\eta}+\boldsymbol{\epsilon})] \\ &= \boldsymbol{\eta}'\boldsymbol{\eta} - 2\boldsymbol{\eta}'\boldsymbol{X}_i(\boldsymbol{X}_i'\boldsymbol{X}_i)^{-1}\boldsymbol{X}_i'(\boldsymbol{\eta}+\boldsymbol{\epsilon}) + (\boldsymbol{\eta}+\boldsymbol{\epsilon})'\boldsymbol{X}_i(\boldsymbol{X}_i'\boldsymbol{X}_i)^{-1}\boldsymbol{X}_i'(\boldsymbol{\eta}+\boldsymbol{\epsilon}),\end{aligned}$$

which has expected value

$$= \boldsymbol{\eta}'\boldsymbol{\eta} - \boldsymbol{\eta}'\boldsymbol{X}_i(\boldsymbol{X}_i'\boldsymbol{X}_i)^{-1}\boldsymbol{X}_i'\boldsymbol{\eta} + \sigma^2.\text{trace}[\boldsymbol{X}_i(\boldsymbol{X}_i'\boldsymbol{X}_i)^{-1}\boldsymbol{X}_i']. \tag{4.4}$$

We note that as trace$(\boldsymbol{AB})$ = trace$(\boldsymbol{BA})$, it can be shown that the trace above has value $p_i$, provided that the columns of $\boldsymbol{X}_i$ have full rank.

Spjøtvoll suggested that the quantity $\boldsymbol{\eta}'\boldsymbol{X}_i(\boldsymbol{X}_i'\boldsymbol{X}_i)^{-1}\boldsymbol{X}_i'\boldsymbol{\eta}$ be used as the measure of goodness of fit of a regression function, with larger values denoting a better fit. Replacing $\boldsymbol{\eta}$ with $\boldsymbol{X\beta}$, we want then to make inferences about the difference

$$\boldsymbol{\beta}'\boldsymbol{X}'[\boldsymbol{X}_1(\boldsymbol{X}_1'\boldsymbol{X}_1)^{-1}\boldsymbol{X}_1' - \boldsymbol{X}_2(\boldsymbol{X}_2'\boldsymbol{X}_2)^{-1}\boldsymbol{X}_2']\boldsymbol{X\beta} = \boldsymbol{\beta}'\boldsymbol{C\beta} \tag{4.5}$$

which is a quadratic form in the unknown $\boldsymbol{\beta}$.

Now if the condition on the left-hand side of (4.3) is satisfied then we can find absolute values for the maximum and minimum of (4.5) as the condition means that $\boldsymbol{\beta}$ must be within the specified closeness of the known $\hat{\boldsymbol{\beta}}$. Only a summary of the results will be given here, more technical detail is given in appendix 4A. Alternatively, the reader can refer to Spj$\phi$tvoll's paper, but this contains a large number of printing errors and the notation differs slightly from ours.

Let $\boldsymbol{P}$ be a matrix such that both

$$\boldsymbol{P}'\boldsymbol{X}'\boldsymbol{X}\boldsymbol{P} = \boldsymbol{I} \tag{4.6}$$

$$\boldsymbol{P}'\boldsymbol{C}\boldsymbol{P} = \boldsymbol{D}, \tag{4.7}$$

where $\boldsymbol{D}$ is a diagonal matrix. Such a matrix always exists, as will be clear from the method for finding $\boldsymbol{P}$ given in Appendix 4A. Let $\gamma = \boldsymbol{P}^{-1}\boldsymbol{\beta}$ and $\hat{\gamma} = \boldsymbol{P}^{-1}\hat{\boldsymbol{\beta}}$ . (N.B. Spj$\phi$tvoll shows $\boldsymbol{P}'$ instead of $\boldsymbol{P}^{-1}$ which is not the same in general). The condition (4.3) is now

$$\text{prob.}\{(\gamma - \hat{\gamma})'(\gamma - \hat{\gamma}) \leq ks^2F\} = 1 - \alpha,$$

while the quadratic form in (4.5) is now just $\sum d_i g_i^2$, where $d_i$, $\gamma_i$ are the $i^{th}$ elements of $\boldsymbol{D}$ and $\gamma$, respectively. If the inequality on the left-hand side of (4.3) is satisfied, then

$$A_1 \leq \boldsymbol{\beta}'\boldsymbol{C}\boldsymbol{\beta} \leq A_2,$$

where

$$A_1 = a\left(\sum \frac{d_i\hat{\gamma_i}^2}{a + d_i} - ks^2F\right)$$

$$A_2 = b\left(\sum \frac{d_i\hat{\gamma_i}^2}{b - d_i} + ks^2F\right),$$

where $a = -min(min\ d_i,\ \lambda_{min})$, $b = max(max\ d_i,\ \lambda_{max})$, and $\lambda_{min}$, $\lambda_{max}$ are the minimum and maximum roots of

$$\sum \frac{d_i\hat{\gamma_i}^2}{(d_i - \lambda)^2} = ks^2F, \tag{4.8}$$

except that $A_1 = 0$ if all of the $d_i$'s are $\geq 0$ and $\sum_{d_i \neq 0} \hat{\gamma_i}^2 \leq ks^2F$, and $A_2 = 0$ if all of the $d_i$'s are $\leq 0$ and $\sum_{d_i \neq 0} \hat{\gamma_i}^2 \leq ks^2F$. These results follow from Forsythe and Golub (1965), or Spj$\phi$tvoll (1972b). As the left-hand side of (4.2.6) tends to $+\infty$ from both sides as $\lambda$ approaches the positive $d_i$'s, and $-\infty$ for the negative $d_i$'s, it is easy to see from a plot of (4.8) that $\lambda_{min}$ is a little smaller than the smallest positive $d_i$. Similarly, $\lambda_{max}$ is a little larger than the largest $d_i$. Any reasonable iterative method finds $\lambda_{min}$ and $\lambda_{max}$ very easily.

If $A_1$ and $A_2$ are both greater than zero, then subset $X_1$ is significantly better than subset $X_2$ at the level used to obtain the value used for $F$. If the final term in (4.4) had been included somehow in the quadratic form $\boldsymbol{\beta}'\boldsymbol{C}\boldsymbol{\beta}$, then we would have been able to conclude that subset $X_1$ gave significantly better predictions for the particular $X$-values used. It does not appear to be easy to do this, though a crude adjustment is simply to add $s^2(p_2 - p_1)$ to both $A_1$ and $A_2$ as though $s^2$ were a perfect, noise-free estimate of $\sigma^2$.

A particularly attractive feature of Spj$\phi$tvoll's method is that it gives simultaneous confidence limits and/or significance tests for any number of comparisons based upon the same data set. If the condition on the left-hand side of (4.3) is satisfied, and note that this condition is unrelated to the subsets being compared, then all of the confidence or significance statements based upon calculated values of $A_1$ and $A_2$ are simultaneously true.

A special case of the Spj$\phi$tvoll test is that in which one subset is completely contained in the other. This is the case, for instance, in both forward selection and backward elimination, and in carrying out the lack-of-fit test in which one 'subset' consists of all $k$ available predictors. It is readily shown (see Appendix 4A), that the $d_i$'s corresponding to variables that are common to both subsets are equal to zero, as are those corresponding to variables that are in neither subset. Thus if there are $p_0$ variables common to both subsets, $(k-p_1-p_2+p_0)$ of the $d_i$'s are equal to zero. If subset $X_2$ is completely contained in subset $X_1$, in which case $p_2 = p_0$, then the remaining $(p_1 - p_2)$ values of $d_i$ are all equal to $+1$, and the corresponding $\gamma_i$'s are just the projections of $Y$ on the $(p_1 - p_2)$ $X$-variables in $X_1$, but not $X_2$, after making them orthogonal to the $p_0$ common variables. (Conversely, if $X_1$ is contained in $X_2$, then there are $(p_2 - p_1)$ values of $d_i$ equal to –1, all other $d_i$'s being equal to zero.) In this case, (4.8) becomes

$$(1-\lambda)^{-2} \sum \hat{\gamma}_i{}^2 \;=\; ks^2 F,$$

so that $\lambda_{max}$ and $\lambda_{min} \;=\; 1 \pm (\sum \hat{\gamma}_i{}^2 / ks^2 F)^{\frac{1}{2}}$. Now $\sum d_i \hat{\gamma}_i{}^2$ is the difference in $RSS$ between fitting subsets $X_1$ and $X_2$. In this case, it simplifies to $\sum \hat{\gamma}_i{}^2$. Denote this difference by $\Delta RSS$, then

$$A_1 \;=\; \max[0, \{\Delta RSS^{\frac{1}{2}} \;-\; (ks^2 F)^{\frac{1}{2}}\}^2]$$

$$A_2 \;=\; \{\Delta RSS^{\frac{1}{2}} \;+\; (ks^2 F)^{\frac{1}{2}}\}^2.$$

Hence, subset $X_1$ (the one with more variables) is significantly better than $X_2$ if $A_1 > 0$; that is, if $\Delta RSS$ is greater than $ks^2 F$. If it had been decided *a priori* to compare subsets $X_1$ and $X_2$, then the usual test would have been to compute the variance ratio $\Delta RSS/(p_1 - p_2)$ divided by $s^2$. Subset $X_1$ would then have been deemed significantly better if $\Delta RSS$ were greater than $(p_1 - p_2)s^2 F$. The replacement of $(p_1 - p_2)$ by $k$ in the required difference in residual sums of squares, $\Delta RSS$, is an indication of the degree of conservatism in the Spj$\phi$tvoll test.

This special case of Spj$\phi$tvoll's test has also been derived, using different arguments, by Aitkin (1974), McCabe (1978), and Tarone (1976). Borowiak (1981) appears to have derived a similar result for the case in which the residual variance is assumed to be known.

The argument used by Aitkin is as follows. If we had decided, *a priori,* that we wanted to compare subset $X_2$ with the full model containing all of the variables in $X$, then we would have used the likelihood-ratio test which gives the variance ratio statistic:

$$F = \frac{(RSS_p - RSS_k)/(k - p)}{RSS_k/(n - k)}, \tag{4.9}$$

where the counts of variables ($p$ and $k$) include one degree of freedom for a constant if one is included in the models. Under the null hypothesis that none of the $(k - p)$ variables excluded from $X_2$ is in the 'true' model, this quantity is distributed as $F(k-p, n-k)$, subject of course to assumptions of independence, normality, and homoscedacity of the residuals from the model. Aitkin then considers the statistic:

$$U(X_2) = (k-p)F. \tag{4.10}$$

The maximum value of $U$ for all possible subsets (but including a constant) is then

$$U_{max} = \frac{RSS_1 - RSS_k}{RSS_k/(n - k)}.$$

Hence, a simultaneous $100\alpha\%$ test for all hypotheses $\beta_2 = 0$ for all subsets $X_2$ is obtained by testing that

$$U(X_2) = (k - 1)F(\alpha, k-1, n-k). \tag{4.11}$$

Subsets satisfying (4.11) are called $R^2$-adequate sets. Aitkin expresses (4.9) in terms of $R^2$ instead of the residual sums of squares, and hence (4.10) can also be so expressed.

The term 'minimal adequate sets' is given to subsets which satisfy (4.11) but which are such that if any variable is removed from the subset, it fails to satisfy the condition. Edwards and Havranek (1987) give an algorithm for deriving minimal adequate sets.

### *4.2.1 Applications of Spj$\phi$tvoll's method*

Spj$\phi$tvoll's method was applied to the best subsets found for the STEAM, DETROIT, and POLLUTE data sets. Table 4.6 shows some of the comparisons for the STEAM data set. From this table, we see that the best-fitting subsets of two variables are significantly better than the best-fitting single variable (number 7) at the 5% level, but that adding further variables produces no further improvement. Referring back to table 3.16, we might anticipate that subsets (4,7) and (7,9) might be significantly worse than the best-three of three variables because of the big difference in $RSS$, but this is not so.

Notice that though we have obtained 90% and 98% confidence levels, we are quoting 5% and 1% significance levels. As the subsets that are being compared have been selected conditional upon their position among the best-fitting subsets, we know that a significant difference can occur in only one tail, and that tail is known before carrying out the test. It is appropriate then that a single-tail test is used.

The comparisons among the single variables in Table 4.6 are interesting. At

Table 4.6 *Spjøtvoll's upper and lower confidence limits for the difference in regression sums of squares (or in RSS) for selected subset comparisons for the STEAM data set*

| Subset $X_1$ | Subset $X_2$ | Diff. in RSS Sub. 2 - Sub. 1 | $\alpha$-level (%) | $A_1$ | $A_2$ |
|---|---|---|---|---|---|
| 7 | 6 | 19.4 | 10 | 3.0 | 40.7 |
| | | | 2 | -1.1 | 47.1 |
| 7 | 5 | 27.25 | 10 | -8.8 | 69.0 |
| 7 | 3 | 31.2 | 10 | 7.1 | 63.7 |
| | | | 2 | 1.3 | 73.4 |
| 7 | 8 | 35.7 | 10 | 10.5 | 70.5 |
| | | | 2 | 4.8 | 81.0 |
| 7 | 1,7 | -9.3 | 10 | -31.7 | -0.21 |
| | | | 2 | -39.5 | 0 |
| 7 | 5,7 | -8.6 | 10 | -30.4 | -0.12 |
| | | | 2 | -38.1 | 0 |
| 7 | 2,7 | -8.4 | 10 | -30.1 | -0.10 |
| | | | 2 | -37.7 | 0 |
| 7 | 4,7 | -2.6 | 10 | -17.7 | 0 |
| 7 | 7,9 | -2.2 | 10 | -16.7 | 0 |
| 1,7 | 5,7 | 0.7 | 10 | -13.6 | 15.5 |
| 1,7 | 2,7 | 0.9 | 10 | -5.1 | 7.5 |
| 1,7 | 4,7 | 6.7 | 10 | -7.8 | 27.0 |
| 1,7 | 7,9 | 7.1 | 10 | -3.4 | 24.5 |
| 1,7 | 4,5,7 | -1.6 | 10 | -15.1 | 10.8 |
| 1,7 | 1,5,7 | -1.2 | 10 | -13.7 | 0 |
| 1,7 | 1,3,5,7,8,9 | -3.6 | 10 | -20.1 | 0 |

Table 4.7 *Correlations with variable $X_7$ of some of the other predictors for the STEAM data set*

| Variable | $X_6$ | $X_5$ | $X_3$ | $X_8$ |
|---|---|---|---|---|
| Correlation | -0.86 | -0.21 | -0.62 | -0.54 |
| Range | 37.7 | 77.8 | 56.6 | 60.0 |

the 5% level, variable 7 fits significantly better than the second, fourth and fifth-best, but not the third-best. This is a situation which will often occur. If there are high correlations, positive or negative, between the variables in the two subsets, they span almost the same space and the upper and lower limits are relatively close together. In this case, the correlations between variable 7 and the others in the best five, together with the range between upper and lower 90% limits, are as shown in Table 4.7.

In the case of the DETROIT data set, the values of $F$ are from the distribution with 12 degrees of freedom for the numerator but only 1 degree of

Table 4.8 *Spjøtvoll's upper and lower confidence limits for the difference in regression sums of squares (or in RSS) for selected subset comparisons for the DETROIT data*

| Subset $X_1$ | Subset $X_2$ | Diff. in RSS Sub. 2 – Sub. 1 | $\alpha$-level (%) | $A_1$ | $A_2$ |
|---|---|---|---|---|---|
| 6 | 1 | 27 | 10 | –176 | 231 |
| 6 | 9 | 65 | 10 | –347 | 476 |
| 6 | 8 | 78 | 10 | –337 | 495 |
| 6 | 7 | 98 | 10 | –317 | 517 |
| 6 | 4,6 | –166 | 10 | –440 | –23 |
| | | | 2 | –2858 | 0 |
| 4,6 | 2,7 | 11 | 10 | –143 | 250 |
| 4,6 | 1,9 | 21 | 10 | –135 | 181 |
| 4,6 | 4,5 | 22 | 10 | –94 | 148 |
| 4,6 | 3,8 | 29 | 10 | –152 | 210 |
| 4,6 | 2,4,11 | –27 | 10 | –148 | 90 |
| 2,4,11 | 4,6,10 | 14 | 10 | –93 | 132 |
| 4,6 | 1,2,4,6,7,11 | –32 | 10 | –190 | 0 |
| 2,4,11 | 1,2,4,6,7,11 | –5 | 10 | –109 | 0 |

freedom for the denominator. The 10% and 2% points in the tail are at 60.7 and 1526 respectively. If we had one year's data more, then these values would be down to 9.41 and 49.4. The huge confidence ellipsoids for the regression coefficients are reflected in the big differences between the upper and lower limits, $A_1$ and $A_2$. Table 4.8 shows the results of a few comparisons of subsets for this data set. Only one of the comparisons in this table yields a significant difference; subset (4,6) fits significantly better at the 5% level than variable 6 alone.

For the POLLUTE data set, there was a moderately large number of degrees of freedom (44) for the residual variance so that much closer confidence limits and more powerful tests are possible. No significant differences at the 5% level were found when comparing subsets of the same size using only the five best-fitting subsets of each size as listed in Table 3.21. In most practical cases, more subsets of each size should be retained; the number was limited to five here to keep down the amount of space consumed by these tables. Table 4.9 shows only the comparisons between the best-fitting subsets of each size.

In Table 4.9, we see that the subset of two variables, 6 and 9, fits just significantly better at the 5% level than variable 9 alone. The upper limit of –32 for this comparison is extremely small as the residual variance estimate for these data is 1220. In view of the term in (4.4), which was left out of the quadratic form used for the comparison of pairs of subsets, we cannot infer that the subset (6,9) will yield better predictions.

Table 4.9 *Spjøtvoll's upper and lower confidence limits for the difference in regression sums of squares (or in RSS) for selected subset comparisons for the POLLUTE data*

| Subset $X_1$ | Subset $X_2$ | Diff. in RSS Sub. 2 – Sub. 1 | $\alpha$-level (%) | $A_1$ | $A_2$ |
|---|---|---|---|---|---|
| 9 | 6,9 | –34,000 | 10 | –131,000 | –32 |
| | | | 2 | –153,000 | 0 |
| 6,9 | 2,6,9 | –17,000 | 10 | –96,000 | 0 |
| 6,9 | 1,2,9,14 | –31,000 | 10 | –134,000 | 51,000 |
| 6,9 | 1,2,6,9,14 | –35,000 | 10 | –134,000 | –86 |
| | | | 2 | –156,000 | 0 |
| 6,9 | 1,2,3,6,9,14 | –39,000 | 10 | –142,000 | –395 |
| | | | 2 | –164,000 | 0 |
| 2,6,9 | 1,2,3,6,9,14 | –22,000 | 10 | –106,000 | 0 |

Code in Fortran for implementing Spjøtvoll's test is in Applied Statistics algorithm AS 273 (Miller (1992a)).

### *4.2.2 Using other confidence ellipsoids*

Spjøtvoll has pointed out that in applying his method, the confidence ellipsoid (4.3) that was used as the starting point can be shrunk by reducing it with one in a smaller number of variables. For instance, if certain variables, such as the dummy variable representing the constant in the model, are to be forced into all models, then those variables can be removed from $X$ and $\beta$. If this leaves $k^*$ variables, then $ks^2F_{\alpha,k,n-k}$ should be replaced with $k^*s^2F_{\alpha,k^*,n-k}$. It must be emphasized that ellipsoids of the form (4.3), but for subsets of variables, are only valid if those subsets had been determined *a priori;* they are not valid if, for instance, they include only those variables which appear in certain best-fitting subsets, as the least-squares estimates of the regression coefficients are then biased estimates of the coefficients for that subset (see Chapter 6 for a detailed discussion of selection bias).

To assist in appreciating the difference between these confidence regions, Figure 4.1 shows a hypothetical confidence ellipse for two regression coefficients, $\beta_1$ and $\beta_2$, and the confidence limits for the same confidence level for one of the coefficients. If the two $X$-variables have been standardized to have unit standard deviation, and the $\boldsymbol{X'X}$-matrix, excluding the rows and columns for the constant in the model is:

$$\boldsymbol{X'X} = \begin{pmatrix} 1 & \rho \\ \rho & 1 \end{pmatrix},$$

then the elliptical confidence regions are given by

$$x^2 + 2\rho xy + y^2 \leq 2s^2F$$

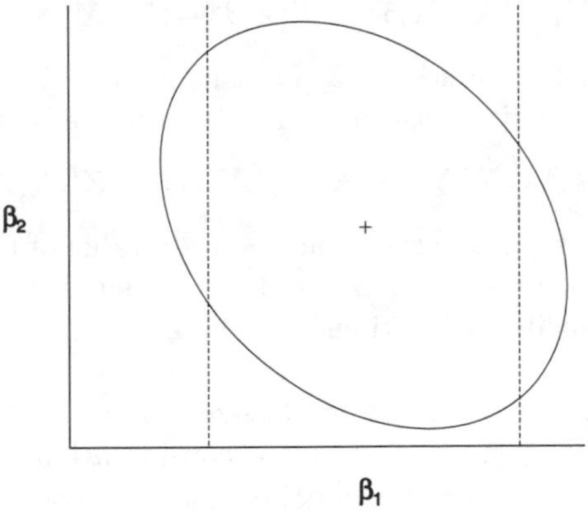


Figure 4.1 *Confidence ellipse and confidence limits for a hypothetical case.*

where $x = (\beta_1 - \hat{\beta}_1)$, $y = (\beta_2 - \hat{\beta}_2)$, $s^2$ is the usual residual variance estimate with $\nu$ degrees of freedom, and $F$ is the appropriate percentage point of the $F$-distribution with 2 and $\nu$ degrees of freedom respectively for the numerator and denominator. The most extreme points on the ellipses are then

$$\hat{\beta}_i \pm (s^2/(1-\rho^2))^{\frac{1}{2}}.(2F)^{\frac{1}{2}},$$

compared with the corresponding confidence limits (i.e. for individual regression coefficients) which are:

$$\hat{\beta}_i \pm (s^2/(1 - \rho^2))^{\frac{1}{2}}.t,$$

where $t$ is the appropriate percentage point from the $t$-distribution. For moderately large numbers of degrees of freedom, $t \approx 2$ and $F \approx 3$ so that the ratio of ranges of regression coefficients in this case (i.e. for two coefficients) is about 1.22:1.

In general, by reducing the number of $\beta$'s, the ranges of those remaining is reduced but at the expense of allowing infinite ranges for the $\beta$'s omitted.

*Goodness-of-fit outside the calibration region*

Unless there is sound scientific justification for a model, it is always hazardous to attempt to use it with estimated coefficients outside the range over which it may have been found to fit tolerably well. If we have sufficient trust in the linear model, which includes all of the predictor variables, then Spjøtvoll's method may be used to test whether one subset of variables can be expected to perform better than another in any specified region of the $X$-space. Let $\boldsymbol{Z}$ be an $m \times k$ array of values of the $X$-variables for which predictions are required. Then, proceeding as before, we have that the sum of squares of deviations of the expected values of the dependent variable from those predicted using subset $X_i$, is

$$[\boldsymbol{Z\beta} - \boldsymbol{Z}_i(\boldsymbol{X}_i'\boldsymbol{X}_i)^{-1}\boldsymbol{X}_i'(\boldsymbol{X\beta}+\boldsymbol{\epsilon})]'[\boldsymbol{Z\beta} - \boldsymbol{Z}_i(\boldsymbol{X}_i'\boldsymbol{X}_i)^{-1}\boldsymbol{X}_i'(\boldsymbol{X\beta}+\boldsymbol{\epsilon})],$$

where $\boldsymbol{Z}_i$ refers to the values in $\boldsymbol{Z}$ for the subset of variables in $X_i$. The equivalent goodness-of-fit measure to that used earlier is then

$$\boldsymbol{\beta}'[-2\boldsymbol{Z}'\boldsymbol{Z}_i(\boldsymbol{X}_i'\boldsymbol{X}_i)^{-1}\boldsymbol{X}_i'\boldsymbol{X} + \boldsymbol{X}'\boldsymbol{X}_i(\boldsymbol{X}_i'\boldsymbol{X}_i)^{-1}\boldsymbol{Z}_i'\boldsymbol{Z}_i(\boldsymbol{X}_i'\boldsymbol{X}_i)^{-1}\boldsymbol{X}_i'\boldsymbol{X}]\boldsymbol{\beta}.$$

The new matrix $\boldsymbol{C}$ is the difference between the value of the expression inside the square brackets for subsets $X_1$ and $X_2$. A small amount of simplification results by replacing the various $\boldsymbol{X}$ and $\boldsymbol{Z}$ matrices by their orthogonal reductions.

Finally in this chapter, we must emphasize that the methods described here will usually eliminate some subsets from consideration but will rarely leave a single subset which is indisputably the best. We must also emphasize that satisfying a significance test and providing good predictions are not synonymous.

# Appendix A
# Spj$\phi$tvoll's method, detailed description

In applying Spj$\phi$tvoll's method, we want to find maximum and minimum values of the quadratic form

$$\boldsymbol{\beta}'\boldsymbol{X}'[\boldsymbol{X}_1(\boldsymbol{X}_1'\boldsymbol{X}_1)^{-1}\boldsymbol{X}_1' - \boldsymbol{X}_2(\boldsymbol{X}_2'\boldsymbol{X}_2)^{-1}\boldsymbol{X}_2']\boldsymbol{X\beta} = \boldsymbol{\beta}'\boldsymbol{C\beta}, \tag{4.5}$$

subject to $\boldsymbol{\beta}$ being close to the known vector of least-squares estimates $\hat{\boldsymbol{\beta}}$, where the degree of closeness is such that

$$(\boldsymbol{\beta} - \hat{\boldsymbol{\beta}})'\boldsymbol{X}'\boldsymbol{X}(\boldsymbol{\beta} - \hat{\boldsymbol{\beta}}) \le ks^2F_{\alpha,k,n-k}.$$

The subscripts on $F$ will be dropped. We will assume that $\boldsymbol{X}'\boldsymbol{X}$ is of full rank. First, we attempt to find a matrix $\boldsymbol{P}$ such that

$$\boldsymbol{P}'\boldsymbol{X}'\boldsymbol{XP} = \boldsymbol{I} \tag{4.6}$$

and

$$\boldsymbol{P}'\boldsymbol{CP} = \boldsymbol{D}, \tag{4.7}$$

where $\boldsymbol{D}$ is a diagonal matrix. If we can do this, then we have transformed the problem into that of finding the maximum and minimum of $\boldsymbol{\gamma}'\boldsymbol{D\gamma}$ subject to

$$(\gamma - \hat{\gamma})'(\gamma - \hat{\gamma}) \le ks^2F,$$

where $\boldsymbol{\gamma} = \boldsymbol{P}^{-1}\boldsymbol{\beta}$, and $\hat{\boldsymbol{\gamma}} = \boldsymbol{P}^{-1}\hat{\boldsymbol{\beta}}$.

First, let us form the Cholesky factorization, $\boldsymbol{X}'\boldsymbol{X} = \boldsymbol{R}'\boldsymbol{R}$, where $\boldsymbol{R}$ is an upper-triangular matrix, though at this stage any nonsingular factorization will suffice. Then $\boldsymbol{P} = \boldsymbol{R}^{-1}$ satisfies (4.6). Now let us find a matrix $\boldsymbol{V}$ whose columns are the normalized eigenvectors of $\boldsymbol{R}^{-T}\boldsymbol{C}\boldsymbol{R}^{-1}$; that is, $\boldsymbol{V}$ satisfies

$$\boldsymbol{V}'(\boldsymbol{R}^{-T}\boldsymbol{C}\boldsymbol{R}^{-1})\boldsymbol{V} = \boldsymbol{D},$$

where $\boldsymbol{V}'\boldsymbol{V} = \boldsymbol{I}$ and $\boldsymbol{D}$ is a diagonal matrix with the eigenvalues of $\boldsymbol{R}^{-T}\boldsymbol{C}\boldsymbol{R}^{-1}$ on its diagonal. The matrix $\boldsymbol{P} = \boldsymbol{R}^{-1}\boldsymbol{V}$ then satisfies (4.6) and (4.7). We have then that $\hat{\gamma} = \boldsymbol{V}'\boldsymbol{R}\boldsymbol{\beta}$. If $\boldsymbol{R}$ were taken as the Cholesky factor of $\boldsymbol{X}'\boldsymbol{X}$, then $\boldsymbol{R}\boldsymbol{\beta}$ is just the vector of projections of the dependent variable on the space spanned by the $X$'s and so would normally be calculated in any least-squares calculations using orthogonal reduction.

In calculating the eigenstructure of the matrix $\boldsymbol{R}^{-T}\boldsymbol{C}\boldsymbol{R}^{-1}$, it is convenient to order the variables as follows (the order is not that used by Spj$\phi$tvoll):

1. The $p_0$ variables which are common to both $X_1$ and $X_2$.
2. The $(p_1 - p_0)$ variables in $X_1$ but not in $X_2$.
3. The $(p_2 - p_0)$ variables in $X_2$ but not in $X_1$.
4. The remaining $(k - p_1 - p_2 + p_0)$ variables in neither $X_1$ nor $X_2$.

In practice, some of the above groups will often be empty.

Then we form the orthogonal reduction

$$\begin{aligned}\boldsymbol{X} &= \boldsymbol{Q}\boldsymbol{R} \\ &= (\boldsymbol{Q}_1, \boldsymbol{Q}_2, \boldsymbol{Q}_3, \boldsymbol{Q}_4)\begin{pmatrix} \boldsymbol{R}_{11} & & & \\ \boldsymbol{R}_{12} & \boldsymbol{R}_{22} & & \\ \boldsymbol{R}_{13} & \boldsymbol{R}_{23} & \boldsymbol{R}_{33} & \\ \boldsymbol{R}_{14} & \boldsymbol{R}_{24} & \boldsymbol{R}_{34} & \boldsymbol{R}_{44}, \end{pmatrix}\end{aligned}$$

where the columns of the $\boldsymbol{Q}$ are orthonormal; that is, $\boldsymbol{Q}'\boldsymbol{Q} = \boldsymbol{I}$. The columns of $\boldsymbol{Q}$ are of length $n$ equal to the number of observations and $\boldsymbol{R}$ is upper triangular with $k$ rows and columns. In practice, this orthogonal reduction will usually be obtained from another orthogonal reduction by re-arranging the order of variables.

Now we can write

$$\begin{aligned}\boldsymbol{X}_1 &= (\boldsymbol{Q}_1, \boldsymbol{Q}_2)\begin{pmatrix} \boldsymbol{R}_{11} & \boldsymbol{R}_{12} \\ & \boldsymbol{R}_{22} \end{pmatrix} \\ \boldsymbol{X}_2 &= (\boldsymbol{Q}_1, \boldsymbol{Q}_2, \boldsymbol{Q}_3)\begin{pmatrix} \boldsymbol{R}_{11} & \boldsymbol{R}_{13} \\ & \boldsymbol{R}_{23} \\ & \boldsymbol{R}_{33} \end{pmatrix}.\end{aligned}$$

This gives us an upper-triangular factorization for $\boldsymbol{X}_1$, which is simply the first $p_1$ rows and columns of $\boldsymbol{R}$. Unfortunately, it does not give us anything simple for the factorization of $\boldsymbol{X}_2$. However, by the use of planar rotations again, we can obtain a triangular factorization for $\boldsymbol{X}_2$. Let $\boldsymbol{P}_0$ be a product of the planar rotations thus required. Then $\boldsymbol{P}_0$ has the form:

$$\boldsymbol{P}_0 = \begin{pmatrix} \boldsymbol{I} & \boldsymbol{P}_1 & \boldsymbol{P}_2 \\ & \boldsymbol{P}_3 & \boldsymbol{P}_4 \end{pmatrix},$$

where $\boldsymbol{I}$ has $p_0$ rows and columns, $\boldsymbol{P}_1$, $\boldsymbol{P}_2$ have $(p_2 - p_0)$ rows, $\boldsymbol{P}_3$, $\boldsymbol{P}_4$ have $(p_1 - p_0)$ rows, and it will later be convenient to split the columns so that $\boldsymbol{P}_1$, $\boldsymbol{P}_3$ have $(p_1 - p_0)$ columns, and of course, $\boldsymbol{P}_0'\boldsymbol{P}_0 = \boldsymbol{I}$. Then

$$\begin{aligned} \boldsymbol{X}_2 &= (\boldsymbol{Q}_1, \boldsymbol{Q}_2, \boldsymbol{Q}_3)\boldsymbol{P}_0'\boldsymbol{P}_0 \begin{pmatrix} \boldsymbol{R}_{11} & \boldsymbol{R}_{13} \\ & \boldsymbol{R}_{23} \\ & \boldsymbol{R}_{33} \end{pmatrix} \\ &= (\boldsymbol{Q}_1, \boldsymbol{Q}_3^*, \boldsymbol{Q}_2^*) \begin{pmatrix} \boldsymbol{R}_{11} & \boldsymbol{R}_{13} \\ & \boldsymbol{R}_{33}^* \\ & \boldsymbol{0} \end{pmatrix} \\ &= (\boldsymbol{Q}_1, \boldsymbol{Q}_3^*) \begin{pmatrix} \boldsymbol{R}_{11} & \boldsymbol{R}_{13} \\ & \boldsymbol{R}_{33}^* \end{pmatrix}. \end{aligned}$$

The matrix $\boldsymbol{P}_0$ can be formed while $\boldsymbol{Q}_2$ is being forced out of the orthogonal reduction, by applying the same planar rotations to another matrix which is initially an identity matrix with $(p_1 - p_0) + (p_2 - p_0)$ rows and columns.

We then find that the matrix of which we want the eigenstructure can be written as

$$\boldsymbol{R}^{-T}\boldsymbol{C}\boldsymbol{R}^{-1} =$$

$$\begin{pmatrix} \boldsymbol{I} & \boldsymbol{0} \\ \boldsymbol{0} & \boldsymbol{I} \\ \boldsymbol{0} & \boldsymbol{0} \\ \boldsymbol{0} & \boldsymbol{0} \end{pmatrix} \begin{pmatrix} \boldsymbol{I} & \boldsymbol{0} & \boldsymbol{0} & \boldsymbol{0} \\ \boldsymbol{0} & \boldsymbol{I} & \boldsymbol{0} & \boldsymbol{0} \end{pmatrix} - \begin{pmatrix} \boldsymbol{I} & \boldsymbol{0} \\ \boldsymbol{0} & \boldsymbol{A} \\ \boldsymbol{0} & \boldsymbol{B} \\ \boldsymbol{0} & \boldsymbol{0} \end{pmatrix} \begin{pmatrix} \boldsymbol{I} & \boldsymbol{0} & \boldsymbol{0} & \boldsymbol{0} \\ \boldsymbol{0} & \boldsymbol{A}' & \boldsymbol{B}' & \boldsymbol{0} \end{pmatrix},$$

where

$$\begin{pmatrix} \boldsymbol{A} \\ \boldsymbol{B} \end{pmatrix} = \begin{pmatrix} \boldsymbol{Q}_2' \\ \boldsymbol{Q}_3' \end{pmatrix} \boldsymbol{Q}_3^*.$$

But

$$\boldsymbol{Q}_3^* = (\boldsymbol{Q}_2, \boldsymbol{Q}_3) \begin{pmatrix} \boldsymbol{P}_1' \\ \boldsymbol{P}_2' \end{pmatrix}$$

and hence $(\boldsymbol{A}', \boldsymbol{B}') = (\boldsymbol{P}_1', \boldsymbol{P}_2')$.

In the general case, the matrix $\boldsymbol{R}^{-T}\boldsymbol{C}\boldsymbol{R}^{-1}$ will have $p_0$ rows of zeroes at the top and $(k - p_1 - p_2 + p_0)$ rows of zeroes at the bottom, and has a zero eigenvalue for each such row. We are left then to find the eigenstructure of the inner matrix:

$$\begin{pmatrix} \boldsymbol{I} & \boldsymbol{0} \\ \boldsymbol{0} & \boldsymbol{0} \end{pmatrix} - \begin{pmatrix} \boldsymbol{P}_1' \\ \boldsymbol{P}_2' \end{pmatrix} (\boldsymbol{P}_1, \boldsymbol{P}_2) = \begin{pmatrix} \boldsymbol{I} - \boldsymbol{P}_1'\boldsymbol{P}_1 & -\boldsymbol{P}_1'\boldsymbol{P}_2 \\ -\boldsymbol{P}_2'\boldsymbol{P}_1 & -\boldsymbol{P}_2'\boldsymbol{P}_2 \end{pmatrix} = \boldsymbol{Z} \text{ say.}$$

We notice that $\boldsymbol{P}_0\boldsymbol{P}_0' = \boldsymbol{I}$ (as well as $\boldsymbol{P}_0'\boldsymbol{P}_0 = \boldsymbol{I}$) and hence $\boldsymbol{P}_1\boldsymbol{P}_1' + \boldsymbol{P}_2\boldsymbol{P}_2' = \boldsymbol{I}$. Using this, we find that

$$\boldsymbol{Z}^2 = \begin{pmatrix} \boldsymbol{I} - \boldsymbol{P}_1'\boldsymbol{P}_1 & \boldsymbol{0} \\ \boldsymbol{0} & \boldsymbol{P}_2'\boldsymbol{P}_2 \end{pmatrix},$$

which means that the $(p_1 - p_0)$ columns on the left-hand side of $\boldsymbol{Z}$ are orthogonal to the last $(p_2 - p_0)$ columns.

Now the eigenvalues of $\boldsymbol{Z}$ satisfy $|\boldsymbol{Z} - d\boldsymbol{I}| = 0$ or

$$\begin{vmatrix} (1-d)\boldsymbol{I} - \boldsymbol{P}_1'\boldsymbol{P}_1 & -\boldsymbol{P}_1'\boldsymbol{P}_2 \\ -\boldsymbol{P}_2'\boldsymbol{P}_1 & -\boldsymbol{P}_2'\boldsymbol{P}_2 - d\boldsymbol{I} \end{vmatrix} = 0.$$

As the determinant of the product of two matrices equals the product of the determinants, we can multiply $(\boldsymbol{Z} - d\boldsymbol{I})$ by any other matrix, being careful not to introduce any more zeroes, and the determinant will still be zero. A convenient matrix to use is $\boldsymbol{Z}$. Multiplying on the right by $\boldsymbol{Z}$ gives

$$\begin{vmatrix} (1-d)\boldsymbol{I} - \boldsymbol{P}_1'\boldsymbol{P}_1 & d\boldsymbol{P}_1'\boldsymbol{P}_2 \\ d\boldsymbol{P}_2'\boldsymbol{P}_1 & (1+d)\boldsymbol{P}_2'\boldsymbol{P}_2 \end{vmatrix} = 0.$$

This matrix consists of the blocks of $\boldsymbol{Z}$ multiplied by different scalars. We now multiply on the right by

$$\begin{pmatrix} \boldsymbol{I} & d\boldsymbol{P}_1'\boldsymbol{P}_2 \\ \boldsymbol{0} & (1-d)\boldsymbol{P}_2'\boldsymbol{P}_2 \end{pmatrix},$$

noting that in the process we introduce $(p_2 - p_0)$ roots equal to 1, then the top right-hand side block is transformed to a matrix of zeroes, and we can write

$$(1-d)^{p_1 - p_0} |\boldsymbol{I} - \boldsymbol{P}_1'\boldsymbol{P}_1| . |\boldsymbol{P}_2'[(1-d^2)\boldsymbol{I} - \boldsymbol{P}_1\boldsymbol{P}_1']\boldsymbol{P}_2| = 0.$$

If the eigenvalues of $\boldsymbol{P}_1\boldsymbol{P}_1'$ are $\lambda_i$, $i = 1, ..., p_2 - p_0$, then

$$d_i = \pm (1 - \lambda_i)^{\frac{1}{2}}.$$

Our present equation has $(p_1 - p_0) + 2(p_2 - p_0)$ roots, of which $(p_2 - p_0)$ equal to 1 were introduced. In general then, if $p_1 > p_2$, there will be $(p_1 - p_2)$ of the $d_i$'s equal to $+1$ and $(p_2 - p_0)$ pairs of $d_i$'s with opposite signs, while if $p_1 < p_2$ there will be $(p_2 - p_1)$ of the $d_i$'s equal to –1 and $(p_1 - p_0)$ pairs of $d_i$'s with opposite signs.

The above theory gives a method for calculating the eigenvalues of $\boldsymbol{Z}$ but not its eigenvectors. The eigenvalues and eigenvectors of $\boldsymbol{Z}$ could be found using any routine for symmetric matrices. This would not utilize the special form of $\boldsymbol{Z}$. Also, though there are very accurate and efficient routines for calculating eigenvalues, they do not necessarily return accurate eigenvectors (see, for instance Wilkinson and Reinsch (1971), page 193). An algorithm for symmetric matrices that is particularly attractive when the eigenvectors are needed is the JK-algorithm of Kaiser (1972). In this, planar rotations are applied to the columns of $\boldsymbol{Z}$ to make them orthogonal. The lengths of the columns are then equal to the modulii of the eigenvalues. In this case, we know that many of the columns are initially orthogonal so that the number of planar rotations is greatly reduced. The JK-algorithm actually finds the eigenvalues and eigenvectors of the square of the matrix of interest. As we know that many of the eigenvalues occur in pairs, the signs of the eigenvalues present no problem. However, when a matrix has multiple eigenvalues, the eigenvectors corresponding to the multiple values are free to range over a hyperplane with dimension equal to the multiplicity. In our case, if $\boldsymbol{x}_1$ and $\boldsymbol{x}_2$

are different eigenvectors corresponding to a double eigenvalue of $Z^2$, then any linear combination of $\boldsymbol{x}_1$ and $\boldsymbol{x}_2$ is also an eigenvector. However, $\boldsymbol{Z}$ does not in general have multiple eigenvalues, so that having found an orthogonal pair $\boldsymbol{x}_1$ and $\boldsymbol{x}_2$, the right linear combinations must be taken to obtain the eigenvectors of $\boldsymbol{Z}$. This can be done by solving the eigen-equation $ZV = VD$ for pairs of columns at a time, where the matrix $\boldsymbol{V}$ of the columns of eigenvectors is known to have pairs of columns of the form $(c\boldsymbol{x}_1 - s\boldsymbol{x}_2)$ and $(s\boldsymbol{x}_1 + c\boldsymbol{x}_2)$ where $c^2 + s^2 = 1$.

CHAPTER 5

# When to stop?

## 5.1 What criterion should we use?

There are many criteria being used in deciding how many variables to include. These can be broadly divided into three classes, namely,

- Prediction criteria
- Likelihood or information criteria
- Maximizing Bayesian posterior probabilities.

The emphasis in this book is upon prediction rather than upon finding the 'right' model, and hence there is far more material here on prediction than on likelihood criteria. Bayesian methods will be treated later in Chapter 7.

Under prediction criteria, we first look at the mean squared error of prediction ($MSEP$) for the case in which the same values of the $X$-variables will be used for prediction as those in the data used for model construction. This is sometimes known as the fixed model or $X$-fixed case. The principal criterion used here is Mallows' $C_p$ and variants of it. We look in detail at the case in which the $X$-predictors are orthogonal. Though this case is not common in practice, with most data being from observational studies rather than from designed experiments, it is useful to examine this case for the insights that it gives into the biases that occur in more general cases.

The random model or $X$-random case assumes that future cases for which predictions will be required will come from the same multivariate normal population as that of the data used for model construction.

The section on prediction criteria concludes with a review of some of the applications of cross-validation to model building. The basic idea here is that part of the data is used for model selection, and the remainder is used for validation. We start with the $PRESS$ statistic, which uses leave-one-out cross-validation, and uses the model chosen using the full data set. We then look at leaving out, say $n_v$ cases at the time, selecting a model or a number of models that fit the remaining $(n - n_v)$ cases well, then finding how well the model or models fit the $n_v$ validation cases. This is repeated many times for different random samples of $n_v$ cases out of the $n$.

Under likelihood and other criteria, including the minimum description length (MDL), we look at the Akaike Information Criterion (AIC) and variations of it such as the Bayesian Information Criterion (BIC) and others.

In the appendix at the end of this chapter, some of the criteria are roughly equated to $F$-to-enter criteria.

## 5.2 Prediction criteria

Suppose we want to minimize prediction errors in some sense, e.g. by minimizing the mean squared error of prediction, then a larger subset will often be appropriate than for the case in which there is to be a trade-off between the future cost of measuring more predictors and the prediction accuracy achieved.

The choice of subset, as well as the size of subset, will also depend upon the region of the $X$-space in which we wish to predict. A fairly common situation is one in which one variable, say $X_1$, is expected to vary or be varied in future over a wider range than in the past. If the range of values of $X_1$ in the calibration sample was very small, then the influence of that variable may have been so small that the selection procedure(s) failed to select it. In such circumstances, it may be desirable to force the variable into the model.

Typically, we would like to minimize with respect to the size of subset, $p$, a sum of the kind

$$\|\boldsymbol{X\beta} - \boldsymbol{X_p\hat{\beta}_p}\|_2^2 = \sum_{i=1}^{N} \left( \sum_{j=1}^{k} x_{ij}\beta_j - \sum_{j=1}^{k} I_j x_{ij}\hat{\beta}_j \right)^2 + \sum_{j=1}^{k} C_j I_j, \qquad (5.1)$$

where $I_j = 1$ if the $j^{th}$ variable is in the subset, and $= 0$ otherwise, $C_j$ is the cost of measuring the $j$th variable, where the matrix $\boldsymbol{X} = \{x_{ij}\}$ of values of the predictor variables is specified, as is the method of obtaining the estimated regression coefficients, $\hat{\beta}_j$, and $N$ is the number of future values to be predicted.

As the future values of $x_{ij}$ for which predictions will be required are often not known, the matrix $\boldsymbol{X}$ is often taken to be identical with the observation matrix, or to be the multivariate normal distribution with covariance matrix equal to the sample covariance matrix for the calibration sample. This sometimes leads to some simple analytic results, provided that unbiased estimates are available for the $\beta_j$'s.

There are few examples in the literature of the use of any other assumed future values for the predictor variables, other than Galpin and Hawkins (1982, 1986). They demonstrate that different subsets should be chosen for different ranges of future $X$-values, though their derivation neglects the bias in the parameter estimates that results from the selection process.

Consider one future prediction for which $\boldsymbol{x}$ is the vector of values of the $k$ predictors. The variance of the predictor $\boldsymbol{x}'\hat{\boldsymbol{\beta}}$, where $\boldsymbol{\beta}$ is the vector of least-squares (LS) regression coefficients, is

$$\sigma^2 \boldsymbol{x}'(\boldsymbol{X}'\boldsymbol{X})^{-1}\boldsymbol{x},$$

where $\sigma^2$ is the residual variance with all of the predictors in the model. If we take the gamble of ignoring the biases due to the selection process and assume the same residual variance for selected subsets, then we could minimize this quadratic form, substituting the appropriate subsets for $\boldsymbol{X}$ and

$\boldsymbol{x}$ to find a suitable subset for prediction at $\boldsymbol{x}$. An algorithm for minimizing such quadratic forms has been given by Ridout (1988). Despite the biases, this could be used to indicate a possible subset for future prediction.

The representation (5.1) can also be used when the objective is to estimate the regression coefficients. This can be treated as equivalent to estimating $\boldsymbol{X\beta}$ when each row of $\boldsymbol{X}$ contains zeroes in all except one position. Other $\boldsymbol{X}$-matrices can similarly be constructed if the purpose is to estimate contrasts. For the simplest contrasts, each row of $\boldsymbol{X}$ contains one element equal to +1, one element equal to −1, and the remainder equal to zero.

Unfortunately, almost all of the available theory assumes that we have unbiased least-squares estimates (i.e. no selection bias) of the regression coefficients. This would apply if separate, independent data are used for model selection and for estimation. For instance, Bendel and Afifi (1977) appear at first glance to have solved the common problem of the choice of stopping rule using LS regression coefficients for the case of one data set for both selection and estimation until one notices the requirement that "the subset is selected without reference to the regression sample". In the derivations of mean squared error of prediction ($MSEP$), which follow later in this chapter, we shall see that in many practical cases, the minimum $MSEP$ is obtained by using all of the available predictors, i.e. with no selection, if no correction is made for competition bias in selection.

The basic criterion we will use will be that of minimizing the $MSEP$, but it will be shown that in practice this often yields the same results as using either a false $F$-to-enter criterion, or a likelihood criterion such as the Akaike Information Criterion.

### *5.2.1 Mean squared errors of prediction (MSEP)*

We will consider two models for the predictor variables which Thompson (1978) described as the fixed and random models:

*(a) Fixed model*

The values of the $X$ variables are fixed or controllable, as, for instance, when the data are from a controlled experiment.

*(b) Random model*

The $X$ variables are random variables. This type of model is relevant to observational data when the $X$ variables cannot be controlled.

Calculations of the $MSEP$ often use the same actual values of the $X$ variables as in the calibration data. This effectively treats them as fixed variables even if they are not. The $MSEP$ is higher for the case of random $X$ variables.

In many practical cases using observational data, there will be a mixture of fixed and random variables. For instance, in studies of ozone concentration in the atmosphere, some of the predictors could be season, day of the week, time of day, location, all of which will be known, i.e. fixed variables, while other predictors, such as meteorological variables and concentrations of pollutants such as nitric oxides and hydrocarbons, will be random variables.

### 5.2.2 *MSEP for the fixed model*

Let $\boldsymbol{X_A}$ denote the $n \times p$ matrix consisting of the $p$ columns of $\boldsymbol{X}$ for the $p$ selected variables. If the prediction equation is to contain a constant or intercept term, one of these columns will be a column of 1's. For convenience, it will be assumed that the columns of $\boldsymbol{X}$ have been ordered so that the first $p$ are those for the selected variables. Let $\boldsymbol{X}$ be partitioned as

$$\boldsymbol{X} = (\boldsymbol{X_A}, \boldsymbol{X_B}),$$

where $\boldsymbol{X_B}$ is an $n \times (k-p)$ matrix, and let $\boldsymbol{b_A}$ be the vector of least-squares (LS) regression coefficients, where

$$\boldsymbol{b_A} = (\boldsymbol{X'_A X_A})^{-1}\boldsymbol{X'_A y},$$

Let $\boldsymbol{x}$ be one vector of values of the predictors for which we want to predict $Y$, and let $\hat{y}(\boldsymbol{x})$ denote the predicted value using the LS regression coefficients, i.e.

$$\hat{y}(\boldsymbol{x}) = \boldsymbol{x'_A b_A}.$$

If the true relationship between $Y$ and the $X$ variables is

$$\boldsymbol{y} = \boldsymbol{X\beta} + \boldsymbol{\epsilon},$$

where the residuals, $\boldsymbol{\epsilon}$, have zero mean and variance $\sigma^2$, then the prediction error is

$$\hat{y}(\boldsymbol{x}) - \boldsymbol{x'\beta} = \boldsymbol{x'_A b_A} - \boldsymbol{x'\beta} \tag{5.2}$$

where $\boldsymbol{x'} = (\boldsymbol{x'_A}, \boldsymbol{x'_B})$, i.e. it includes the values of the other $(k-p)$ predictors that were not selected. The prediction error given by (5.2) can be regarded as having the following components:

1. a sampling error in $\boldsymbol{b_A}$,
2. a bias in $\boldsymbol{b_A}$, the selection bias, if the same data were used both to select the model and to estimate the regression coefficients, and
3. a bias due to the omission of the other $(k-p)$ predictors.

If there is no selection bias, which would be the case if independent data had been used to select the model, then

$$\begin{aligned} E(\boldsymbol{b_A}) &= \boldsymbol{\gamma_A}, \text{ say} \\ &= (\boldsymbol{X'_A X_A})^{-1}\boldsymbol{X'_A X\beta} \\ &= (\boldsymbol{X'_A X_A})^{-1}\boldsymbol{X'_A}(\boldsymbol{X_A}, \boldsymbol{X_B})\begin{pmatrix}\boldsymbol{\beta_A} \\ \boldsymbol{\beta_B}\end{pmatrix} \\ &= \boldsymbol{\beta_A} + (\boldsymbol{X'_A X_A})^{-1}\boldsymbol{X'_A X_B \beta_B}; \end{aligned} \tag{5.3}$$

that is, $\boldsymbol{\beta_A}$ is augmented by the regression of $\boldsymbol{X_B \beta_B}$ on the $p$ selected variables.

Now let us rewrite the prediction error (5.2) using (5.3) above as:

$$\begin{aligned}
\hat{y}(\boldsymbol{x}) - \boldsymbol{x}'\boldsymbol{\beta} &= \boldsymbol{x}'_{\boldsymbol{A}}\{[\boldsymbol{b_A} - E(\boldsymbol{b_A})] + [E(\boldsymbol{b_A}) - \gamma_{\boldsymbol{A}}] + [\gamma_{\boldsymbol{A}} - \boldsymbol{\beta_A}]\} \\
&\quad - \boldsymbol{x}'_{\boldsymbol{B}}\boldsymbol{\beta_B} \\
&= \boldsymbol{x}'_{\boldsymbol{A}}\{(1) + (2) + (\boldsymbol{X}'_{\boldsymbol{A}}\boldsymbol{X_A})^{-1}\boldsymbol{X}'_{\boldsymbol{A}}\boldsymbol{X_B}\boldsymbol{\beta_B}\} - \boldsymbol{x}'\boldsymbol{\beta_B} \\
&= \boldsymbol{x}'_{\boldsymbol{A}}\{(1) + (2)\} + [\boldsymbol{x}'_{\boldsymbol{A}}(\boldsymbol{X}'_{\boldsymbol{A}}\boldsymbol{X_A})^{-1}\boldsymbol{X}'_{\boldsymbol{A}}\boldsymbol{X_B} - \boldsymbol{x}'_{\boldsymbol{B}}]\boldsymbol{\beta_B} \\
&= \boldsymbol{x}'_{\boldsymbol{A}}\{(1) + (2)\} + (3),
\end{aligned}$$

where the messy expression for (3) is for the projection of the $Y$ variable on that part of the $\boldsymbol{X_B}$ predictors which is orthogonal to $\boldsymbol{X_A}$.

The expected squared error of prediction is then

$$\begin{aligned}
E[\hat{y}(\boldsymbol{x}) - \boldsymbol{x}'\boldsymbol{\beta}]^2 &= \boldsymbol{x}'_{\boldsymbol{A}}\{\boldsymbol{V}(\boldsymbol{b_A}) + [E(\boldsymbol{b_A}) - \gamma_{\boldsymbol{A}}][E(\boldsymbol{b_A}) - \gamma_{\boldsymbol{A}}]'\}\boldsymbol{x_A} \\
&\quad + (omission\ bias)^2 \\
&= \boldsymbol{x}'_{\boldsymbol{A}}\{\boldsymbol{V}(\boldsymbol{b_A}) + (sel.bias)(sel.bias)'\}\boldsymbol{x_A} \\
&\quad + (omission\ bias)^2, \qquad (5.4)
\end{aligned}$$

where 'sel.bias' denotes the selection bias, and $V(\boldsymbol{b_A})$ is the covariance matrix of the $p$ elements of $\boldsymbol{b_A}$ about their (biased) expected values.

Note that equation (5.4) still holds if $\boldsymbol{b_A}$ has been estimated using a method other than least squares.

To derive the $MSEP$ from (5.4), we need to supply a set of $\boldsymbol{x}$-vectors over which to average. A simple, well-known result can be obtained if we make the following choice of $\boldsymbol{x}$-vectors and the following assumptions about $\boldsymbol{b_A}$:

1. The future $\boldsymbol{x}$-vectors will be the same as those in the $\boldsymbol{X}$-matrix used for the estimation of the regression coefficients.
2. There is no selection bias.
3. $\boldsymbol{V}(\boldsymbol{b_A}) = \sigma^2(\boldsymbol{X}'_{\boldsymbol{A}}\boldsymbol{X_A})^{-1}$.

The second and third conditions apply for LS regression coefficients when the subset has been chosen independently of the data used for the estimation of $\boldsymbol{b_A}$.

Subject to these conditions, we have

$$E(RSS_p) = \sum_{i=1}^{n}(omission\ bias)_i^2 + (n-p)\sigma^2, \qquad (5.5)$$

where $(omission\ bias)_i$ is the bias in estimating the $i^{th}$ observation caused by omitting the $(k-p)$ predictors. We can obtain an estimate of the sum of squares of these omission biases if we replace the left-hand side of (5.5) with the sample $RSS_p$. Note that when the same data are used for both selection and estimation, the sample value of $RSS_p$ is liable to be artificially low for the selected subset.

Now we need to find the sum of the $\boldsymbol{x}'_{\boldsymbol{A}}\boldsymbol{V}(\boldsymbol{b_A})\boldsymbol{x_A}$ terms in (5.4). That is, we need to calculate

$$\sigma^2 \sum_{i=1}^{n} \boldsymbol{x}_i'(\boldsymbol{X}_A'\boldsymbol{X}_A)^{-1}\boldsymbol{x}_i,$$

where $\boldsymbol{x}_i'$ is the $i^{th}$ row of $\boldsymbol{X}_A$. By examining which terms are being multiplied, we see that this is the sum of diagonal elements, i.e. the trace of a matrix product, and that

$$\begin{aligned} \sigma^2 \sum_{i=1}^{n} \boldsymbol{x}_i'(\boldsymbol{X}_A'\boldsymbol{X}_A)^{-1}\boldsymbol{x}_i &= \sigma^2 \text{ trace}\{\boldsymbol{X}_A(\boldsymbol{X}_A'\boldsymbol{X}_A)^{-1}\boldsymbol{X}_A'\} \\ &= \sigma^2 \text{ trace}\{(\boldsymbol{X}_A'\boldsymbol{X}_A)^{-1}\boldsymbol{X}_A'\boldsymbol{X}_A\} \\ &= \sigma^2 \text{ trace}\{\boldsymbol{I}_{p\times p}\} \\ &= p\sigma^2. \end{aligned} \quad (5.6)$$

It is assumed that $\boldsymbol{X}_A'\boldsymbol{X}_A$ is of full rank (why select a subset with redundant variables?). Here we have used the property that for any pair of matrices $\boldsymbol{P}$, $\boldsymbol{Q}$ with appropriate dimensions,

$$\text{trace } \boldsymbol{PQ} = \text{trace } \boldsymbol{QP}.$$

Finally, summing the terms (5.4) over the $n$ future observations and using (5.5) and (5.6), we have that the sum of squared errors

$$\begin{aligned} &\approx p\sigma^2 + RSS_p - (n-p)\sigma^2 && (5.7) \\ &= RSS_p - (n-2p)\sigma^2. && (5.8) \end{aligned}$$

If we divide through by $\sigma^2$, we obtain the well-known Mallows' $C_p$ statistic (Mallows, 1973):

$$C_p = \frac{RSS_p}{\sigma^2} - (n-2p). \quad (5.9)$$

In practice, $\sigma^2$ is replaced with the unbiased estimate

$$\hat{\sigma}^2 = \frac{RSS_k}{(n-k)};$$

that is, the residual variance for the full model.

The mean squared error of prediction ($MSEP$) is defined as

$$E(y - \hat{y}(\boldsymbol{x}))^2,$$

where $y$ is a future value of the $Y$-variable. As we have so far been looking at differences between $\hat{y}(\boldsymbol{x})$ and its expected value, we must now add on an extra term for the future difference between the expected value and the actual value. Hence the $MSEP$ is obtained from (5.8) by dividing through by $n$ and then adding an extra $\sigma^2$. Finally, for the fixed-variables model with unbiased LS-estimates of regression coefficients,

$$MSEP \approx (RSS_p + 2p\sigma^2)/n. \quad (5.10)$$

Minimizing Mallows' $C_p$ has been widely used as a criterion in subset selection despite the requirement for unbiased regression coefficients. Most of the applications have been to situations with predictors that are random variables, whereas the derivation requires that the $X$'s be fixed or controllable variables.

Mallows himself warns that minimizing $C_p$ can lead to the selection of a subset which gives an $MSEP$, using LS regression coefficients, which is much worse than if there were no selection at all and all predictors are used. His demonstration uses $k$ orthogonal predictors; a similar derivation follows. First though, we present a shorter, simpler derivation of the $MSEP$ and hence of Mallows' $C_p$, based upon an orthogonal projection approach.

Write an orthogonal reduction of $\boldsymbol{X}$ as

$$\begin{aligned} \boldsymbol{X} &= (\boldsymbol{X_A},\ \boldsymbol{X_B}) \\ &= (\boldsymbol{Q_A},\ \boldsymbol{Q_B}) \begin{pmatrix} \boldsymbol{R_A} & \boldsymbol{R_{AB}} \\ & \boldsymbol{R_B} \end{pmatrix}, \end{aligned}$$

where $\boldsymbol{Q_A}$, $\boldsymbol{Q_B}$ are orthogonal matrices, and $\boldsymbol{R_A}$, $\boldsymbol{R_B}$ are upper triangular (though $\boldsymbol{R_{AB}}$ is not). Let the vector of orthogonal projections be

$$\begin{pmatrix} \boldsymbol{Q}'_A \\ \boldsymbol{Q}'_B \end{pmatrix} \boldsymbol{y} = \begin{pmatrix} \boldsymbol{t_A} \\ \boldsymbol{t_B} \end{pmatrix}$$

with expected values $\boldsymbol{\tau_A}$, $\boldsymbol{\tau_B}$ for the vectors $\boldsymbol{t_A}$, $\boldsymbol{t_B}$. The LS-regression coefficients for the selected variables are obtained by solving

$$\boldsymbol{R_A b_A} = \boldsymbol{t_A}.$$

Hence the prediction errors, if our future $\boldsymbol{X}$ is exactly the same as the $\boldsymbol{X}$ used for estimation, are given by

$$\begin{aligned} \boldsymbol{X\beta} - \boldsymbol{X_A b_A} &= \boldsymbol{QR\beta} - \boldsymbol{Q_A R_A b_A} \\ &= \boldsymbol{Q\tau} - \boldsymbol{Q_A t_A}. \end{aligned}$$

Hence, because of the orthogonality of the columns of $\boldsymbol{Q}$, the sum of squared errors

$$\begin{aligned} &= E(\boldsymbol{X\beta} - \boldsymbol{X_A b_A})'(\boldsymbol{X\beta} - \boldsymbol{X_A b_A}) \\ &= \sum_{i=1}^{p} var\ t_{Ai} + \sum_{i=p+1}^{k} \tau_{Bi}^2, \end{aligned} \tag{5.11}$$

where $t_{Ai}$, $\tau_{Bi}$ are the $i^{th}$ elements of $\boldsymbol{t_A}$ and $\boldsymbol{\tau_B}$ respectively. As

$$E(RSS_p) = \sum_{i=p+1}^{k} \tau_{Bi}^2 + (n-p)\sigma^2,$$

and the projections $t_{Ai}$ have variance $\sigma^2$, we have that the sum of squared errors

$$= p\sigma^2 + E(RSS_p) - (n-p)\sigma^2.$$

Replacing the expected value of $RSS_p$ with its sample value gives formula (5.8).

*Modifications of Mallows' $C_p$*

Let us look in a little more detail at the components of equation (5.7). This quantity is an estimate of the total squared error of prediction ($TSEP$) of the response variable, $Y$, for $n$ cases with the same values of the $X$ variables as in the calibration data.

$$TSEP = p\sigma^2 + RSS_p - (n-p)\sigma^2,$$

and after substituting $\hat{\sigma}^2 = RSS_k/(n-k)$, we have an estimated $TSEP$:

$$\hat{TSEP} = p\hat{\sigma}^2 + RSS_p - (n-p)\hat{\sigma}^2$$

Now split $RSS_p$ into two parts comprising the regression sum of squares on the $(k-p)$ rejected predictors and the residual with all $k$ predictors in the model; that is,

$$\begin{aligned} \hat{TSEP} &= p\hat{\sigma}^2 + (RSS_p - RSS_k) - (p-k)\hat{\sigma}^2 \\ &\quad + RSS_k - (n-k)\hat{\sigma}^2 \\ &= p\hat{\sigma}^2 + (RSS_p - RSS_k) - (p-k)\hat{\sigma}^2 \qquad (5.12) \end{aligned}$$

The first term above, $p\hat{\sigma}^2$, is contributed by the uncertainty in the regression coefficients, and neglects the bias in these coefficients that is introduced by the variable selection process. At the present time, we have very little idea of the size of these biases or of how to reduce them. There is more on this subject in the next chapter. In many cases, this quantity should be several times larger.

The remainder of 5.12 is an estimate of the squared error due to omitting the last $(k-p)$ variables. It will usually be too small, and in many cases will be negative, which seems strange for the estimate of a squared quantity. We have that

$$RSS_p - RSS_k = \sum_{i=p+1}^{k} t_i^2,$$

that is, it is the sum of squares of the last $(k-p)$ projections. Notice that

$$E(t_i^2) = \tau_i^2 + \sigma^2,$$

where the $\tau_i$'s are the expected values of the projections. Hence $RSS_p - RSS_k - (p-k)\sigma^2$ is an estimate of $\sum_{i=p+1}^{k} \tau_i^2$, and it is this estimate which is often negative.

Instead of trying to obtain an unbiased estimated of this regression sum of squares, which is what Mallows did, let us look at trying to minimize the squared error loss. We will do this by separately shrinking each squared projection in turn.

Suppose that the $i^{th}$ projection, $t_i$, has an unknown expected value equal to $\tau_i$, and variance $\sigma^2$. Let $C_i$ be the shrinkage factor, between 0 and 1, then

Table 5.1 *Shrinkage of projections applied to the STEAM data*

| | Const | $X_7$ | $X_5$ | $X_4$ | $X_1$ | $X_2$ | $X_3$ | $X_6$ | $X_8$ | $X_9$ |
|---|---|---|---|---|---|---|---|---|---|---|
| $t_i$ | 47.12 | −6.75 | 2.93 | 1.51 | 0.74 | −0.51 | −0.33 | −0.41 | −0.72 | −0.93 |
| $\sqrt{C_i}t_i$ | 47.10 | −6.63 | 2.69 | 1.17 | 0.39 | −0.21 | −0.10 | −0.14 | −0.38 | −0.57 |

the squared error loss is

$$E(C_i t_i^2 - \tau_i^2)^2 = C_i^2(\tau_i^4 + 6\tau_i^2\sigma^2 + 3\sigma^4) \; - \; 2C_i\tau_i^2(\tau_i^2 + \sigma^2) \; + \; \tau_i^4$$

where the '3' on the right-hand side is the kurtosis for the normal distribution. The value of $C_i$ which minimizes this is

$$C_i = \frac{(\tau_i^2/\sigma^2)^2 \; + \; (\tau_i^2/\sigma^2)}{(\tau_i^2/\sigma^2)^2 \; + \; 6(\tau_i^2/\sigma^2) \; + \; 3}. \tag{5.13}$$

Unfortunately, this involves the unknown $\tau_i$. Let us simply substitute the value of $t_i$. Substituting for $\sigma^2$ as well, we obtain our estimate of each $\tau_i^2$ as

$$\hat{\tau_i^2} = t_i^2 \; \frac{(t_i^2/\hat{\sigma}^2)^2 \; + \; (t_i^2/\hat{\sigma}^2)}{(t_i^2/\hat{\sigma}^2)^2 \; + \; 6(t_i^2/\hat{\sigma}^2) \; + \; 3}, \tag{5.14}$$

The above shrinkage factor was derived in Miller (2000).

Our modified $C_p$ is then obtained by dividing the estimated total sum of squares by $\hat{\sigma}^2$ giving

$$\text{Modified } C_p \; = \; p \; + \; \sum_{p+1}^{k} C_i(t_i^2/\hat{\sigma}^2). \tag{5.15}$$

Let us look at this shrinkage method applied to the STEAM data. We have seen that variables numbered 4, 5 and 7 seem to provide a good fit to the data. Let us put these three variables at the start, together with the constant term. Table 5.1 shows the projections with the variables in the order shown. The residual sum of squares with all 9 predictors in the model is 4.869 with 15 degrees of freedom, giving an estimate of the residual variance of 4.869/15 = 0.3246. The residual standard deviation is the square root of this, or about 0.57, which has units of pounds of steam. We see that none of the last six projections is large (say more than double) compared with this estimate of $\sigma$.

The last row of Table 5.1 shows the projections after applying the shrinkage just described. As the shrinkage factors, $C_i$, apply to the squared projections, the projections in the table have been scaled by $\sqrt{C_i}$. We see that there is very little shrinkage of the large projections. A Taylor series expansion shows that for large $t_i^2$, the shrinkage of $t_i^2$ is by approximately $5\sigma^2$, while for small $t_i^2$, the shrinkage is to approximately $t_i^4/(3\sigma^2)$.

Figure 5.1 shows Mallows' $C_p$ and the modified $C_p$ calculated using (5.15) for all subsets of variables from the STEAM data, but truncated at 40. If we look at the smallest values of $C_p$ and the modified $C_p$, we see that Mallows'

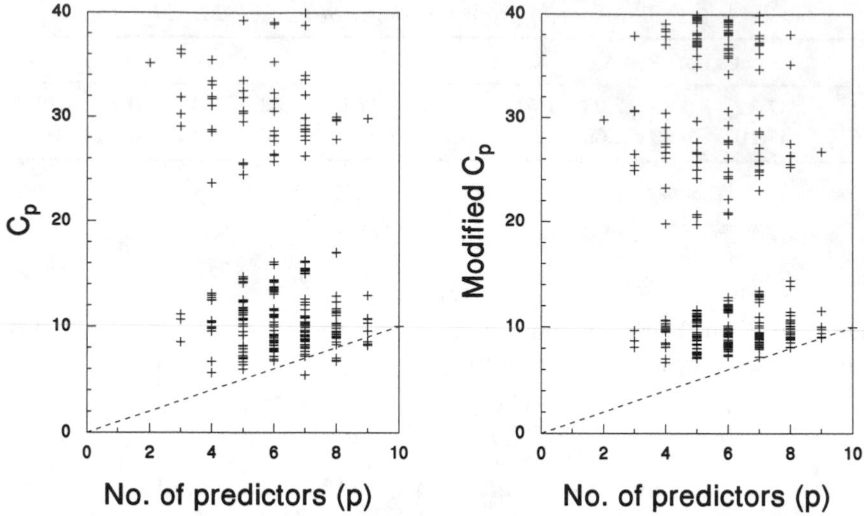


Figure 5.1 *Mallows' $C_p$ and a modified $C_p$ for subsets of the STEAM data.*

Table 5.2 *Demonstration of the range of values of the modified $C_p$ over the order of the rejected variables, using the STEAM data*

| p | Selected variables | Mallows' $C_p$ | Modified $C_p$ |
|---|---|---|---|
| 2 | 7 | 35.14 | (26.27, 33.06) |
| 3 | 1 7 | 8.51 | ( 7.53, 9.73) |
| 4 | 4 5 7 | 5.62 | ( 5.94, 6.64) |
| 5 | 1 4 5 7 | 5.96 | ( 6.50, 7.16) |
| 6 | 1 2 5 7 9 | 6.75 | ( 7.23, 7.37) |
| 7 | 1 3 5 7 8 9 | 5.41 | ( 7.16, 7.18) |
| 8 | 1 3 4 5 7 8 9 | 6.71 | ( 8.07, 8.08) |

$C_p$ reaches a low value at $p = 4$, and then fluctuates about this level and has its minimum at $p = 7$, whereas the modified $C_p$ starts to climb slowly from its minimum at $p = 4$. At $p =$ k (10 for the STEAM data), that is, when all of the variables are included, both $C_p$ and the modified form are equal to $p$. The value of the modified $C_p$ cannot fall below $p$, whereas Mallows' $C_p$ often does. The broken line in both parts of Figure 5.1 is $C_p = p$.

The values of the projections depend upon the order of the variables. Thus, the value of this modified $C_p$ depends upon the order of the rejected variables. Table 5.2 shows the range of values of the modified $C_p$ over all combinations

Table 5.3 *Demonstration of the range of values of the modified $C_p$ over the order of the rejected variables, using the POLLUTE data*

| p | Selected variables | Mallows' $C_p$ | Modified $C_p$ |
|---|---|---|---|
| 2 | 9 | 53.59 | (38.67, 50.63) |
| 3 | 6 9 | 23.70 | (20.83, 28.14) |
| 4 | 2 6 9 | 15.53 | (13.07, 16.67) |
| 5 | 1 2 9 14 | 6.68 | ( 8.22, 9.83) |
| 6 | 1 2 6 9 14 | 4.98 | ( 8.11, 9.23) |
| 7 | 1 2 3 6 9 14 | 6.71 | ( 8.21, 8.97) |
| 8 | 1 2 3 5 6 9 14 | 7.42 | ( 8.70, 8.97) |
| 9 | 1-6 9 14 | 9.42 | ( 9.63, 9.89) |
| 10 | 1-6 9 12 13 | 11.33 | (10.16, 10.23) |
| 11 | 1-6 8 9 12 13 | 12.81 | (11.02, 11.04) |
| 12 | 1-6 8 9 12-14 | 14.28 | (12.01, 12.01) |
| 13 | 1-9 12-14 | 13.65 | (13.00, 13.00) |

of the rejected variables, using the STEAM data set. Notice that the range rapidly decreases as $p$ increases.

Table 5.3 shows similar results for the POLLUTE data set. In this case, Mallows' $C_p$ shows a clear minimum at $p = 6$ (constant + five variables), whereas the modified $C_p$ shows very little difference between $p = 5$, 6 or 7.

The above modification to Mallows' $C_p$ is not very satisfactory because of its dependence upon the order of the omitted variables. It has been included here in the hope that it will stimulate further research.

Apart from a scaling factor, $\sigma^2$, the problem of estimating $\sum \tau_i^2$ for the omitted variables is the same as that of estimating the non-centrality parameter of the non-central chi-squared distribution with $k-p$ degrees of freedom. This could be done using maximum likelihood, see e.g. Anderson (1981) and Spruill (1986). This is feasible but messy, involving an iterative solution to an equation requiring the calculation of modified Bessel functions.

Further alternatives have been suggested by Neff and Strawderman (1976), but only for the case in which our $(k-p)$ is at least 5. A simple alternative is to estimate $\sum \tau_i^2$ as $(RSS_p - RSS_k) - (k-p)\hat{\sigma}^2$ when this is positive, and zero otherwise. Numerical results contained in Saxena and Alam (1982) show that in terms of mean squared error, this is a better estimate than the maximum likelihood estimator over a wide range of numbers of degrees of freedom and values of the noncentrality parameter. Using this estimator, our modified Mallows' $C_p$ estimator becomes

$$C_p^* = (2p-k) + max\left(\frac{RSS_p - RSS_k}{\hat{\sigma}^2}, (k-p)\right), \qquad (5.16)$$

If this $C_p^*$ is used for the STEAM data, the only change to the left-hand part of Figure 5.1 is that those points that lie below the broken line $C_p = p$ are pulled up to that line.

A further modifications to Mallows' $C_p$ has been suggested by Gilmour (1996). In each of the versions of $C_p$ above, we need an estimate of $1/\sigma^2$, not of $\sigma^2$. If we use $\hat{\sigma}^2 = RSS_k/(n-k)$, then the expected value of its reciprocal is $(n-k)/(n-k-2) \times 1/\sigma^2$. Thus, Gilmour suggests that the term in $1/\sigma^2$ in the expression for $C_p$ should be reduced by multiplying by $(n-k-2)/(n-k)$. This of course requires that $(n-k-2) > 0$.

In general, the expected value of a nonlinear function of $\hat{\sigma}^2$ will not be an unbiased estimate of the same function of $\sigma$. One important case is in the estimation of a population standard deviation, in which case,

$$s = \frac{\text{Sum of squares}}{\text{Deg. of freedom} - 0.5}$$

is approximately unbiased. (This can be derived using a Taylor series expansion.)

A further variation on Mallows' $C_p$ has been proposed by Ronchetti and Staudte (1994) for use in robust regression.

*Further discussion of Mallows' $C_p$*

Mallows (1995) derived the asymptotic form of the curve of minimum $C_p$ plotted against $p$ when the predictors are orthogonal. He assumes that all of the true regression coefficients, the $\beta$'s, are nonzero, or equivalently that all of the $\tau$'s are nonzero, but that some of them, $k'$ out of the $k$, are 'small'. He lets $k$ and $k' \to \infty$ as the sample size, $n \to \infty$, then assumes either that the small $\beta$'s are uniformly distributed between $-\tau$ and $\tau$ with density $\lambda = k'/(2\tau)$, or that they are normally distributed with zero mean and standard deviation $\tau'$, giving a density at zero of $\lambda = k'/(\tau'/\sqrt{2\pi})$. Using either density, he derives the asymptotic form

$$C_p = k + \frac{(k-p)^3}{12\lambda^2} - 2(k-p),$$

for $k - k' \leq p \leq k$. This approximates the shape of the relationship observed in practice, with a suitable value for the unknown parameter $\lambda$.

Gilmour (1996) suggested that instead of minimizing $C_p$, hypothesis testing should be applied to the differences between consecutive smallest values of $C_p$, or rather between successive values of his slightly modified $C_p$. Let us suppose that the model with $p$ parameters with the smallest value of $C_p$ contains all of the variables with nonzero regression coefficients. There are $(k-p)$ remaining variables. Let us calculate the $F$-ratio, $(RSS_p - RSS_{p+1})/s^2$, for any one of these variables chosen at random to add next, where $s^2 = RSS_k/(n-k)$. Gilmour argues that this $F$-ratio has the $F$-distribution with 1 and $(n-k)$ degrees of freedom for the numerator and denominator, respectively. By looking at only the largest of these $F$-ratios, we are looking at the largest of $(k-p)$ $F$-ratios, which have independent numerators but the same denominator. He

Table 5.4 *Coefficients in rational polynomial which approximates Gilmour's percentage points for the distribution of the maximum F-ratio*

| Coeff. | $\alpha = 10\%$ | $\alpha = 5\%$ | $\alpha = 1\%$ |
|---|---|---|---|
| $p_{00}$ | –256.7 | 152.9 | 2241. |
| $p_{10}$ | 182.6 | 116.7 | 1204. |
| $p_{01}$ | 334.3 | 226.7 | 4671. |
| $p_{20}$ | 2.706 | 3.841 | 6.635 |
| $p_{11}$ | 490.79 | 334.73 | 1449.1 |
| $p_{02}$ | 71.85 | 56.73 | 281.3 |
| $q_{00}$ | –262.3 | –99.5 | –448. |
| $q_{10}$ | 186.1 | 80.7 | 286. |
| $q_{01}$ | –93.22 | –79.2 | –350. |
| $q_{11}$ | 56.73 | 34.32 | 122.6 |
| $q_{02}$ | 1.98 | 1.22 | 2.89 |

presents a table of these maximum $F$-ratios, which can be used to test for a significant drop in $C_p$ when another variable is added.

The values in Gilmour's table for the upper percentage points of the maximum $F$-ratio are very well approximated by the ratio of two quadratics:

$$\max F(\nu,\ k-p,\ \alpha) =$$

$$\frac{p_{00} + p_{10}\nu + p_{01}(k-p) + p_{20}\nu^2 + p_{11}\nu(k-p) + p_{22}(k-p)^2}{q_{00} + q_{10}\nu + q_{01}(k-p) + \nu^2 + q_{11}\nu(k-p) + q_{22}(k-p)^2},$$

where $\nu = (n-k)$ and $(k-p)$ are as above, and $100\alpha\%$ is the upper tail area or significance level. The values of the $p$ and $q$ coefficients are given in Table 5.4. These values have been obtained by least-squares fitting to the logarithms of the percentage points given in Gilmour's Table 2, except that accurate values read from tables of the $F$-distribution were used for the case $(k-p) = 1$ including values for $(n-k) = 3$ and $(n-k) = \infty$.

In practice, the argument above is not quite correct. The subset of $p$ variables, which gives the used value of $C_p$, may already include one or more variables for which the true regression coefficients are zero. Thus, the $(k-p)$ remaining variables may be from $(k-p+1)$ or $(k-p+2)$ which have zero regression coefficients. Thus, the largest of the $(k-p)$ $F$-ratios is the second or third largest of slightly more than $(k-p)$ values with the largest one or two already selected. The argument used to derive the distribution of the maximum $F$-ratio is only correct if the variables already selected have nonzero regression coefficients.

The effect of using Gilmour's suggestion, and testing at, say the 5% level of significance, is that the reduction in $C_p$ is often not significant before the minimum value of $C_p$ is reached, so that smaller subsets are often selected.

*The case of orthogonal predictors*

Now let us look at the case of orthogonal predictors with estimation and selection from the same data. Adding variable $X_i$ to the subset of selected variables reduces the residual sum of squares by $t_i^2$; hence the approximate $MSEP$, or equivalently Mallows' $C_p$, is minimized by including all of those variables for which $t_i^2 > 2\hat{\sigma}^2$. Hence, if the $t_i$'s are normally distributed with expected values $\tau_i$ and, for convenience, $\sigma = 1$, then

$$E(t_i \mid variable\ X_i\ is\ selected) =$$

$$\frac{\int_{-\infty}^{-\sqrt{2}} t\,\phi(t-\tau_i)dt + \int_{\sqrt{2}}^{\infty} t\,\phi(t-\tau_i)dt}{\Phi(-\sqrt{2}-\tau_i) + 1 - \Phi(\sqrt{2}-\tau_i)},$$

where $\phi$ and $\Phi$ are the density function and distribution function, respectively, for the standard normal distribution. Here, for simplicity, it is assumed that $\sigma = 1$; a more rigorous derivation would use the t-distribution instead of the normal.

The true sum of squared errors in this case is then given by

$$\sum_{i=1}^{p}(selection\ bias)_i^2 + \sum_{i=1}^{p} var(t_{Ai}) + \sum_{i=p+1}^{k} \tau_{Bi}^2, \tag{5.17}$$

where the variance of the projections of selected variables is no longer equal to $\sigma^2$. Table 5.5 shows the contributions to the sum of squared errors according to whether a variable is selected, using the first two terms of (5.17), or rejected, using last term of (5.17), as a function of the expected projection, $\tau_i$.

In practice, we do not know the expected values of the projections, so where should we apply the cut-off? To answer this question, let us look at the sum of squared errors for a mixture of $\tau_i$'s. Let $\tau_1 = 10$, so that the first variable will almost always be selected, and let $\tau_i = 10\alpha^{i-1}$ for some $\alpha$ between 0 and 1. Let the residual standard deviation, $\sigma$, $= 1$. Let $C$ be the cut-off value such that variable $X_i$ is selected if $|t_i| \geq C$, and rejected otherwise. For variable $X_i$, the expected contribution to the sum of squared errors is

$$prob(|t_i| \geq C).E[(t_i - \tau_i)^2 \text{ given } |t_i| \geq C] + prob(|t_i| < C).\tau_i^2$$

$$= \int_{-\infty}^{-C} (t-\tau_i)^2\phi(t-\tau_i)dt + \int_{C}^{\infty} (t-\tau_i)^2\phi(t-\tau_i)dt$$

$$+ \tau_i^2 \int_{-C}^{C} \phi(t-\tau_i)dt, \tag{5.18}$$

This quantity is shown as function $m(\tau)$ in Figure 4 of Mallows (1973), and is shown here in Figure 5.2.

We see from Figure 5.2 that if the true but unknown projections are less than about 0.8 in absolute value, then we should reject that variable. In this case, the higher the value of $C$ the better. But if the unknown true projec-

Table 5.5 *Expected values of contributions to the sum of squared errors against expected projections for orthogonal predictors, using Mallows' $C_p$ as the stopping rule*

| $\tau_i$ | Contribution if selected | Contribution if rejected | Wtd. average from (6.2.1.12) |
|---|---|---|---|
| 0.2 | 3.46 | 0.04 | 0.61 |
| 0.4 | 3.01 | 0.16 | 0.70 |
| 0.6 | 2.48 | 0.36 | 0.84 |
| 0.8 | 1.99 | 0.64 | 1.02 |
| 1.0 | 1.59 | 1.00 | 1.20 |
| 1.2 | 1.28 | 1.44 | 1.37 |
| 1.4 | 1.06 | 1.96 | 1.51 |
| 1.6 | 0.90 | 2.56 | 1.60 |
| 1.8 | 0.79 | 3.24 | 1.65 |
| 2.0 | 0.73 | 4.00 | 1.64 |
| 2.5 | 0.72 | 6.25 | 1.49 |
| 3.0 | 0.81 | 9.00 | 1.27 |
| 3.5 | 0.90 | 12.25 | 1.11 |
| 4.0 | 0.96 | 16.00 | 1.04 |
| 4.5 | 0.989 | 20.25 | 1.009 |
| 5.0 | 0.998 | 25.00 | 1.002 |

tions are greater than about 1.1 in absolute value, then a large value of $C$ is undesirable.

Figures 5.3 and 5.4 show the error sums of squares for the mixture of $\tau_i$'s described above for $\alpha = 0.8$ and 0.6, respectively. In Figure 5.3, for $k = 10$ available predictors, the smallest true projection is $\tau_{10} = 1.34$, so that there are no very small projections. In this case, the use of any cut-off is undesirable. Even in the case of $k = 20$, when the smallest true projection is $\tau_{20} = 0.144$, there are sufficient true projections between 1 and 10 that nothing is gained by using any cut-off. For larger values of $k$ or smaller values of $\alpha$ (see Figure 5.4), when there are sufficiently many true small projections, moderate gains in reducing the error sum of squares can be obtained by using a cut-off in the vicinity of 2.0 standard deviations, as opposed to the $\sqrt{2}$ for Mallows' $C_p$.

Looking back to equation (5.17), we see that the first term in the sum of squared errors is the square of the selection bias. If we can halve this bias, then its contribution to the squared error will be divided by four. We can anticipate that the likelihood estimation method described in section 5.4 will thus give squared errors closer to the broken lines than to the solid ones in figures 5.3 and 5.4, and hence that a stopping rule, such as minimizing Mallows' $C_p$ or an $F$-to-enter of about 2.0, will produce something close to the optimal linear subset predictor.

The conclusions of this section are that for prediction when future values

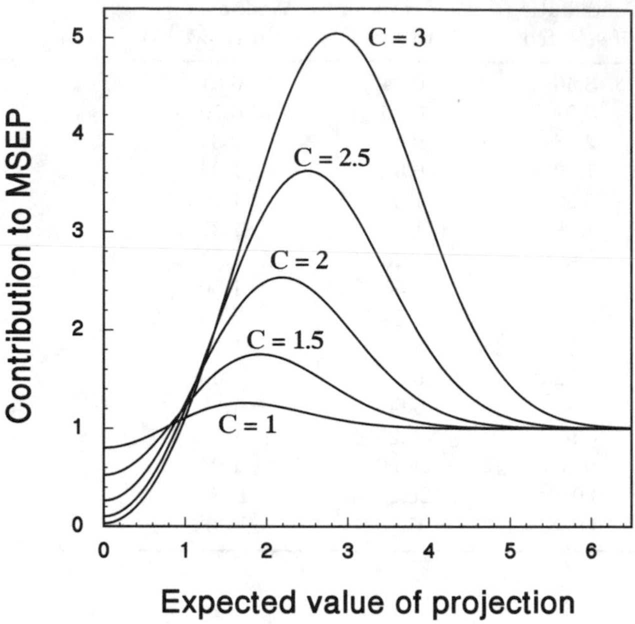


Figure 5.2 . *Expected contributions to the MSEP for the case of orthogonal predictors, against the expected value of a projection,* $\tau$*, for a range of cut-off values,* $C$.

of the predictor variables will take the same values as those in the calibration sample:

- Using biased LS-regression coefficients estimated from the same data as were used to select the model, a cut-off value of about 2.0 standard deviations for the absolute value of the LS-projections (roughly equivalent to an $F$-to-enter of 4.0) is about optimal when an appreciable fraction of the true (but unknown!) projections are less than about 0.8 standard deviations in absolute value; otherwise, no selection should be used.
- It is desirable to try to reduce or eliminate the bias in the LS-regression coefficients. If this can be achieved, then using a cut-off of between 1.5 and 2.0 standard deviations for the sample LS-projections may be about optimal.

The results described in this section have been for orthogonal predictors. In this case, the contribution of a variable to the sum of squared errors is independent of the other variables that have been selected, so there is no competition bias. The results should be applied with caution in cases where the

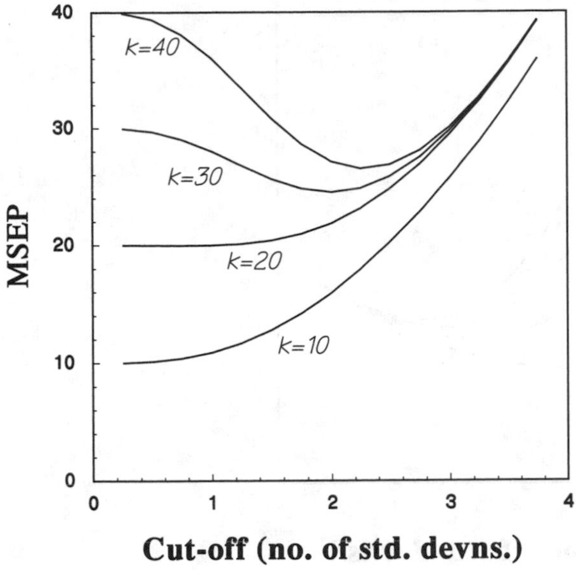


Figure 5.3 *MSEP against cut-off for* $k = 10, 20, 30$ *or* $40$ *predictors, and* $\tau_i = 10\alpha^{i-1}$ *with* $\alpha = 0.8$.

predictors in the calibration data are not orthogonal. It should be emphasized that these results are for an assumed pattern of values of the true projections; there is no evidence that this pattern is realistic.

*The Risk Inflation Criterion (RIC)*

Foster and George (1994) introduced the Risk Inflation Criterion (RIC). They defined 'predictive risk' as

$$R(\boldsymbol{\beta}, \hat{\boldsymbol{\beta}}) = E_{\boldsymbol{\beta}}|\boldsymbol{X}\hat{\boldsymbol{\beta}} - \boldsymbol{X}\boldsymbol{\beta}|^2,$$

where the elements of $\hat{\boldsymbol{\beta}}$ corresponding to variables that are not selected are set equal to zero, and $E_{\boldsymbol{\beta}}$ indicates that the expected value is a function of the unknown vector $\boldsymbol{\beta}$ of 'true' regression coefficients. This last point is made as the notation $E_x$ is often used to mean that the expectation is taken with respect to the subscripted variable, $x$. They first derive this risk for the case of orthogonal predictors, deriving equation (5.18) of this book (giving appropriate references to Mallows (1973) and to the first edition of this book). $\Pi$ in the notation of Foster and George equals the square of our cut-off $C$. By finding the value of each regression coefficient $\beta_j$ for which the risk is the

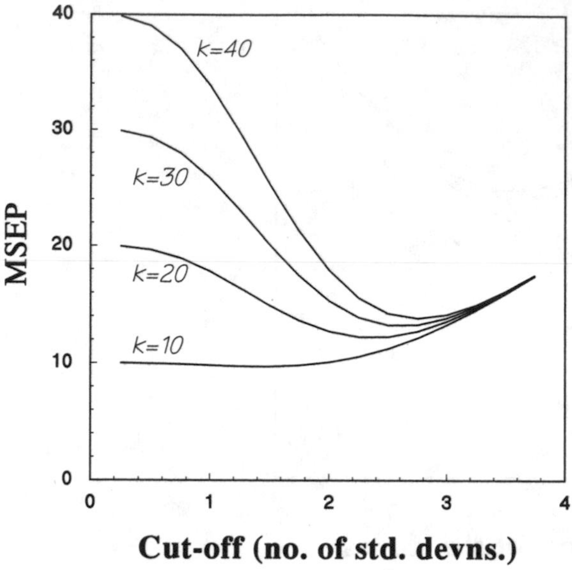


Figure 5.4 *MSEP against cut-off for k = 10, 20, 30 or 40 predictors, and* $\tau_i = 10\alpha^{i-1}$ *with* $\alpha = 0.6$.

worst, they show that a good cut-off rule is to use

$$C^2 = 2\ln(k), \tag{5.19}$$

$C^2$ is equal to the $F$-to-enter in the simple forward, backward and Efroymson stepwise selection procedures. Thus, with say 10 available predictors, we should use an $F$-to-enter of 4.6; for 50, we should use 7.8, and for 200 we should use 10.6.

The RIC contrasts strikingly with many of the other stopping criteria, in that it is *not* a function of the sample size $n$. (N.B. There is a small dependency upon $n$ through the number of degrees of freedom of the $F$-distribution, but this is very small for large $n$.)

In a later paper, George and Foster (2000), give a heuristic 'derivation' or discussion of the RIC in terms of the largest order statistic of the squares of $k$ independent samples from the $t$-distribution. Based upon this idea, the number of variables falsely selected, that is, variables for which the true value of $\beta = 0$, will very slowly decrease as $k$ tends to $\infty$. Table 5.6 shows the probability that all of the false variables will be rejected as a function of the number, $k$, of false variables. The probabilities in this table are $\Phi(z)^k$, where $\Phi(z)$ is the area under the normal curve up to a cut-off $z = \sqrt{2\ln k}$. This is the probability if we have a large number of residual degrees of freedom.

Table 5.6 *Probability that all false variables are rejected using the RIC criterion, against the number of predictors, k, when they are false*

| $k$ | $\sqrt{2\ln k}$ | Prob. |
|---|---|---|
| 10 | 2.146 | 0.8516 |
| 20 | 2.448 | 0.8657 |
| 50 | 2.797 | 0.8789 |
| 100 | 3.035 | 0.8866 |
| 200 | 3.255 | 0.8929 |
| 500 | 3.526 | 0.8997 |
| 1000 | 3.717 | 0.9041 |
| 2000 | 3.899 | 0.9079 |
| 5000 | 4.127 | 0.9123 |
| 10000 | 4.292 | 0.9152 |

Similar tables can be constructed using the $t$-distribution for small numbers of degrees of freedom.

How does this compare with Mallows' $C_p$, and other criteria to be introduced later, such as Akaike's AIC, the BIC, the Hannan and Quinn criteria? From the first few lines of Table 5.6, we see that the RIC is far more ruthless than Mallows' $C_p$ and the AIC at rejecting variables that should be rejected. This also applies to the BIC and Hannan and Quinn criteria for moderate sample sizes, until $n$ is quite large, so that the $\log(n)$ and $\log(\log(n))$ terms in these criteria exceed the value of the $\sqrt{2\ln k}$ term in the RIC. Conversely, this means that the RIC is more reluctant to include variables with nonzero but small regression coefficients until the sample size is much larger. Where the expected value of a least-squares projection is nonzero, that projection increases proportional to $\sqrt{n}$, so that the sample projection will ultimately exceed the cut-off value in Table 5.6.

For a fixed number of predictors, $k$, the RIC is not quite asymptotically consistent. Asymptotically, it will always include all of the variables that should be included, but there will always be a small probability, of the order of perhaps 10%, that it will include one or two variables that should be rejected.

For the case of nonorthogonal predictors, Foster and George (1994) argue that the RIC is still valid.

George and Foster give credit to Donoho and Johnstone (1994), who independently discovered the same criterion in the context of wavelet modelling.

### *5.2.3 MSEP for the random model*

In this case, the omission bias in the fixed model becomes additional residual variation. However, in practice, the columns of the $X$-matrix will usually be far from orthogonal, and there will often be considerable competition between variables for selection.

Suppose that

$$Y = \boldsymbol{X_A}\boldsymbol{\beta_A} + \boldsymbol{\epsilon},$$

where the residuals, $\boldsymbol{\epsilon}$, have zero expected value and $E(\epsilon^2) = \sigma_A^2$, i.e. the residual variance is a function of the subset. Assume for the moment that we have unbiased estimates $\boldsymbol{b_A}$ of $\boldsymbol{\beta_A}$ and $s_A^2$ of $\sigma_A^2$.

For one future prediction, the prediction error is

$$\hat{y}(\boldsymbol{x_A}) - \boldsymbol{x}'_{\boldsymbol{A}}\boldsymbol{\beta_A} = \boldsymbol{x}'_{\boldsymbol{A}}(\boldsymbol{b_A} - \boldsymbol{\beta_A}).$$

Hence, the predictions are unbiased with variance

$$\boldsymbol{x}'_{\boldsymbol{A}}\boldsymbol{V}(\boldsymbol{b_A})\boldsymbol{x_A},$$

where $\boldsymbol{V}(\boldsymbol{b_A})$ is the covariance matrix of $\boldsymbol{b_A}$.

If we use independent data for selection and estimation, which would give us unbiased LS-regression coefficients, then the covariance of the regression coefficients is

$$\boldsymbol{V}(\boldsymbol{b_A}) = \sigma_A^2(\boldsymbol{X}'_{\boldsymbol{A}}\boldsymbol{X_A})^{-1},$$

where $\boldsymbol{X_A}$ is the subset of the $\boldsymbol{X}$-matrix used for estimation. If we now average the squared prediction errors over $\boldsymbol{x_A}$'s comprising the rows of $\boldsymbol{X_A}$, we derive

$$\begin{aligned} MSEP &= \frac{\sigma_A^2}{n}\sum_{i=1}^{n}\boldsymbol{x}'_{\boldsymbol{i}}(\boldsymbol{X}'_{\boldsymbol{A}}\boldsymbol{X_A})^{-1}\boldsymbol{x_i} + \sigma_A^2 \\ &= \sigma_A^2 . \frac{n+p}{n} \qquad (5.20) \end{aligned}$$

from (5.6), where $p$ is the rank of $\boldsymbol{X}'_{\boldsymbol{A}}\boldsymbol{X_A}$. If we replace $\sigma_A^2$ with its sample estimate, then the estimated $MSEP$ is

$$\approx \frac{RSS_p}{n-p}.\frac{n+p}{n} \qquad (5.21)$$

This expression for the $MSEP$ usually gives numerical values that are almost the same as those from the fixed-variables model, provided that the sample size, $n$, is large compared with the number of predictors, $p$. If we replace the $\sigma^2$ in (5.8) by $RSS_p/(n-p)$, then we obtain (5.21).

At the minimum $MSEP$ we have

$$\frac{RSS_p(n+p)}{(n-p)n} \leq \frac{RSS_{p+1}(n+p+1)}{(n-p-1)n}.$$

A little rearrangement shows that at the minimum

$$(n+p)(n-p-1)(RSS_p - RSS_{p+1}) \leq 2n.RSS_{p+1}$$

or

$$\frac{RSS_p - RSS_{p+1}}{RSS_{p+1}/(n-p-1)} \leq \frac{2n}{n+p}. \qquad (5.22)$$

The left-hand side of (5.22) is the usual '$F$-to-enter' statistic, so that when the $MSEP$ is minimized, the '$F$-to-enter' for the next larger subset is less than $2n/(n+p)$, or a little less than 2 if $n \gg p$.

For the random model, it is unreasonable to assume that future $\boldsymbol{x}$'s will take the same values as in the calibration sample. It may be reasonable though to assume that the future $\boldsymbol{x}$'s will be sampled from the same distribution as the $\boldsymbol{x}$'s in the calibration sample. It may be reasonable in some circumstances to assume that future $\boldsymbol{x}$'s will be sampled from the same distribution as the $\boldsymbol{x}$'s in the calibration sample. The $MSEP$ for future $\boldsymbol{x}$'s is

$$MSEP \;=\; \sigma_A^2 \;+\; E[\boldsymbol{x}'_{\boldsymbol{A}}\boldsymbol{M_2}(\boldsymbol{b_A} - \boldsymbol{\beta_A})\boldsymbol{x_A}],$$

where $\boldsymbol{M_2}(\boldsymbol{b_A} - \boldsymbol{\beta_A})$ is the matrix of second moments of $\boldsymbol{b_A}$ about $\boldsymbol{\beta_A}$. The expectation has to be taken over both the future $\boldsymbol{x}$'s and $\boldsymbol{M_2}$.

Again, let us assume that we have unbiased estimates $b_A$ of $\beta_A$ with covariance matrix

$$\boldsymbol{V}(\boldsymbol{b_A}) \;=\; \sigma_A^2(\boldsymbol{X}'_{\boldsymbol{A}}\boldsymbol{X_A})^{-1},$$

as would be the case if independent data had been used for model selection and parameter estimation. We have then that

$$MSEP \;=\; \sigma_A^2 \;+\; \sigma_A^2 E[\boldsymbol{x}'_{\boldsymbol{A}}(\boldsymbol{X}'_{\boldsymbol{A}}\boldsymbol{X_A})^{-1}\boldsymbol{x_A}], \tag{5.23}$$

In circumstances in which the future $\boldsymbol{x}$'s are already known, substitution in (5.23) gives the $MSEP$. Galpin and Hawkins (1982) do precisely that, and show that in some cases quite different subsets should be chosen for different $\boldsymbol{x}$'s to give the best predictions in the $MSEP$ sense.

A simple general formula for the $MSEP$ can be obtained if we assume

- that a constant is being fitted in the model, and
- that the calibration sample and the future $\boldsymbol{x}$'s are independently sampled from the same multivariate normal distribution.

From (5.23), we have

$$MSEP \;=\; \sigma_A^2\{1 \;+\; (1/n) \;+\; E[\boldsymbol{x}'^*_{\boldsymbol{A}}(\boldsymbol{X}'^*_{\boldsymbol{A}}\boldsymbol{X}^*_{\boldsymbol{A}})^{-1}\boldsymbol{x}^*_{\boldsymbol{A}}]\},$$

where the asterisks indicate that the sample mean of the calibration data has been removed from each variable. The $(1/n)$ is for the variance of the mean. The vectors are now of length $(p-1)$.

If the $X$-variables are sampled from a multivariate normal distribution with covariance matrix $\boldsymbol{\Sigma_A}$, then the variance of the future $\boldsymbol{x}^*_{\boldsymbol{A}}$'s, after allowing for the removal of the sample mean, is

$$[1 \;+\; (1/n)]\boldsymbol{\Sigma_A}.$$

An estimate of $\boldsymbol{\Sigma_A}$ is provided by

$$\boldsymbol{V} \;=\; (\boldsymbol{X}'^*_{\boldsymbol{A}}\boldsymbol{X}^*_{\boldsymbol{A}})/(n-1).$$

Hence, we can write

$$\boldsymbol{x}'^*_{\boldsymbol{A}}(\boldsymbol{X}'^*_{\boldsymbol{A}}\boldsymbol{X}^*_{\boldsymbol{A}})^{-1}\boldsymbol{x}^*_{\boldsymbol{A}} \;=\; \frac{n+1}{n(n-1)}\boldsymbol{t}'\boldsymbol{V}^{-1}\boldsymbol{t}, \tag{5.24}$$

where $\boldsymbol{t}$ is a vector of statistics with zero mean and covariance matrix $\boldsymbol{\Sigma_A}$.

Now $\boldsymbol{t}'\boldsymbol{V}^{-1}\boldsymbol{t}$ is Hotelling's $T^2$-statistic (see any standard text on multivariate analysis, e.g. Morrison (1967, pages 117-124) or Press (1972, pages 123-126)). The quantity

$$(n-p+1)T^2/((p-1)(n-1))$$

is known to have an $F$-distribution, with $(p-1)$ and $(n-p+1)$ degrees of freedom for the numerator and denominator, respectively. The expected value of $F(\nu_1, \nu_2)$ is $\nu_2/(\nu_2-2)$, where $\nu_1, \nu_2$ are the numbers of degrees of freedom, and hence the expected value of (5.24) is

$$\frac{n+1}{n(n-1)} . \frac{(n-1)(p-1)}{n-p+1} . \frac{n-p+1}{n-p-1} = \frac{(n+1)(p-1)}{n(n-p-1)} .$$

Finally, we derive

$$MSEP = \sigma_A^2\{1 + (1/n) + \frac{(n+1)(p-1)}{n(n-p-1)}\}. \tag{5.25}$$

This result is due to Stein (1960), though the derivation given here is that of Bendel (1973). Notice that the $p$ variables include the constant in the count, in line with the usual derivation of Mallows' $C_p$, which is the equivalent result for the fixed-variables case. Other authors, including Thompson (1978) and Bendel, quote the result without the constant included in the count for $p$.

For large $n$ and $p \ll n$, the last term in (5.25) is approximately $(p-1)/n$, so that the numerical value of (5.25) is nearly the same as that of (5.20). Notice that as $p/n$ increases, the last term of (5.25) increases rapidly.

If we replace $\sigma_A^2$ with the estimate $RSS_p/(n-p)$, then the $MSEP$ becomes

$$\frac{RSS_p}{n-p} . \frac{(n+1)(n-2)}{n(n-p-1)} , \tag{5.26}$$

Minimizing this estimated $MSEP$ with respect to $p$ is then equivalent to minimizing $RSS_p/((n-p)(n-p-1))$. At the minimum we have that

$$\frac{RSS_p}{(n-p)(n-p-1)} < \frac{RSS_{p+1}}{(n-p-1)(n-p-2)}$$

or, after some rearrangement, that

$$\frac{RSS_p - RSS_{p+1}}{RSS_{p+1}/(n-p-1)} < \frac{2(n-p-1)}{n-p-2} .$$

That is, at the minimum the '$F$-to-enter' statistic is less than a quantity that is just greater than 2. In practice, minimizing the estimated $MSEP$ given by (5.26) often selects the same size of subset as minimizing Mallows' $C_p$.

A somewhat surprising feature of all the formulae for the $MSEP$ that we have derived is that they do not involve the $\boldsymbol{X}$-matrix. That is, the estimated $MSEP$ is the same whether the $X$-variables are highly correlated or almost independent. The basic reason for this independence of $\boldsymbol{X}$ is that the same pattern of correlations has been assumed for future $\boldsymbol{x}$'s. These correlations are important if either

- the future $\boldsymbol{x}$'s will have a different pattern from those used for calibration, or
- the same data are used for both model selection and estimation of the regression coefficients.

### 5.2.4 A simulation with random predictors

In most practical cases, the same data are used for both model selection and for estimation. To examine the effect of competition bias in inflating the $MSEP$, a simulation experiment was carried out. The experiment was a $5 \times 2^3$ complete factorial with replication. The four factors were as follows:

- A. Number of available predictors, $k = 10(10)50$, plus a constant.
- B. Sample size, $n$, either small ($n = 2k$) or moderate ($n = 4k$).
- C. Ill-conditioning of $\boldsymbol{X}'\boldsymbol{X}$, either low or moderate.
- D. Size of smallest LS projection, either small or moderate.

To obtain the desired amount of ill-conditioning, a method similar to that of Bendel (1973) was employed. A diagonal matrix was constructed with diagonal elements decreasing geometrically except for a small constant, $\delta$, added to prevent them from becoming too small. The initial values of the diagonal elements were $\lambda_1 = k(1-\alpha)(1-\delta)/(1-\alpha^k) + \delta$, and $\lambda_i = \alpha(\lambda_{i-1} - \delta) + \delta$ for $i = 2, ..., k$. $\delta$ was arbitrarily set equal to 0.001. The $\lambda$'s are then of course the eigenvalues of this matrix, and are chosen to sum to $k$, which is the trace of a $k \times k$ correlation matrix. Similarity transformations of the kind

$$\boldsymbol{A}_{r+1} = \boldsymbol{P}\boldsymbol{A}_r\boldsymbol{P}^{-1}$$

preserve the eigenvalues. Random planar rotations were applied to each pair of rows and columns. A further $(k-1)$ similarity transformations were then used to produce 1's on the diagonal. Thus, if the diagonal block for rows $i$ and $i+1$ is

$$\begin{pmatrix} w & x \\ x & z \end{pmatrix},$$

then the required similarity transformation is

$$\begin{pmatrix} c & s \\ -s & c \end{pmatrix}\begin{pmatrix} w & x \\ x & z \end{pmatrix}\begin{pmatrix} c & -s \\ s & c \end{pmatrix} = \begin{pmatrix} 1 & x^* \\ x^* & z^* \end{pmatrix},$$

where the tangent ($t = s/c$) satisfies

$$t^2(1-z) - 2tx + (1-w) = 0;$$

that is,

$$t = \frac{x \pm \{x^2 - (1-z)(1-w)\}^{1/2}}{(1-z)}.$$

In the rare event that the roots of this quadratic were not real, rows $i$ and $i+2$, or $i$ and $i+3$ if necessary, were used. As the diagonal elements must average 1.0, it is always possible to find a diagonal element $z$ on the opposite side of 1.0 from $w$ so that the product $(1 - z)(1 - w)$ is negative and hence the roots are real.

The matrix so generated was used as the covariance matrix, $\mathbf{\Sigma}$, of a multivariate normal distribution. If we form the Cholesky factorization

$$\mathbf{\Sigma} = \boldsymbol{L}\boldsymbol{L}',$$

where $\boldsymbol{L}$ is a lower-triangular matrix, then a single sample can be generated from this distribution using

$$\boldsymbol{x} = \boldsymbol{L}\boldsymbol{\epsilon},$$

where $\boldsymbol{\epsilon}$ is a vector of elements $\epsilon_i$ which are sampled independently from the standard normal distribution. This follows as the covariance matrix of the elements of $\boldsymbol{x}$ is

$$\begin{aligned} E(\boldsymbol{x}\boldsymbol{x}') &= E(\boldsymbol{L}\boldsymbol{\epsilon}\boldsymbol{\epsilon}'\boldsymbol{L}') \\ &= \boldsymbol{L}\boldsymbol{L}' \\ &= \mathbf{\Sigma}. \end{aligned}$$

The projections of the $Y$-variable were chosen so that their expected values, $\tau_i = n^{1/2}\gamma^{i-1}$ for $i = 1, ..., k$. The expected values of the regression coefficients for these projections were then obtained by solving

$$\boldsymbol{L}'\boldsymbol{\beta} = \boldsymbol{\tau},$$

and each value $y$ of $Y$ was generated as

$$y = \boldsymbol{x}'\boldsymbol{\beta} + \epsilon,$$

where the residuals, $\boldsymbol{\epsilon}$, were sampled from the standard normal distribution.

To decide upon suitable values to use for $\alpha$, the eigenvalues (principal components) of some correlation matrices were examined. Table 5.7 shows the eigenvalues for the DETROIT, LONGLEY, POLLUTE and CLOUDS data sets. A crude way of choosing a value for $\alpha$ for each data set is to fit the largest and smallest (nonzero) eigenvalues, ignoring $\delta$. If we call these values $\lambda_{max}$ and $\lambda_{min}$, then the fitted value of $\alpha$ is $(\lambda_{min}/\lambda_{max})^{1/(r-1)}$, where $r$ is the rank. Table 5.8 shows these values, together with similar values calculated for the data sets examined by Bendel (1973, page 91).

Based upon Table 5.8, the values chosen for $\alpha$ and $\gamma$ in the simulations were as in Table 5.9.

A high value for $\alpha$ meant that the $X$-predictors were not highly correlated, while a low value meant that they were. A high value for $\gamma$ meant that the smallest projection was of the order of the residual standard deviation ($\sigma = 1$) when $n = 2k$, and this led to the selection of large subsets. The smaller values of $\gamma$ meant that a moderate number of the expected projections were very large compared with the noise in the sample projections, and this led to the selection of much smaller subsets.

Table 5.7 *Eigenvalues (principal components) of some correlation matrices*

| Data set | | | |
|---|---|---|---|
| DETROIT | LONGLEY | POLLUTE | CLOUDS |
| 2.83 | 2.15 | 2.13 | 2.85 |
| 1.38 | 1.08 | 1.66 | 2.26 |
| 0.90 | 0.45 | 1.43 | 1.81 |
| 0.32 | 0.12 | 1.16 | 1.37 |
| 0.26 | 0.051 | 1.11 | 1.05 |
| 0.18 | 0.019 | 0.98 | 0.45 |
| 0.14 | | 0.78 | 0.40 |
| 0.12 | | 0.69 | 0.31 |
| 0.075 | | 0.61 | 0.22 |
| 0.050 | | 0.47 | 0.11 |
| 0.020 | | 0.41 | 0.093 |
| | | 0.36 | 0.060 |
| | | 0.34 | 0.021 |
| | | 0.21 | |
| | | 0.070 | |

Table 5.8 *Ratios of smallest to largest eigenvalues of correlation matrices. Data for the last eight data sets have been calculated from the table on page 91 of Bendel (1973)*

| Data set | Rank(r) | $\lambda_1$ | $\lambda_r$ | $(\lambda_1/\lambda_r)^{1/(r-1)}$ |
|---|---|---|---|---|
| DETROIT | 11 | 2.83 | 0.020 | 0.37 |
| LONGLEY | 6 | 2.15 | 0.019 | 0.15 |
| POLLUTE | 15 | 2.13 | 0.070 | 0.61 |
| CLOUDS | 13 | 2.85 | 0.021 | 0.44 |
| AFI | 13 | 3.9 | 0.02 | 0.64 |
| CMA | 16 | 4.2 | 0.14 | 0.80 |
| CRD | 11 | 3.2 | 0.11 | 0.71 |
| CWE | 19 | 2.1 | 0.27 | 0.89 |
| MCC | 15 | 2.4 | 0.10 | 0.80 |
| MEY | 23 | 8.8 | 0.05 | 0.79 |
| ROC[a] | 12 | 3.5 | 0.36 | 0.81 |
| ROC[b] | 12 | 7.7 | 0.17 | 0.71 |

Table 5.9 *Values of $\alpha$ and $\gamma$ used in simulations*

| No. of predictors ($k$) | 10 | 20 | 30 | 40 | 50 |
|---|---|---|---|---|---|
| High $\alpha$ or $\gamma$ | 0.75 | 0.80 | 0.85 | 0.90 | 0.95 |
| Low $\alpha$ or $\gamma$ | 0.45 | 0.55 | 0.70 | 0.75 | 0.80 |

N.B. The number of predictors $k$ shown above excludes the constant that was fitted.

Table 5.10 *Sample means and standard deviations (in brackets) of sizes of selected subsets using minimum estimated $MSEP$ as the stopping rule. The constant has been excluded from the count of variables in the table below*

| $k$ | $n$ | Large $\alpha$ | | Small $\alpha$ | |
|---|---|---|---|---|---|
| | | Large $\gamma$ | Small $\gamma$ | Large $\gamma$ | Small $\gamma$ |
| 10 | 20 | 4.9 (1.0) | 4.4 (1.6) | 4.8 (1.6) | 4.9 (2.2) |
| | 40 | 6.3 (1.3) | 4.7 (0.8) | 5.9 (1.1) | 4.0 (0.9) |
| 20 | 40 | 10.1 (2.4) | 6.7 (1.9) | 8.3 (2.4) | 7.2 (2.9) |
| | 80 | 10.1 (2.5) | 7.2 (1.9) | 9.5 (2.4) | 6.6 (3.1) |
| 30 | 60 | 14.1 (2.7) | 10.6 (2.9) | 12.8 (2.3) | 12.1 (5.7) |
| | 120 | 15.8 (2.4) | 11.6 (2.5) | 15.1 (2.8) | 10.5 (2.6) |
| 40 | 80 | 20.9 (2.8) | 14.5 (2.1) | 22.0 (4.4) | 13.9 (3.0) |
| | 160 | 23.2 (2.1) | 13.4 (3.1) | 23.3 (2.2) | 13.4 (3.7) |
| 50 | 100 | 33.9 (3.3) | 18.0 (3.1) | 35.0 (3.0) | 19.9 (4.2) |
| | 200 | 40.3 (2.6) | 18.3 (4.1) | 37.0 (4.1) | 17.7 (3.6) |

The method used to find subsets which fitted well was sequential replacement. This was used as a compromise in speed between the widely-used Efroymson algorithm, and the use of the exhaustive search algorithm.

Using 10 replicates, that is, 10 artificial data sets for each case, and minimizing the estimated $MSEP$ given by (5.26) as the stopping rule, the sample means and standard deviations of the sizes of selected subsets were as given in Table 5.10. It is interesting to note that the degree of ill-conditioning of the $\mathbf{\Sigma}$-matrix, as indicated by $\alpha$, had very little effect upon the size of subset selected.

For comparison, Mallows' $C_p$ was also used as a stopping rule, though its derivation is only for the fixed predictors case. In 69% of cases, it selected the same size of subset; in 1% of cases it selected a larger subset; in the remaining 30% of cases, it selected a smaller subset. When the true $MSEP$'s were compared, those selected using Mallows' $C_p$ were smaller in 21% of cases and larger in 10% than the true values when the estimated $MSEP$ was minimized.

Table 5.11 *Average MSEP's at the minimum of the estimated MSEP: (a) Average estimated MSEP for LS regression coefficients; (b) Average true MSEP for LS regression coefficients; and (c) Average MSEP for unbiased LS regression coefficients. Each average is based upon 10 replications*

| $k$ | $n$ | Large $\alpha$ | | | | | | Small $\alpha$ | | | | | |
|---|---|---|---|---|---|---|---|---|---|---|---|---|---|
| | | Large $\gamma$ | | | Small $\gamma$ | | | Large $\gamma$ | | | Small $\gamma$ | | |
| | | (a) | (b) | (c) | (a) | (b) | (c) | (a) | (b) | (c) | (a) | (b) | (c) |
| 10 | 20 | 1.32 | 2.22 | 1.96 | 1.05 | 1.77 | 1.53 | 1.20 | 2.03 | 1.79 | 0.98 | 2.20 | 1.47 |
| | 40 | 1.13 | 1.58 | 1.45 | 1.05 | 1.34 | 1.26 | 1.16 | 1.44 | 1.44 | 1.01 | 1.22 | 1.22 |
| 20 | 40 | 1.17 | 2.00 | 1.82 | 0.98 | 1.54 | 1.36 | 1.10 | 1.63 | 1.56 | 0.92 | 1.66 | 1.37 |
| | 80 | 1.00 | 1.39 | 1.29 | 1.01 | 1.26 | 1.13 | 1.09 | 1.32 | 1.29 | 0.95 | 1.22 | 1.16 |
| 30 | 60 | 1.14 | 1.83 | 1.56 | 1.00 | 1.67 | 1.44 | 1.08 | 1.68 | 1.66 | 0.84 | 1.65 | 1.41 |
| | 120 | 1.06 | 1.29 | 1.28 | 1.04 | 1.24 | 1.16 | 1.07 | 1.29 | 1.26 | 1.00 | 1.24 | 1.16 |
| 40 | 80 | 1.08 | 1.86 | 1.74 | 0.87 | 1.56 | 1.30 | 1.08 | 1.95 | 1.67 | 0.96 | 1.54 | 1.33 |
| | 160 | 1.14 | 1.36 | 1.31 | 1.04 | 1.22 | 1.14 | 1.06 | 1.35 | 1.28 | 1.05 | 1.23 | 1.14 |
| 50 | 100 | 1.28 | 2.44 | 2.32 | 0.93 | 1.65 | 1.34 | 1.17 | 2.20 | 2.00 | 0.97 | 1.73 | 1.40 |
| | 200 | 1.22 | 1.44 | 1.45 | 0.97 | 1.26 | 1.16 | 1.20 | 1.48 | 1.43 | 1.01 | 1.21 | 1.15 |

Table 5.11 shows the average values of three different $MSEP$'s when the stopping rule used was that of minimizing the estimated $MSEP$. The first of these, labelled (a) in the table, is the estimated $MSEP$ given by (5.26). The second is the true $MSEP$. As we know the true population values of the regression coefficients, the error in a future prediction for a known vector, $\boldsymbol{x_A}$, of values of the predictors in the selected subset A is

$$\boldsymbol{x}'_{\boldsymbol{A}}(\boldsymbol{b_A} - \boldsymbol{\beta_A}) \; + \; \boldsymbol{\eta},$$

where $\boldsymbol{b_A}$, $\boldsymbol{\beta_A}$ are the vectors of estimated and population regression coefficients, and $\boldsymbol{\eta}$ is a vector of residuals with standard deviation $\sigma_A$. Future $\boldsymbol{x_A}$'s can be generated as

$$\boldsymbol{x_A} \; = \; \boldsymbol{L_A}\boldsymbol{\epsilon},$$

where $\boldsymbol{L_A}$ is the lower-triangular Cholesky factorization of those rows and columns of $\boldsymbol{\Sigma}$ relating to variables in subset A, and $\boldsymbol{\epsilon}$ is a vector of elements sampled from the standard normal distribution. Hence the prediction error can be written as

$$\boldsymbol{\epsilon}'\boldsymbol{L}'_{\boldsymbol{A}}(\boldsymbol{b_A} - \boldsymbol{\beta_A}) \; + \; \boldsymbol{\eta}$$

and hence the true $MSEP$ is

$$(\boldsymbol{b_A} - \boldsymbol{\beta_A})'\boldsymbol{L_A}\boldsymbol{L}'_{\boldsymbol{A}}(\boldsymbol{b_A} - \boldsymbol{\beta_A}) \; + \; \sigma_A^2. \tag{5.27}$$

This quantity is shown as (b) in Table 5.11.

The third $MSEP$ shown in Table 5.11, as (c), is that for unbiased LS regression coefficients, such as would be obtained from an independent set of data from the same population with the same sample size. This $MSEP$ has been calculated using (5.25) with the known population variance $\sigma_A^2$.

A number of important hypotheses are suggested by Table 5.11. First we notice that when the sample size is only double the number of available predictors (excluding the constant), the estimated $MSEP$ is an underestimate (column (a) v. column (b)). The true $MSEP$ using LS regression coefficients is between 20% and 60% larger than that estimated. However, when the sample size is four times the number of available predictors, there is no evidence of any underestimation of the $MSEP$. There are two contributory factors to this. First, the larger sample size allows much better discrimination between subsets. This means that with the larger sample sizes, there is less competition between close subsets and hence, less bias in the regression coefficients. This will be shown more clearly in a later table. The other contributory factor is the increased number of degrees of freedom for the residual sum of squares. The artificially high regression sum of squares for the best-fitting subset has a relatively smaller effect in depressing the residual sum of squares when $n = 4k$ than when $n = 2k$.

The $MSEP$'s in column (c), which are those which would be obtained using an independent set of $n$ observations to estimate the regression coefficients, are usually slightly smaller than the true $MSEP$'s using the biased LS regression coefficients (column (b)). Hence, for situations similar to those simulated here, it is much better to use all of the data to select the model and to use the same data to obtain biased LS regression coefficients for the selected model, than to split the data into equal parts and use one part for model selection and the other for estimation. The penalty is that if the first path is chosen, the estimate of the $MSEP$ is optimistically small.

Figure 5.5 shows the (true) $MSEP$'s using LS regression coefficients from an independent data set, for two cases from the simulations. Both cases are for 40 available predictors and sample size = 80. Both were the first sets generated for $\alpha = 0.90$ (low correlations between predictors), but one is for $\gamma = 0.90$ (all projections moderately large), while the other is for $\gamma = 0.75$ (many small projections).

The curves of $MSEP$ v. $p$ are usually fairly flat near the minimum, so that the exact choice of stopping point is often not very critical.

Probably the most frequently asked question with respect to subset selection in regression is 'What stopping rule should be used?'. Table 5.12 was drawn up to illustrate the answer. If we generalize Mallows' $C_p$ to

$$C_p(M) = \frac{RSS_p}{\hat{\sigma}^2} - (n - M.p), \tag{5.28}$$

where $M = 2$ for Mallows' $C_p$, then minimizing this quantity gives a range of stopping rules. Large values of $M$ will lead to the selection of small subsets, and vice-versa. $M$ can be regarded as a penalty or cost for each additional variable in the selected subset.

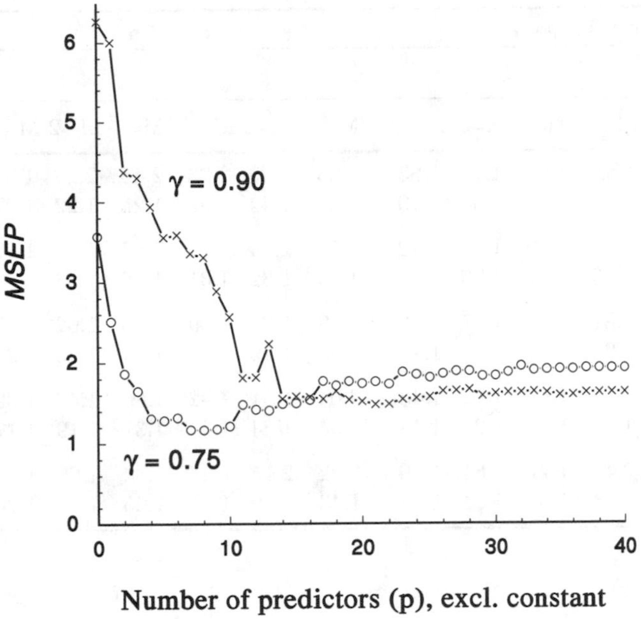


Figure 5.5 *True MSEP using LS for simulated data for k = 40 available predictors and a sample size of 80 observations. See text for further description.*

Table 5.12 shows the true $MSEP$'s using this stopping rule and LS regression coefficients for $M = 1$, 2 and 3. The conclusion is clear. For large $\gamma$, we should use a small $M$ (selecting a large subset), while for small $\gamma$, we should use a large $M$ (selecting a small subset). Thus, the answer to the question, 'What stopping rule should we use?', does not have one answer. In some circumstances, it is preferable to include many predictors, in others we should select a small number.

In practice, nothing equivalent to the $\gamma$ of the simulations will be available, so the finding from Table 5.12 is of limited value. Suppose the rule of minimizing the estimated $MSEP$ is used, with the $MSEP$ falsely estimated from (5.26). If this selects a small subset, with $p < k/2$ say, it suggests that there may be many predictors which make little or no real contribution after the more important predictors have been included. This corresponds to the small $\gamma$ case, and suggests that we should then use the modified Mallows' $C_p$ with $M =$ say 3. On the other hand, if minimizing (5.26) selects a large subset, with $p > k/2$ say, then we should use $M =$ say 1. It appears that it is undesirable to pick a subset that is close to half the number of available predictors. In

Table 5.12 *Average true MSEP's using LS regression coefficients and the generalized Mallows' $C_p$ as the stopping rule for M = 1, 2 and 3 (M = 2 for Mallows' $C_p$)*

| k | n | Large α | | | | | | Small α | | | | | |
|---|---|---|---|---|---|---|---|---|---|---|---|---|---|
| | | Large γ | | | Small γ | | | Large γ | | | Small γ | | |
| | | M=1 | M=2 | M=3 | M=1 | M=2 | M=3 | M=1 | M=2 | M=3 | M=1 | M=2 | M=3 |
| 10 | 20 | 2.23 | 2.14 | 2.19 | 1.83 | 1.79 | 1.83 | 2.23 | 2.03 | 2.22 | 2.45 | 2.20 | 1.87 |
| | 40 | 1.53 | 1.59 | 1.61 | 1.31 | 1.30 | 1.29 | 1.45 | 1.44 | 1.51 | 1.26 | 1.22 | 1.22 |
| 20 | 40 | 2.06 | 2.05 | 1.82 | 1.62 | 1.45 | 1.40 | 1.73 | 1.64 | 1.71 | 1.87 | 1.52 | 1.46 |
| | 80 | 1.43 | 1.40 | 1.36 | 1.27 | 1.26 | 1.22 | 1.35 | 1.32 | 1.31 | 1.26 | 1.23 | 1.22 |
| 30 | 60 | 1.84 | 1.73 | 1.81 | 1.87 | 1.57 | 1.46 | 1.89 | 1.68 | 1.69 | 1.73 | 1.62 | 1.40 |
| | 120 | 1.30 | 1.28 | 1.27 | 1.27 | 1.22 | 1.19 | 1.33 | 1.29 | 1.24 | 1.34 | 1.23 | 1.20 |
| 40 | 80 | 1.95 | 1.84 | 1.79 | 1.67 | 1.52 | 1.43 | 2.04 | 1.91 | 1.92 | 1.86 | 1.47 | 1.39 |
| | 160 | 1.35 | 1.36 | 1.37 | 1.27 | 1.20 | 1.18 | 1.35 | 1.34 | 1.35 | 1.31 | 1.19 | 1.17 |
| 50 | 100 | 2.27 | 2.44 | 2.42 | 1.77 | 1.61 | 1.60 | 2.22 | 2.15 | 2.27 | 1.86 | 1.69 | 1.54 |
| | 200 | 1.39 | 1.43 | 1.48 | 1.31 | 1.24 | 1.22 | 1.43 | 1.49 | 1.51 | 1.25 | 1.21 | 1.20 |

such cases, the number of alternative subsets of the same size is a maximum, and we can anticipate large competition biases. It is of course possible to construct artificial examples in which the best subset for prediction is of exactly half the number of available predictors, and doubtless such cases will occur sometimes in real-life examples.

If we use all of the available predictors without considering subset selection, then the regression coefficients are unbiased. This is sometimes a better strategy. Table 5.13 shows how often, out of the 10 replications in each case, the $MSEP$ using all the available predictors was better than or equal to that for the selected subset using biased LS regression coefficients for that subset. Using all the predictors was nearly always as good as or only slightly worse for large $\gamma$, but rarely so for small $\gamma$. Note that in a few cases, particularly for $k = 10$, the selected subset was that of all $k$ predictors.

Table 5.14 shows the estimated average bias of the LS regression coefficients after standardization and adjustment for sign. If $b_{Ai}$ and $\beta_{Ai}$ are the sample regression coefficient and its expected value, respectively, for variable $X_i$ in subset A, then the standardized and sign-adjusted difference used was

$$\frac{b_{Ai} \;-\; \beta_{Ai}}{s_{LS}(b_{Ai})}.sign(\beta_{Ai}),$$

where the estimated standard errors, $s_{LS}(b_{Ai})$, are those for LS regression when the model has been chosen *a priori*, that is they are the square roots of

Table 5.13 *Frequencies, out of 10, for which the MSEP using all available predictors (using LS regression coefficients) was smaller or equal to the true MSEP for the subset selected by minimizing the estimated MSEP*

| $k$ | $n$ | Large $\alpha$ | | Small $\alpha$ | |
|---|---|---|---|---|---|
| | | Large $\gamma$ | Small $\gamma$ | Large $\gamma$ | Small $\gamma$ |
| 10 | 20 | 6 | 2 | 2 | 2 |
| | 40 | 6 | 5 | 6 | 3 |
| 20 | 40 | 5 | 1 | 0 | 1 |
| | 80 | 5 | 5 | 4 | 3 |
| 30 | 60 | 3 | 2 | 1 | 2 |
| | 120 | 4 | 3 | 5 | 2 |
| 40 | 80 | 3 | 0 | 2 | 2 |
| | 160 | 5 | 3 | 5 | 2 |
| 50 | 100 | 6 | 2 | 4 | 2 |
| | 200 | 8 | 3 | 7 | 2 |

Table 5.14 *Average estimated bias and sample standard deviation (in brackets) of LS regression coefficients after standardization and sign adjustment, for selected subsets*

| $k$ | $n$ | Large $\alpha$ | | Small $\alpha$ | |
|---|---|---|---|---|---|
| | | Large $\gamma$ | Small $\gamma$ | Large $\gamma$ | Small $\gamma$ |
| 10 | 20 | 0.26 (0.69) | 0.72 (0.70) | 0.36 (0.52) | 0.23 (0.22) |
| | 40 | 0.31 (0.36) | 0.72 (0.25) | 0.49 (0.26) | 0.38 (0.26) |
| 20 | 40 | 0.38 (0.72) | 0.72 (0.50) | 1.03 (0.26) | 1.17 (0.89) |
| | 80 | 0.55 (0.64) | 0.37 (0.24) | 0.58 (0.16) | 1.29 (1.44) |
| 30 | 60 | 0.58 (0.52) | 0.52 (0.83) | 0.92 (0.27) | 0.65 (0.93) |
| | 120 | 0.31 (0.46) | 0.57 (0.24) | 0.36 (0.31) | 0.66 (0.58) |
| 40 | 80 | 0.54 (0.55) | 0.46 (0.60) | 0.88 (0.34) | 0.68 (0.45) |
| | 160 | 0.48 (0.56) | 0.58 (0.35) | 0.67 (0.41) | 0.64 (0.44) |
| 50 | 100 | 0.49 (0.81) | 0.47 (0.50) | 1.03 (0.53) | 0.57 (0.77) |
| | 200 | 0.17 (0.40) | 0.24 (0.50) | 0.38 (0.31) | 0.41 (0.46) |

Table 5.15 *Analysis of variance of estimated biases of LS regression coefficients*

| Factor | Sum of squares | Deg. of freedom | Mean square | F-ratio |
|---|---|---|---|---|
| $\alpha$ | 0.388 | 1 | 0.388 | 8.68** |
| $\gamma$ | 0.041 | 1 | 0.041 | 0.92 |
| $k$ | 0.538 | 4 | 0.135 | 3.01* |
| $n$ | 0.156 | 1 | 0.156 | 3.49 |
| Residual | 1.431 | 32 | 0.045 | |
| Total | 2.554 | 39 | | |

* Significant at the 5% level, ** Significant at the 1% level.

the diagonal elements of

$$s_{LS}^2(\boldsymbol{b_A}) = \sigma_A^2.diag(\boldsymbol{X'_A X_A})^{-1},$$

and $\sigma_A^2$ is estimated from the residual sum of squares, $RSS_A$, for subset A in the usual way, i.e.

$$\sigma_A^2 = RSS_A/(n-p),$$

where $p$ is the number of variables, including the constant, in subset A.

The quantities shown in Table 5.14 are the averages of all the standardized differences between sample and population regression coefficients for the selected variables except the constant in the model. Thus for the first entry, the numbers of regression coefficients used for the 10 replicates were 6, 3, 4, 4, 5, 10, 7, 4, 7 and 4.

We notice first that the averages in Table 5.14 are positive. The overall average of the bias estimates in the table is 0.57 of a standard error. As each subset selected would have contained a number of 'dominant' variables for which the bias would have been very small, a substantial proportion of the biases for other variables were well in excess of one standard error.

Table 5.15 contains a simple analysis of variance of the average estimated biases, with only main effects fitted and the interactions used to estimate the residual variance. The only effects that are significant (other than the constant in the model) are those related to $\alpha$ and $k$. The average estimated bias for large $\alpha$ is 0.47 of a standard error, while that for small $\alpha$ is 0.67 of a standard error. With small $\alpha$, some of the $X$-predictors are highly correlated so that there is considerable competition amongst the predictors for selection. The average estimated biases are 0.43, 0.76, 0.57, 0.62 and 0.47 standard errors for the five values of $k$.

Another important feature of the simulation results in Table 5.14 is that most of the standard deviations are substantially less than 1.

The low variance of the biased LS regression coefficients is the basic explanation for the relatively good $MSEP$'s in the columns labelled (b) in Table 5.11 compared with the $MSEP$'s for unbiased LS regression coefficients in the columns labelled (c). If we refer back to formula (5.4), we see that an important part of the $MSEP$ is

$$\boldsymbol{V}(\boldsymbol{b_A}) \;+\; (sel.bias)(sel.bias)',$$

which is the second moment matrix of the regression coefficients. In the standardized units used for Table 5.14, for unbiased LS regression coefficients from an independent data set, the diagonal elements of $\boldsymbol{b_A}$ are all equal to 1.0, while the selection bias is zero. Thus the second moments for unbiased LS regression coefficients are all equal to 1.0. Using the estimated biases in Table 5.14 and their sample standard deviations, only 6 of the 40 estimated second moments exceed 1.0, while many are less than 0.5. This means that, in terms of the mean squared error of prediction, using LS estimates from an independent data set will usually yield worse predictions than those obtained by using the biased LS estimates from the model selection data set.

Thus, subset selection is very much like ridge regression or James and Stein/ Sclove regression in trading bias in the parameter estimates for a reduced variance. However, the regression coefficients from subset selection are biased in the direction of being too large, while those from the shrinkage estimators are too small.

How well do we do if we use a subset selection procedure and then apply some form of shrinkage to the subset of selected variables? Table 5.16 shows the $MSEP$'s for shrinkage using the Sclove (1968) estimator and for ridge regression using the Lawless and Wang (1976) shrinkage parameter. The Sclove shrinkage always reduces the $MSEP$ by a few percent, but ridge regression can be disastrous.

As explained in section 3.9, ridge regression largely ignores the correlation between the dependent variable and the eigenvectors associated with the smaller principal components. In the simulations performed here, there was no attempt to associate the dependent variable with the larger principal components. Thus, these simulations have generated some of the the kind of data on which Fearn (1983) warned that ridge regression performs badly. The cases with small values of $\alpha$ had many small eigenvalues, and the performance of ridge regression in these cases was particularly poor. Such cases are probably more typical of the physical sciences where the predictor variables may be similar quantities measured at different times or locations, or the variables may be constructed by taking polynomials, logarithms, cross-products, etc., of a small set of original variables.

## 5.3 Cross-validation and the *PRESS* statistic

The *PRESS* (prediction sum of squares) statistic is a cross-validation statistic suggested by Allen (1974) for model selection. Cross-validation means that a small part of the data, often only one case, is set aside and then predicted from the rest of the data. This is done repeatedly, setting aside a different part of the data and predicting it from the remainder.

Table 5.16 *MSEP using LS regression coefficients compared with those for shrunken estimates for the selected subset using the Sclove estimator and the Lawless-Wang ridge estimator*

| $k$ | $n$ | $\alpha$ | Large $\gamma$ | | | Small $\gamma$ | | |
|---|---|---|---|---|---|---|---|---|
| | | | LS | Sclove | Ridge | LS | Sclove | Ridge |
| 10 | 20 | 0.75 | 1.004 | 0.986 | 0.956 | 0.924 | 0.901 | 0.878 |
| | 40 | | 1.011 | 1.005 | 1.030 | 0.925 | 0.917 | 0.903 |
| | 20 | 0.45 | 0.828 | 0.816 | 1.150 | 0.954 | 0.941 | 2.206 |
| | 40 | | 0.943 | 0.938 | 2.510 | 0.834 | 0.831 | 0.857 |
| 20 | 40 | 0.80 | 0.906 | 0.875 | 0.947 | 0.742 | 0.723 | 0.679 |
| | 80 | | 0.965 | 0.954 | 1.026 | 0.942 | 0.936 | 0.941 |
| | 40 | 0.55 | 0.754 | 0.729 | 77.2 | 0.844 | 0.826 | 12.4 |
| | 80 | | 0.900 | 0.892 | 17.2 | 0.865 | 0.854 | 22.1 |
| 30 | 60 | 0.85 | 0.905 | 0.873 | 0.869 | 0.833 | 0.799 | 0.879 |
| | 120 | | 0.901 | 0.894 | 0.932 | 0.899 | 0.892 | 0.901 |
| | 60 | 0.70 | 0.781 | 0.759 | 20.7 | 0.852 | 0.833 | 15.2 |
| | 120 | | 0.902 | 0.901 | 29.0 | 0.910 | 0.903 | 17.4 |
| 40 | 80 | 0.90 | 0.891 | 0.869 | 0.885 | 0.822 | 0.800 | 0.775 |
| | 160 | | 0.921 | 0.914 | 0.917 | 0.892 | 0.884 | 0.881 |
| | 80 | 0.75 | 0.936 | 0.915 | 56.9 | 0.805 | 0.782 | 75.3 |
| | 160 | | 0.942 | 0.938 | 54.3 | 0.896 | 0.889 | 38.1 |
| 50 | 100 | 0.95 | 1.033 | 1.003 | 0.983 | 0.826 | 0.813 | 0.811 |
| | 200 | | 0.978 | 0.971 | 0.980 | 0.909 | 0.903 | 0.901 |
| | 100 | 0.80 | 0.991 | 0.967 | 57.9 | 0.883 | 0.856 | 26.8 |
| | 200 | | 0.995 | 0.986 | 46.3 | 0.891 | 0.885 | 29.8 |

In calculating the *PRESS* statistic for a given set of $p$ predictors, each observation, $y_i$, in turn, of the dependent variable is predicted from the other $(n-1)$ cases using least-squares regression. If $\hat{y}_{ip}$ denotes the predicted value for $y_i$, then the *PRESS* statistic for a particular subset of $p$ predictors is defined as

$$PRESS_p = \sum_{i=1}^{n}(y_i - \hat{y}_{ip})^2. \tag{5.29}$$

In calculating (5.29), a different set of regression coefficients is calculated for each case, but the subset of predictor variables remains the same.

The idea of performing $n$ multiple regressions for every subset in which we are interested can be very daunting. Fortunately, there is a mathematical result that can reduce the amount of computation substantially. We use the well-known formula for a rank one update of the inverse of a nonsingular matrix $\boldsymbol{A}$:

$$(\boldsymbol{A}+\boldsymbol{x}\boldsymbol{x}')^{-1} = \boldsymbol{A}^{-1} - \boldsymbol{A}^{-1}\boldsymbol{x}(1 + \boldsymbol{x}'\boldsymbol{A}^{-1}\boldsymbol{x})^{-1}\boldsymbol{x}'\boldsymbol{A}^{-1}, \qquad (5.30)$$

where $\boldsymbol{x}$ is a vector of new 'data'. In our case, the matrix $\boldsymbol{A}$ is $\boldsymbol{X}_p'\boldsymbol{X}_p$, where $\boldsymbol{X}_p$ is that part of $\boldsymbol{X}$ containing the $p$ predictors of interest. However, we want to remove $\boldsymbol{x}$, not add it. There is a corresponding formula for downdating:

$$(\boldsymbol{A}-\boldsymbol{x}\boldsymbol{x}')^{-1} = \boldsymbol{A}^{-1} + \boldsymbol{A}^{-1}\boldsymbol{x}(1 - \boldsymbol{x}'\boldsymbol{A}^{-1}\boldsymbol{x})^{-1}\boldsymbol{x}'\boldsymbol{A}^{-1}. \qquad (5.31)$$

Neither (5.30) nor (5.31) should ever be used for computational purposes, though the update formula is often used in Kalman filter calculations. If the matrix $\boldsymbol{A}$ is ill-conditioned, the accuracy of the calculations can be very poor, particularly for the downdating. Updating or downdating the Cholesky factorization of $\boldsymbol{A}$ is far more accurate. Unless $\boldsymbol{A}^{-1}$ is required after every update, rather than quantities derived using it, updating or downdating the Cholesky factorization will usually be faster and more accurate.

If $\boldsymbol{b}_{ip}$ is the vector of LS regression coefficients based upon the $(n-1)$ cases excluding the $i^{th}$, then

$$\begin{aligned}
\hat{y}_{ip} &= \boldsymbol{x}_i'\boldsymbol{b}_{ip} \\
&= \boldsymbol{x}_i'(\boldsymbol{A}-\boldsymbol{x}_i\boldsymbol{x}_i')^{-1}(\boldsymbol{X}_p'\boldsymbol{y} - \boldsymbol{x}_i'y_i) \\
&= \boldsymbol{x}_i'(\boldsymbol{A}^{-1} + \boldsymbol{A}^{-1}\boldsymbol{x}_i d_i^{-1}\boldsymbol{x}_i'\boldsymbol{A}^{-1})(\boldsymbol{X}_p'\boldsymbol{y} - \boldsymbol{x}_i'y_i),
\end{aligned}$$

where the scalar $d_i = 1 - \boldsymbol{x}_i'\boldsymbol{A}^{-1}\boldsymbol{x}_i$.

If $\boldsymbol{b}_p$ denotes the vector of LS regression coefficients when all the observations are used, i.e.

$$\boldsymbol{b}_p = \boldsymbol{A}^{-1}\boldsymbol{X}_p'\boldsymbol{y},$$

then we find that

$$\begin{aligned}
y_i - \hat{y}_{ip} &= (y_i - \boldsymbol{x}_i'\boldsymbol{b}_p)/d_i \\
&= e_i/d_i, \qquad (5.32)
\end{aligned}$$

where $e_i$ is the LS residual using all the observations.

The $d_i$'s can easily be calculated from the Cholesky factorization. Thus, if

$$\boldsymbol{A} = \boldsymbol{X}_p'\boldsymbol{X}_p = \boldsymbol{R}_p'\boldsymbol{R}_p,$$

where $\boldsymbol{R}_p$ is upper triangular, then

$$\begin{aligned}
d_i &= 1 - \boldsymbol{x}_i'\boldsymbol{R}^{-1}(\boldsymbol{x}_i'\boldsymbol{R}^{-1})' \\
&= 1 - \boldsymbol{z}_i'\boldsymbol{z}_i,
\end{aligned}$$

where $\boldsymbol{z}_i' = \boldsymbol{x}_i'\boldsymbol{R}^{-1}$.

Finally, from (5.29) and (5.32) we derive

$$PRESS_p = \sum_{i=1}^{n}(e_i/d_i)^2. \qquad (5.33)$$

Table 5.17 *Estimated MSEP's for the best-fitting subsets of each size for the STEAM data set*

| No. of predictors | 0 | 1 | 2 | 3 | 4 |
|---|---|---|---|---|---|
| Estimated $MSEP$ | 2.765 | 0.861 | 0.462 | 0.418 | 0.428 |
| No. of predictors | 5 | 6 | 7 | 8 | 9 |
| Estimated $MSEP$ | 0.448 | 0.416 | 0.448 | 0.492 | 0.555 |

The $d_i$'s above are the diagonal elements of $(\boldsymbol{I} - \boldsymbol{V})$, where $\boldsymbol{V}$ is the projection operator, sometimes also known as the 'hat' matrix:

$$\boldsymbol{V} = \boldsymbol{X_p}(\boldsymbol{X'_p}\boldsymbol{X_p})^{-1}\boldsymbol{X'_p}.$$

It was shown in (5.6) that the sum of the diagonal elements of this matrix equals $p$. Hence, the average value of the $d_i$'s is $(1 - p/n)$. If $n \gg p$, then

$$\begin{aligned} PRESS_p &\approx \sum_{i=1}^{n} e_i^2/(1-p/n)^2 \\ &= RSS_p/(1-p/n)^2. \end{aligned}$$

Hence,

$$PRESS_p \approx RSS_p \frac{n^2}{(n-p)^2}.$$

After dividing by the sample size, $n$, this is very similar to formula (5.26) for the (false) estimated $MSEP$, so that we can expect that minimizing the $PRESS$ statistic with respect to $p$ will often pick the same size of subset as minimizing the estimated $MSEP$.

Notice that here the residual, $e_i$, is divided by $d_i$. The quantity $e_i/(s\sqrt{d_i})$ is variously known as a 'standardized' or 'studentized' residual, where $s$ is the estimate of the residual standard deviation.

Using the $PRESS$ statistic is not using true cross-validation as we do not repeat the same procedure on each of the sets of $(n-1)$ observations as was applied to the full set of $n$ observations. That is we do not repeat the subset selection procedure each time a different observation is left out. If we do this, then the selected subsets will not always be the same as those selected using the full data set. To illustrate this, let us look at the STEAM and POLLUTE data sets again.

The STEAM data set contained $k = 9$ predictors and $n = 25$ cases. Table 5.17 shows the estimated $MSEP$'s for the best-fitting subsets of each size. The number of predictors shown in this table excludes the constant which was fitted in all subsets.

Notice that the best-fitting subset of 3 predictors (variables numbered 4, 5, 7) gave almost the same estimated $MSEP$ as the best-fitting subset of 6 variables (numbers 1, 3, 5, 7, 8, 9).

Table 5.18 *Subsets chosen from the STEAM data set leaving out one case at a time and minimizing the estimated MSEP*

| Selected subset (variable nos.) | Frequency |
|---|---|
| 1 3 5 7 8 9 | 5 |
| 4 5 7 | 18 |
| 1 2 5 7 | 1 |
| 4 5 7 8 | 1 |

Table 5.19 *Estimated MSEP's for best-fitting subsets of each size for the POLLUTE data set*

| No. of pred. | 0 | 1 | 2 | 3 | 4 | 5 | 6 | 7 |
|---|---|---|---|---|---|---|---|---|
| Est'd MSEP | 3934 | 2385 | 1844 | 1577 | 1373 | 1332 | 1295 | 1298 |
| No. of pred. | 8 | 9 | 10 | 11 | 12 | 13 | 14 | 15 |
| Est'd MSEP | 1327 | 1332 | 1359 | 1409 | 1465 | 1530 | 1599 | 1673 |

Omitting one observation at a time and repeating the exercise of finding the best-fitting subset, followed by applying the stopping rule of minimizing the estimated $MSEP$, gave the frequencies of selection shown in Table 5.18.

The value of the $PRESS$ statistic for this data set was 11.09, for the best subset of 6 variables. The cross-validation sum of squares, using true leave-one-out cross-validation, was 16.41. Dividing by $n$ to give the average squared prediction error gives 0.444 for $PRESS$, which is close to the estimated $MSEP$ of 0.416, but much smaller than the value of 0.656 for true cross-validation.

Repeating the exercise with the POLLUTE data set ($k = 15,\ n = 60$), the estimated $MSEP$'s were as in Table 5.19, using all the data set.

Leaving out one case at a time and then minimizing the estimated $MSEP$ selected the subsets shown in Table 5.20. The value of the $PRESS$ statistic for the best-fitting subset of 6 variables (plus a constant) was 79,490 compared with a value of 116,673 for the true cross-validation sum of squares of prediction errors. If we divide 79,490 by the number of cases (60), we obtain 1325, which is close to the minimum estimated $MSEP$ using all of the data.

Simulation studies by Breiman and Spector (1992) also showed that the use of the $PRESS$-statistic, which they called partial cross-validation, results in extremely misleading estimates of the $MSEP$ compared with full cross-validation, that is, with selecting the subsets afresh each time that a case is omitted.

If the $n$ observations are independent samples from the same population, then the $i^{th}$ observation is independent of the other $(n-1)$ that were used

Table 5.20 *Subsets chosen from the POLLUTE data set leaving out one case at a time and minimizing the estimated MSEP*

| Selected subset (variable nos.) | | | | | | | | | Frequency |
|---|---|---|---|---|---|---|---|---|---|
| 1 | 2 | 3 | 6 | 9 | 14 | | | | 46 |
| 1 | 2 | 3 | 5 | 6 | 9 | 14 | | | 8 |
| 1 | 2 | 6 | 9 | 14 | | | | | 1 |
| 1 | 3 | 8 | 9 | 10 | 14 | | | | 1 |
| 1 | 2 | 3 | 8 | 9 | 10 | 14 | | | 1 |
| 1 | 2 | 3 | 5 | 6 | 9 | 12 | 13 | | 1 |
| 1 | 2 | 3 | 6 | 8 | 9 | 12 | 13 | | 1 |
| 1 | 2 | 3 | 4 | 5 | 6 | 9 | 12 | 13 | 1 |

to predict its value. Hence, $(y_i - \hat{y}_{ip})^2$ is an unbiased estimate of the squared error for the $i$th case. Therefore, the true cross-validation sum of squares, divided by the sample size, gives an unbiased estimate of the $MSEP$ for our procedure based upon $(n-1)$ cases. We can expect that the true $MSEP$ for $n$ cases will be very slightly smaller than this.

Notice that consecutive values of $(y_i - \hat{y}_{ip})^2$, though unbiased, will be correlated with each other.

A forward validation procedure for predicting each $y_i$, using only the previous cases, has been proposed by Hjorth (1982). This attempts to allow for the change in sample size, and the span in the $X$-variables by using generalized cross-validation.

So far in this brief section, only 'one out at a time' cross-validation has been considered. For a review of the statistical literature on this subject, see Stone (1978). Efron (1982), Efron and Tibshirani (1993), and Hjorth (1994) provide books completely devoted to this subject, or with major sections on cross-validation.

Leaving out one case at a time, subset selection usually picks the same subsets as when the full data set is used. It seems sensible to leave out somewhat more, perhaps 10% or 20%. We want to omit a sufficient number of cases so that different subsets will quite often be selected. What fraction should be left out? If we leave out a large number of cases, say 50%, then the biases in the regression coefficients and estimated residual variance due to the selection process may be very different from those for the full data set. Simulation results by Breiman and Spector (1992) suggest that about 20% is a sensible fraction. Given that we have $n$ cases, one way to do this is as follows:

- Generate a random ordering of the integers 1, 2, ..., n.
- For i = 1:5,
- leave out those cases identified by the $i^{th}$ fifth of the numbers in the random ordering.

Table 5.21 *Most-frequently selected subsets using 'leave out 20%' cross-validation on the POLLUTE data set, using the Efroymson stepwise procedure*

| Model variables | Frequency selected | MSEP |
|---|---|---|
| 2 6 9 14 | 648 | 1325 |
| 2 5 6 9 | 901 | 1548 |
| 2 6 9 | 332 | 1613 |
| 1 2 6 9 14 | 412 | 1683 |
| 1 2 3 6 9 14 | 163 | 1759 |
| 2 4 6 9 14 | 381 | 1762 |
| 1 2 9 14 | 2648 | 1768 |
| 1 2 3 9 14 | 171 | 1851 |

Table 5.22 *Cross-validation leaving out 20% applied to the POLLUTE data set, using the Efroymson stepwise procedure to select subsets. Means and standard deviations of regression coefficients from 10,000 replicates*

| Var. | Subset (1,2,9,14) Frequency = 2648 | | Different subset Frequency = 7352 | | All cases Frequency = 10,000 | |
|---|---|---|---|---|---|---|
| | Mean | Std. devn | Mean | Std. devn | Mean | Std. devn |
| X(1) | 1.95 | 0.25 | 2.15 | 0.33 | 2.10 | 0.32 |
| X(2) | –1.93 | 0.28 | –1.67 | 0.41 | –1.74 | 0.40 |
| X(9) | 4.05 | 0.40 | 4.04 | 0.45 | 4.04 | 0.44 |
| X(14) | 0.340 | 0.031 | 0.330 | 0.028 | 0.332 | 0.029 |

- Perform the chosen subset selection procedure, identify a best subset, estimate the regression coefficients for this model, and fit it to the omitted cases.

This process can be repeated many times. Notice that in this way, each case is left out of the construction sample in exactly 20% of runs.

Leaving out 20% cross-validation was applied to the POLLUTE data set, using the Efroymson stepwise procedure to pick the subset each time. As $n = 60$ for this data set, 12 cases were left out each time. The usual default value of 4.0 was used for $F$-to-enter and $F$-to-drop. Using 10,000 replications, that is, 2000 blocks of 5, 153 different models were selected! All except one of these models included variable $X(9)$. Table 5.21 shows the selected subsets with the smallest MSEPs estimated by cross-validation. The subset (1, 2, 9, 14), which was chosen the most often, is that which the Efroymson algorithm selects using the full data set. Notice that replacing variable $X(1)$ with $X(6)$ gives a substantially smaller MSEP. As the residual variance estimate fitting all variables to the full data set is 1220, the MSEP of 1325 is remarkably small.

Table 5.22 shows the means and standard deviations of the regression coefficients for the subset (1, 2, 9, 14), both for the replicates in which that subset was selected and for the remainder. The regression coefficients for variables $X(1)$, $X(2)$ and $X(14)$ are significantly different. In the case of $X(14)$ in which the difference 'looks small', the standard error of the the difference of the regression coefficients is

$$\sqrt{\frac{0.03067)^2}{2648} + \frac{0.02792)^2}{7352}} = 0.00068,$$

so that the difference 0.3397 − 0.3297 is about 15 standard errors. This illustrates that if the same data are used to select a model and to estimate the regression coefficients using ordinary least squares, then those regression coefficients will be biased.

Cross-validation can give us an estimate of the true prediction errors from the subset selection process that we have chosen to use, though it is really an estimate of the prediction errors for sample size $= 0.8n$ if we leave out 20%. If we have some method for shrinking regression coefficients, such as the garrote or the lasso discussed earlier, then for each subset of 80% of the data, we can use a range of values of some parameter describing the amount of shrinkage and find its optimum value. This idea will be explored in more detail in the next chapter.

Thall et al. (1992) give another example of the application of cross-validation in variable selection. They divide the data set into $K$ equal or nearly equal subsets $V_1$, $V_2$, ..., $V_K$ ('V' for verification) and their complementary subsets $E_1$, $E_2$, ..., $E_K$ ('E' for evaluation). They then run backward elimination using a 'significance' level $\alpha$ to delete variables. For each $\alpha$, they find the cross-validation sum of squares of prediction errors, using least-squares regression coefficients, and hence find the value of $\alpha$ which minimizes the prediction errors. However, they do nothing to shrink the regression coefficients of the selected variables.

An important theoretical paper on cross-validation is that of Shao (1993). Shao shows that the number of cases left out for validation, $n_v$, should increase as the sample size, $n$, increases so that $n_v/n \to 1$ as $n \to \infty$. In particular, he argues that if this is not the case, then the selected models will be too large; they will include all of the variables that should be included, but also a finite number for which the true regression coefficients are zero.

Shao shows that using leaving one-at-a-time cross-validation does not converge upon the correct model as the sample size tends to infinity. It will include too many variables, assuming that there are some variables for which the true regression coefficients are zero.

Shao divides the set of models into two groups. Assuming that some, but not all, the true regression coefficients are zero, his category I contains all models that exclude at least one variable with a nonzero regression coefficient. His category II contains all models that include all of the variables with nonzero regression coefficients, though all but one of its models contains some variables with zero regression coefficients.

Recalling from Chapter 2 that we can expect least squares projections to be proportional to $\sqrt{n}$ provided their expected values are nonzero, nearly all stopping rules ($F$-to-enter, Mallows' $C_p$, the RIC, the AIC, etc.) will lead to models in his category II as $n$ becomes large enough. That is, they will err in the direction of including too many variables, rather than too few.

Using the rule $n_v/n \rightarrow 1$ as $n \rightarrow \infty$, Shao shows that cross-validation will converge upon the correct model provided that the models compared include the correct model.

Shao uses $n_c = n^{0.75}$, where $n_c$ is the size of the construction set. In his paper, the simulation study has sample size $n = 40$, which gives $n_c = 15.9$, say 16. The idea of using less than half the data for model construction, and the majority for validation conflicts with usual practice. A rule $n_c = n^{0.9}$ would probably be more acceptable to most practitioners.

Shao also introduced what he called the Balanced Incomplete $CV(n_v)$ design, or BICV. This is an experimental design for cross-validation runs, in which not only is each case $i$ left out of the construction sample exactly the same number of times, but each pair $(i, j)$ is left out exactly the same number of times. This is certainly a desirable feature of the design of cross-validation runs to reduce variability.

Altman and Leger (1997) have extended Shao's results slightly, looking at estimators which are not necessarily differentiable.

Cross-validation has been applied to robust regression by Ronchetti et al. (1997). They use Shao's rule for deciding the sizes of model construction and validation subsets of the data. The regression coefficients are calculated using M-estimators, that is, by using a kind of weighted least-squares method. If the absolute value of the residual, $e_i = y_i - \boldsymbol{x}_i'\boldsymbol{b}$, is less than $1.345\hat{\sigma}$, where $\boldsymbol{b}$ is the current vector of estimated regression coefficients and $\hat{\sigma}$ is the current estimate of the residual standard deviation, then full weight is given to this case; otherwise, its residual is taken to be $1.345\hat{\sigma}$ with the sign of the actual residual. A robust estimate of $\hat{\sigma}$ is used.

## 5.4 Bootstrapping

According to Efron and Tibshirani (1995, page 5), the phrase *to pull oneself up by one's own bootstraps* comes from the 18th century adventures of Baron Munchausen, written by Rudolph Raspe. In one adventure, the baron fell into a deep lake and was in danger of drowning when he realized the brilliant idea of pulling himself to the surface using his own bootstraps!

In data analysis, it is used as follows. Suppose we have a sample of say 100 cases. We have carried out some kind of statistical analysis on the full sample and estimated some statistic $\theta$. For instance, at the start of Chapter 5 of Hjorth (1994), the median is used as the example. We then treat the sample as the complete population, and sample *with replacement* from that population. That is, we take another sample of size 100 with the first case chosen to be equally likely to be any one of the original 100; the second case is also equally likely to be any of the original 100 and could be the same one chosen for the first sample, etc. Thus some of the original 100 will be chosen 2, 3 or 4 times, while others will be left out. The statistic $\theta$ is then calculated for this sample. The exercise is then repeated many times, and an empirical distribution is found for the statistic.

An attractive feature of the bootstrap is that no assumptions about the distribution of the random variable(s) has been made; any inferences are made entirely on the data in the sample.

The bootstrap is closely related to cross-validation. The probability that any particular case will be left out is $(1 - 1/n)^n$, which is approximately $e^{-1} = 0.3679$. Thus, about 36-37% of cases are left out. This is more than is typically left out in cross-validation. The probability that any case is included $r$ times is the binomial probability:

$$p_r = {}^nC_r(1/n)^r(1 - 1/n)^{n-r},$$

which is closely approximated by the Poisson probability:

$$p_r = \frac{e^{-1}}{r!}.$$

Thus, for moderately large $n$, $p_0 = 0.3679$, $p_1 = 0.3679$, $p_2 = 0.1839$, $p_3 = 0.0613$, $p_4 = 0.0153$, and it is rare that any case occurs 5 or more times.

How can the idea be applied to regression subset selection? There are two different methods:

1. Bootstrapping residuals. In this method, the full model is fitted to the data and the vector, $\boldsymbol{e}$, of least-squares residuals is calculated. These residuals are then studentized (standardized) using

$$e_i^* = \frac{e_i}{\sqrt{(1 - d_i)}},$$

   where the $d_i$ are the diagonal elements of the 'hat' matrix, $\boldsymbol{X}(\boldsymbol{X}'X)^{-1}\boldsymbol{X}'$. The original $\boldsymbol{X}$ matrix is used, but corresponding to each case, $X_i$, the $Y$-value is the fitted value of $Y$ from the full model plus one of the $e_i^*$'s sampled with replacement.

2. Bootstrapping the $(X_i, y_i)$-pairs. The $(X, y)$-pairs are sampled with replacement. This can be seen as the equivalent of the case of random predictors, whereas bootstrapping can be viewed as equivalent to the case of fixed predictors.

Let us take the POLLUTE data set again and use the Efroymson stepwise procedure with $F$-to-enter = 4.0, and $F$-to delete = 4.0. We have taken 10,000 bootstrap samples of size 60 from the data set and performed stepwise regression on each sample. The stepwise procedure picks the subset (1, 2, 9, 14) using the full data set. For each bootstrap replicate, the regression coefficients were calculated for this subset, whether or not it was the selected

Table 5.23 *Bootstrap of standardized residuals applied to the POLLUTE data set, using the Efroymson stepwise procedure to select subsets. Means and standard deviations of regression coefficients from 10,000 replicates*

| Var. | Subset (1,2,9,14) Frequency = 954 | | Different subset Frequency = 9046 | | All cases Frequency = 10,000 | |
|---|---|---|---|---|---|---|
| | Mean | Std. devn | Mean | Std. devn | Mean | Std. devn |
| X(1) | 2.19 | 0.48 | 2.03 | 0.54 | 2.05 | 0.53 |
| X(2) | –1.89 | 0.43 | –1.77 | 0.54 | –1.78 | 0.53 |
| X(9) | 4.06 | 0.62 | 4.08 | 0.69 | 4.08 | 0.68 |
| X(14) | 0.348 | 0.070 | 0.328 | 0.079 | 0.330 | 0.078 |

subset. The ordinary least-squares regression coefficients for the chosen subset using the full data set are:

```
X(1)   X(2)   X(9)   X(14)
2.06  -1.77   4.08   0.331
```

Table 5.23 shows the regression coefficients when boostrapping from bootstrapping the residuals. We see that the subset selected from the full data set was only chosen in about 9.5% of the bootstrap replications.

Notice that the average regression coefficients over all 10,000 replicates are very close to the least-squares coefficents for these variables using the original data, which is as they should be. However, the regression coefficients when subset (1, 2, 9, 14) was selected were significantly larger than those when other subsets were selected, except for variable $X(9)$. In the absence of any better method for estimating regression coefficients when the same data are used for model selection and for estimating the regression coefficients (see the next chapter), this suggests that we should use as our regression coefficients for model (1, 2, 9, 14):

```
X(1)   X(2)   X(9)   X(14)
1.92  -1.66   4.06   0.313
```

The weakness of this method is that the selection process has already picked the subset which fits best for the full data set. If the sample size, $n$, is only a little greater than the number of available predictors, $k$, then the fitted values, $\boldsymbol{X}\hat{\beta}$, will also be biased away from the true but unknown values $\boldsymbol{X}\beta$ in a direction which favours the model that has been selected. This bias is partially but not completely overcome by adding the bootstrap residuals to the fitted values. If $n \gg k$, then there is far more smoothing of the fitted values, which should be closer to the true $\boldsymbol{X}\beta$, and the bias will be very small. Unfortunately, we have no theory to quantify this bias.

Table 5.24 shows the regression coefficients when boostrapping $(X, y)$-pairs is used. We see that the subset selected from the full data set was only chosen in about 8.5% of the bootstrap replications.

Notice that the averages of the regression coefficients in the next to last

Table 5.24 *Bootstrap of $(X, y)$-pairs applied to the POLLUTE data set, using the Efroymson stepwise procedure to select subsets. Means and standard deviations of regression coefficients from 10,000 replicates*

| Var. | Subset (1,2,9,14) Frequency = 855 | | Different subset Frequency = 9145 | | All cases Frequency = 10,000 | |
|---|---|---|---|---|---|---|
| | Mean | Std. devn | Mean | Std. devn | Mean | Std. devn |
| X(1) | 1.89 | 0.43 | 2.18 | 0.58 | 2.16 | 0.57 |
| X(2) | –2.02 | 0.46 | –1.70 | 0.68 | –1.72 | 0.67 |
| X(9) | 4.02 | 0.63 | 4.00 | 0.79 | 4.00 | 0.77 |
| X(14) | 0.336 | 0.051 | 0.336 | 0.057 | 0.336 | 0.057 |

column are significantly different from the ordinary least-squares coefficients for the full data set. There is no reason why the mean values should converge to the OLS values for the full data set. One of the dangers when bootstrapping pairs is that the some of bootstrap samples may be rank deficient. For instance, if we have, say 40 predictors but only 60 cases, the average rank of the bootstrap sample will be about 63% (100% - 37%) of 60, or about 38.

From this we see that the average values of the regression coefficients are significantly different in samples in which the variables are selected from those samples in which they are not selected, for variables X(1) and X(2), but not for X(9) or X(14).

For a detailed description of bootstrap methods, the reader is referred to Efron and Tibshirani (1993). For details of its application to subset selection in regression, see Hjorth (1994). For the asymptotic properties of bootstrap estimators see Shao (1996, 1997).

## 5.5 Likelihood and information-based stopping rules

We have so far looked at obtaining the best predictions. An alternative family of criteria are those based upon likelihood or information measures.

Let us suppose that the likelihood of a particular value $y$ of the $Y$-variable is $f(y|\boldsymbol{x}, \boldsymbol{\theta})$, where $\boldsymbol{x}$ is a vector of given values of the predictor variables and $\boldsymbol{\theta}$ is a vector of parameter values. For instance, if $Y$ is normally distributed about the regression line $\boldsymbol{X\beta} = \boldsymbol{\beta_0} + \boldsymbol{\beta_1 X_1} + \boldsymbol{\beta_2 X_2} + ...$ with standard deviation $\sigma$, then the likelihood is

$$f(y|\boldsymbol{x}, \boldsymbol{\theta}) = \frac{e^{-(y-\mu)^2/2\sigma^2}}{\sqrt{2\pi\sigma^2}},$$

where the mean $(\mu) = \boldsymbol{\beta_0} + \boldsymbol{\beta_1 X_1} + \boldsymbol{\beta_2 X_2} + ....$ In this case, the vector of parameter values consists of the regression coefficients and one additional parameter which is the standard deviation.

Likelihoods can similarly be derived for other distributions, such as a gamma distribution or even a discrete distribution such as the Poisson distribution.

The likelihood for a sample of $n$ independent observations is simply the product of the likelihoods for each individual observation.

Maximizing the likelihood (ML), or in practice maximizing its logarithm, is an important method of estimation. It frequently leads to usable estimates in difficult situations, for instance in the case of truncated data where values of a variable greater than a certain size cannot be observed. The maximum likelihood estimates of parameters are often biased, and this must be checked in every case. For instance, in the case of the normal distribution, the ML-estimate of the sample variance ($\sigma^2$) is equal to the sum of squares of residuals divided by the sample size ($n$). An unbiased estimate can be obtained by dividing by $(n-1)$ instead. Taking the square root gives an estimate of the standard deviation, but this is biased using either $n$ or $(n-1)$ as the divisor; if $(n-1.5)$ is used as the divisor, the estimate of the standard deviation is approximately unbiased.

Let us look at how maximum likelihood can be used for model selection. Suppose that we have models $M_j$ for $j = 1, 2, \ldots, J$ to choose from. Let $f(y|\boldsymbol{X}, \boldsymbol{\theta}, M_j)$ be the likelihood for a sample of $n$ values, $y_1, y_2, \ \ldots, y_n$ given the $n \times k$ matrix of values of $k$ predictor variables, a vector of parameters $\boldsymbol{\theta}$ and the model $M_j$. The model may be nonlinear in the parameters, and for the moment, the distribution of the $Y$-values may be any distribution. In most cases, the number of parameters in $\boldsymbol{\theta}$ will vary from model to model. The function $f$ may be a discrete or continuous function of $X$; it will usually be continuous in $\boldsymbol{\theta}$ and will be a discrete function of the model number, $j$.

We could try to find the model $M_j$ and the vector of parameter values $\boldsymbol{\theta}$ that maximizes this likelihood. In many cases, this picks one of the most complex model, with a large number of parameters. Let us look, for instance, at the case in which the $Y$ are normally distributed and the models are all linear but in different subsets of the $k$ predictors. The likelihood in this case is

$$\prod_{i=1}^{n}(2\pi\sigma_j^2)^{-1/2} exp\{-(y_i - \sum_{l \in M_j} \beta_l x_{il})^2/(2\sigma_j^2).\}$$

In practice, it is almost always the natural logarithm, $L$, of the likelihood which is maximized. This is

$$L \ = \ - \ (n/2)\ln(2\pi\sigma_j^2) \ - \ RSS_j/(2\sigma_j^2),$$

where $RSS_j$ is the residual sum of squares for model $M_j$. This is maximized by taking

$$\hat{\sigma_j^2} \ = \ RSS_j/n,$$

where the usual least-squares regression coefficients are used in calculating the residual sum of squares ($RSS_j$). Hence, the maximum value of the log-likelihood, $L_j$, for model $M_j$ is

$$L_j \ = \ -(n/2)\ln(2\pi\hat{\sigma_j^2}) \ - \ (n/2), \qquad (5.34)$$

Over all models, $L$ is maximized by choosing that model which has the smallest residual variance estimate, $\hat{\sigma}_j^2$. If the models are those selected by forward selection, the residual variance estimate decreases each time a new variable is added until the reduction in the residual sum of squares from adding another variable is less than $\hat{\sigma}_j^2$. Thus, for forward selection and linear, normally distributed models, maximizing the likelihood is equivalent to using an $F$-to-enter value of 1.0 instead of more traditional values in the region of 4.0.

Suppose that we have 100 predictors, of which 20 have no predictive value (though we do not know this). The projections for these 20 variables will be normally distributed with zero mean and standard deviation equal to the residual standard deviation, $\sigma$. From tables of the normal distribution, we can expect about 31.7% of them to be either greater than $\sigma$ or less than $-\sigma$. Hence, about 6 or 7 of the 20 useless predictors will probably be included in the model selected by maximum likelihood. At the same time, a few of the remaining 80 predictors may be rejected.

As the sample size, $n$, increases, the projections of the 'real' predictors increase proportionally to $\sqrt{n}$, so that eventually all such variables will be included, but the proportion of 'false' variables included will remain at about 31.7% unless we add a penalty function to the likelihood. (The figure 31.7% is for forward selection; if an exhaustive search procedure is used, and the predictors are correlated, the figure is likely to be much larger.)

Akaike (1973, 1977) suggested that a penalty of $p$ should be deducted from the log-likelihood, where $p_j$ is the number of parameters estimated in model $M_j$. Akaike's Information Criterion (AIC) is minus twice the log-likelihood, and because of the change of sign, it is minimized rather than maximized. It is

$$AIC(j) \;=\; -2(L_j \;-\; p_j). \tag{5.35}$$

The AIC has been widely used as a stopping rule in the field of time series analysis in much the same way that Mallows' $C_p$ is used as the practical solution to exactly the same mathematical problem in selecting subsets of regression variables.

Using (5.34), the AIC can be rewritten as

$$\begin{aligned} AIC(j) &= n\ln(2\pi\hat{\sigma}_j^2) \;+\; n \;+\; 2p_j \\ &= \text{constant} \;+\; n\ln(\hat{\sigma}_j^2) \;+\; 2p_j. \end{aligned}$$

Some authors divide this expression by $n$, and the author has seen the exponential of this expression named as the AIC.

Several modifications of the AIC have been proposed (e.g. Risannen (1978), Schwarz (1978), and Hannan and Quinn (1979)). These all have the form $-2(L_j - p.f(n))$, where $f(n)$ is some slowly increasing function of the sample size, $n$, such as $\ln n$ or $\ln(\ln n)$.

The Schwarz criterion, often known as the Bayesian Information Criterion (BIC), is to maximize

$$BIC(j) \;=\; L_j \;-\; (1/2)p_j\ln(n), \tag{5.36}$$

Risannen's criterion is to maximize

$$RC(j) = nL_j + \sum_{i=1}^{p_j} \ln\left(\theta_i^2 \frac{\partial^2 L_j}{\partial \theta_i^2}\right) + (p_j + 1)\ln(n+2) + \ldots, \quad (5.37)$$

where the $\theta_i$ are the parameters in the model. This criterion was derived for fitting autoregressive, moving average models, and the final term above was $2\ln(r+1)(s+1)$, where $r$ and $s$ were the orders of the autoregressive and moving average parts of the model, and $p_j = r + s$.

Hannan and Quinn (1979) consider the case of selecting time series models in which the number of parameters in the model, $k$ in our notation, increases with sample size. For this situation, they derive the criterion of maximizing

$$HQ(j) = n\ln(\hat{\sigma}_j^2) + 2p_j c\ln(\ln n) \quad (5.38)$$

for some $c > 1$, though in their simulations they use the value $c = 1$. The justification for the double logarithm comes from pages 292–294 of Hannan (1970). Note that this is in Chapter V of Hannan's book, not Chapter VI as stated in the Hannan and Quinn paper.

Asymptotic comparisons of some of these criteria have been made by Stone (1977, 1979), who also showed the asymptotic equivalence of cross-validation and the AIC. These asymptotic results give no indication of the behaviour for finite sample sizes, and contain no discussion of the bias in parameter estimates that result from model selection.

The AIC has often been used as the stopping rule for selecting ARIMA models, where selection is not only between models with different numbers of parameters but also between many models of the same size. In this situation, there is competition bias, and this does not appear to have been considered in the time series literature. The BIC and the Hannan-Quinn criterion have better asymptotic properties. If there are variables that should be omitted, and hopefully there are, or why should we be looking at subset selection, then with the AIC the expected number of such variables included does not tend to zero as the sample size increases.

Hurvich and Tsai (1989) looked at the small-sample properties of the likelihood function and, based upon the expected values of estimators of parameters, derived the following modified AIC:

$$AIC_c = AIC + \frac{2p(p+1)}{(n-p-1)}. \quad (5.39)$$

Compared with the AIC, this gives a sharper cut-off, that is, it tends to select smaller subsets when the number of predictors, $p$, is large compared with the sample size, $n$. For large $n$, it behaves in a similar way to the AIC. This adjustment to the AIC is based upon small sample properties of the estimates assuming that the models have been chosen *a priori*. It does not allow for the bias in selecting the most extreme subset of each size, which is the basis of the RIC criterion described earlier in this chapter. Simulation results in the paper by Hurvich and Tsai show its superior performance, compared with a variety of other criteria, at selecting the correct model size with small sample sizes and small numbers of predictors.

The AIC, and the various modifications of it, can be applied in situations in which normality is not assumed, in which case the maximum likelihood fitting procedure may not be equivalent to least-squares fitting. Ronchetti (1985) has proposed a variation on the AIC using ideas from robust regression.

Burnham and Anderson (1998) use the Kullback-Liebler distance (or information difference/loss). Let $f(\boldsymbol{y}|\boldsymbol{\theta}, \boldsymbol{X})$ be the true but unknown probability density for the values of a $Y$-variable, given the matrix, $\boldsymbol{X}$ of values of the predictors, and a set of parameters $\boldsymbol{\theta}$. Kullback and Leibler (1951), and later Kullback (1959), define the K-L distance between the true model $f$ and an approximation to it, $g$, as

$$I(f, g) = \int f(y) \ln \left( \frac{f(y)}{g(y)} \right) dy.$$

Notice that this 'distance' or 'information loss' is directional; it is not symmetric. As the logarithm of a ratio is equal to the difference of the two separate logarithms, this can be rewritten as

$$I(f, g) = \int f(y) \ln(f(y)) dy - \int g(y) \ln(g(y)) dy.$$

When comparing many models with $f$, the first term on the right-hand side above remains constant, so that we can write

$$I(f, g) = \text{constant} - \int g(y) \ln(g(y)) dy.$$

Burnham and Anderson show how the K-L distance can be used to derive the AIC criterion for model selection, and various variations on it.

### 5.5.1 *Minimum description length (MDL)*

While likelihood theory has developed in the field of statistics, there have been parallel and similar developments in the field of information theory. There has been some 'cross-fertilization', for instance, in the papers by Rissanen (1987) and by Wallace and Freeman (1987), read to the Royal Statistical Society together with the published discussion that follows those papers in the printed version. This is a useful source of references to previous literature in this field. See also the more recent review paper by Hansen and Yu (2001).

Rissanen derives a new stopping criterion, which is to minimize

$$\frac{n}{2} \ln RSS_p + \frac{1}{2} \ln | \boldsymbol{X}_p' \boldsymbol{X}_p |$$

over all prospective models and over all sizes $p$, where $\boldsymbol{X}_p$ is the $n \times p$ matrix of values of the predictors in the current model. The determinant is most easily calculated from the Cholesky factorization $\boldsymbol{X}_p' \boldsymbol{X}_p = \boldsymbol{R}_p' \boldsymbol{R}_p$. As the determinant of a product of two matrices equals the product of the two separate determinants, we can simplify the above expression for computational purposes to

$$\frac{n}{2} \ln RSS_p \; + \; \ln \mid \boldsymbol{R_p} \mid, \tag{5.40}$$

where the determinant of a triangular matrix is simply the product of its diagonal elements. We note that, provided that $\boldsymbol{X'_p X_p}$ is of full rank, we can expect the second term above to increase as $\ln n$, as $n$ increases. It will also increase with the number of variables $p$.

Bryant and Cordero-Brana (2000) have derived alternative stopping criteria based upon the MDL. They define the MDL for a class, $\mathcal{M}$, of models such as the class of all linear subsets of variables, as

$$MDL(Y \mid \mathcal{M}, g) \;=\; -\ln f(Y \mid \hat{\boldsymbol{\beta}}) \;-\; \ln g(\hat{\boldsymbol{\beta}}),$$

where $Y$ is the data, $f$ is the (normal) probability density, $\boldsymbol{\beta}$ is the vector of regression coefficients, and $g$ is the length of message required to 'encode' the optimum values of the regression coefficients. Thus, the term involving $g$ is a penalty for adding a large number of parameters (variables). They consider some examples involving discrete distributions before looking at subset selection in regression.

For the case of linear models with a normally distributed $Y$-variable, they first standardize the least-squares regression coefficients, $\hat{\boldsymbol{\beta}}$. Let

$$\hat{\gamma} \;=\; \frac{(\boldsymbol{X'X})^{1/2}\hat{\boldsymbol{\beta}}}{\hat{\sigma}\sqrt{n}},$$

where they define the square root of $\boldsymbol{X'X}$ as any matrix $\boldsymbol{A}$ such that $\boldsymbol{A'A} = \boldsymbol{X'X}$. This definition is quite often used in the engineering literature, whereas in mathematical literature, the square root is defined as any matrix satisfying $\boldsymbol{AA} = \boldsymbol{X'X}$, that is, $\boldsymbol{A}$ is multiplied by itself, not by its transpose. This engineering definition of a matrix square root enables the Cholesky factorization, $\boldsymbol{R'R} = \boldsymbol{X'X}$, to be used as a square root, where $\boldsymbol{R}$ is an upper triangular matrix. Using the Cholesky factorization shows, we have then that

$$\hat{\gamma} \;=\; \frac{\boldsymbol{R}\hat{\boldsymbol{\beta}}}{\hat{\sigma}\sqrt{n}}$$

and $\boldsymbol{R}\hat{\boldsymbol{\beta}} = \boldsymbol{t}$, which is the vector of least-squares projections.

The Bryant and Cordero-Brana criterion is to minimize the quantity

$$\begin{aligned} MDL \;=\; & (n/2)\ln(\hat{\sigma}^2) \;-\; \ln\Gamma((n-p)/2) \;+\; p\ln(R_0/\sqrt{\pi}) \\ & + \ln\ln(\sigma_{max}/\sigma_{min})^2 \;+\; G(n), \end{aligned} \tag{5.41}$$

where

$$G(n) \;=\; \frac{n}{2}\left[\frac{n-1}{n}\ln n \;+\; \ln\pi\right].$$

and where the argument $R_0$ is the number of standard deviations of range of the standardized regression coefficients over which their MDL is integrated, and $\sigma_{min} < \hat{\sigma} < \sigma_{max}$ for all the models considered. The last two terms in (5.41) are thus constant over the models considered.

The stopping rule that Bryant and Cordero-Brana derive is equivalent to including a variable in the chosen model if its least-squares regression coefficient, $\hat{\beta}_j$ divided by the usual *a priori* estimate of its standard error is greater than

$$\sqrt{(n-p)\left\{\left[\frac{\Gamma((n-p+1)/2)R_0}{\Gamma((n-p)/2)\sqrt{\pi}}\right]^{2/n} - 1\right\}}. \tag{5.42}$$

They give tables for $R_0 = 10,\ 20$ *and* $50$, which show that that the criterion is only weakly dependent upon this strange parameter. They also give the approximation to (5.42)

$$\sqrt{\ln(n) \; - \; 1.838 \; + \; 2\ln(R_0)}$$

for $p > 10, n \leq 1000$ and $R_0 \leq 500$. Notice that the number of variables, $p$, does not appear in this approximation. Thus, the cut-off for the projections increases as the square root of $\ln n$. If we square this to apply it to sums of squares, then the cut-off increases as $\ln n$, which is the same as the Schwarz criterion.

Using a range of $\pm 5$ standard deviations ($R_0 = 10$), and a sample size $n = 50$, with $k = 10$ predictors, the value of the cut-off (5.42) is 2.35 standard deviations. For a sample size $n = 200$, this increases to 2.79 standard deviations. Squaring the numbers of standard deviations in the tables of Bryant and Cordero-Brana shows that their criterion is usually equivalent to using $F$-to-enter values between about 3 and 10, depending upon the values of $n$, $p$ and $R_0$.

# Appendix A
# Approximate equivalence of some stopping rules

In section 5.2.3, a generalized form of Mallows' $C_p$

$$C_p(M) \;=\; \frac{RSS_p}{\hat{\sigma}^2} \;-\; (n - Mp), \tag{A.1}$$

was introduced. There are many different stopping rules in use, and most of them can be shown to yield a similar stopping point to minimizing $C_p(M)$ for some value of $M$, usually a value close to 2.0, which is that for Mallows' $C_p$.

## A.1 *F*-to-enter

If $C_p(M)$ is a minimum at $p = m$, then $C_{m+1}(M) \geq C_m(M)$, where $C_m(M)$ and $C_{m+1}(M)$ are for the best-fitting subsets of $m$ and $(m+1)$ variables that have been found. Substitution from (A.1) and a little rearrangement shows that

$$\frac{RSS_m - RSS_{m+1}}{\hat{\sigma}^2} \leq M. \tag{A.2}$$

The left-hand side of (A.2) is approximately the $F$-to-enter statistic. The difference is that in $C_p(M)$, $\hat{\sigma}^2$ is defined as

$$\frac{RSS_k}{n-k},$$

that is, it is the residual variance estimate with all the available predictors in the model, whereas the denominator of the $F$-to-enter statistic is

$$\frac{RSS_{m+1}}{n-m-1}.$$

These quantities will often be very similar provided that $n \gg k$. Hence, at the minimum of $C_p(M)$, the $F$-to-enter statistic will usually be less than $M$.

If the $F$-to-enter statistic is used as the stopping rule when progressively increasing subset sizes, stopping as soon as it falls below $M$, then we can stop earlier than if we use the rule of minimizing $C_p(M)$. Unless the $X$-predictors are orthogonal, the quantity $C_p(M)$ may have several local minima, and the $F$-to-enter stopping rule tends to stop at the first. If $k$ is small, say of the order of 10-20, then stopping when the $F$-to-enter first drops below $M$ will usually find the minimum of $C_p(M)$, but for larger $k$, it will often pick a smaller subset.

## A.2 Adjusted $R^2$ or Fisher's A-statistic

The adjusted $R^2$-statistic (adjusted for degrees of freedom) is usually defined as

$$A_p = 1 - (1 - R_p^2)\frac{n-1}{n-p}, \tag{A.3}$$

where

$$R_p^2 = 1 - (RSS_p/RSS_1).$$

This is an appropriate definition when a constant is being fitted. When no constant is being fitted, $R_p^2$ is often re-defined as

$$R_p^2 = 1 - (RSS_p/RSS_0)$$

where $RSS_0$ means just the total sum of squares of the values of the $Y$-variable without subtraction of the mean. In this case, it is appropriate to replace the $(n-1)$ in (A.3) with $n$.

Maximizing the adjusted $R^2$-statistic can be shown to be identical to minimizing the quantity

$$s_p^2 = \frac{RSS_p}{(n-p)},$$

with respect to $p$, whichever of the two definitions for $A_p$ is used. This quantity, i.e. $s_p^2$, is the residual mean square for the $p$-variable subset.

A little rearrangement shows that at the minimum of $s_p^2$ we have

$$\frac{RSS_m \ - \ RSS_{m+1}}{RSS_{m+1}/(n-m-1)} \leq 1,$$

that is, the $F$-to-enter statistic has a value not greater than 1. Hence, maximizing the adjusted $R^2$-statistic is approximately equivalent to minimizing $C_p(M)$ with $M = 1$.

## A.3 Akaike's information criterion (AIC)

The AIC is to minimize $-2(L_p - p)$, where $L_p$ is the log-likelihood of a model with $p$ parameters.

In the linear regression situation, the crude log-likelihood, neglecting the issue of model selection, is

$$\begin{aligned} L_p &= -(n/2)\ln(2\pi\sigma_p^2) - \frac{1}{2\sigma_p^2}\sum_{i=1}^{n}(y_i - \sum_{j=1}^{p}\beta_j x_{ij})^2 \\ &= -(n/2)\ln(2\pi\sigma_p^2) - RSS_p/(2\sigma_p^2). \end{aligned}$$

As the maximum likelihood estimate for $\sigma_p^2$ is

$$\hat{\sigma}_p^2 = RSS_p/n,$$

where the division is by $n$, not the more usual $(n-p)$, we have that

$$L_p - p.f(n) = constant - (n/2)\ln(RSS_p) - p.f(n).$$

Changing signs and taking exponentials we see that minimizing these modified AIC's is equivalent to minimizing

$$RSS_p.exp[(2p/n)f(n)].$$

At the minimum, with respect to $p$, we have that

$$RSS_m.exp[(2m/n)f(n)] \leq RSS_{m+1}.exp[((2m+2)/n)f(n)]$$

or that

$$RSS_m \leq RSS_{m+1}.exp[2f(n)/n].$$

A little rearrangement then gives the $F$-to-enter statistic at the minimum,

$$\frac{RSS_m \ - \ RSS_{m+1}}{RSS_{m+1}/(n-m-1)} \leq (n-m-1)(e^{(2/n)f(n)} - 1).$$

Provided that $(2/n)f(n)$ is small, the right-hand side above is approximately

$$2\left(1 - \frac{m+1}{n}\right)f(n).$$

Thus using the AIC in its original form, i.e. with $f(n) = 1$, is equivalent to minimizing $C_p(M)$ with $M$ a little less than 2.

If the estimate used for $\hat{\sigma}_p^2$ is

$$\hat{\sigma}_p^2 \;=\; RSS_p/(n-p),$$

then maximizing $(L_p \;-\; p.f(n))$ is equivalent to minimizing

$$RSS_p.exp[(2p/n)f(n) \;-\; \ln(n-p) \;-\; p/n].$$

Continuing as before, we find that at the minimum the $F$-to-enter statistic for the AIC is not greater than a quantity approximated by

$$2\left(1 \;-\; \frac{m+1}{n}\right).$$

Thus, minimizing the AIC tends to select slightly larger subsets than minimizing Mallows' $C_p$.

CHAPTER 6

# Estimation of regression coefficients

## 6.1 Selection bias

The paper by Zhang (1992) starts with the statement:

> Standard statistical inferences are often carried out based on a model that is determined by a data-driven selection criterion. Such procedures, however, are both logically unsound and practically misleading.

This is a forceful way of stating the problem of overfitting that occurs in all model building processes, and for which we have no general solution. For an extensive discussion of the inferential problems associated with data-driven model building, see the paper by Chatfield (1995) and the discussion which follows it. The reader may, however, be disappointed by the lack of solutions presented there. The reader may be similarly disappointed by this chapter.

In this chapter, we look at some of the ways of estimating regression coefficients for a subset of variables when the data to be used for estimation are the same as those that were used to select the model. Most of these methods are based upon the biased least-squares regression coefficients, and require an estimate of the selection bias or depend upon properties of the selection bias. Such methods will be discussed in section 5.3, while selection bias in a very simple case will be discussed in section 5.2.

The term 'selection bias' was introduced in Chapter 1 but was not precisely defined at that stage. If the true relationship between $Y$ and the predictor variables is

$$Y = X_A\beta_A + X_B\beta_B + \epsilon,$$

where $X = (X_A, X_B)$ is a subdivision of the complete set of variables into two subsets $A$ and $B$, and where the residuals have zero expected value, then the unconditional expected value of the vector, $\boldsymbol{b_A}$, of least-squares regression coefficients for subset $A$ is

$$E(\boldsymbol{b_A}) = \boldsymbol{\beta_A} + (\boldsymbol{X'_A X_A})^{-1}\boldsymbol{X'_A X_B \beta_B}. \tag{6.1}$$

The second term on the right-hand side of (6.1) is what we have called the omission bias. Expression (6.1) is valid when the subset A has been chosen independently of the data. The definition of selection bias is then as the difference between the expected values when the data satisfy the conditions necessary for the selection of the subset A, and the unconditional expected values, that is,

$$\text{selection bias} = E(\boldsymbol{b_A} \mid \text{Subset } A \text{ selected}) - E(\boldsymbol{b_A}).$$

The extent of the bias is therefore dependent upon the selection procedure (e.g. forward selection, sequential replacement, exhaustive search, etc.), and is also a function of the stopping rule. It can usually be anticipated that the more extensive the search for the chosen model, the more extreme the data values must be to satisfy the conditions and hence, the greater the selection bias. Thus, in an exhaustive search, the subset is compared with all other subsets so that larger biases can be expected than for forward selection in which a much smaller number of comparisons is made.

An early paper on selection bias is that of Lane and Dietrich (1976). They carried out simulation studies using only 6 independent predictors, all of which had nonzero real regression coefficients, and found that for their smallest sample size ($n$=20), two of the six sample regression coefficients averaged about double their true values in the cases in which those variables were selected.

There is a large literature on the subject of pretest estimation, though in most cases studied only one or two tests are applied in selecting the model as contrasted with the large number usually carried out in subset selection. For the relatively small number (usually only two) of alternatives considered, explicit expressions can be derived for the biases in the regression coefficients, and hence the effect of these biases on prediction errors can be derived. There is almost no consideration of alternative estimates of regression coefficients other than ridge regression and the James-Stein/ Sclove estimators. Useful references on this topic are the surveys by Bancroft and Han (1977) and Wallace (1977), and the book by Judge and Bock (1978).

Though much of this chapter is concerned with understanding and trying to reduce bias, it should not be construed that bias elimination is a necessary or even a desirable objective. We have already discussed some biased estimators such as ridge estimators. Biased estimators are a standard part of the statistician's toolkit. How many readers use unbiased estimates of standard deviations? (N.B. The usual estimator $s = \{\sum(x - \bar{x})^2/(n - 1)\}^{\frac{1}{2}}$ is biased, though most of the bias can be removed by using $(n - 1.5)$ instead of the $(n - 1)$). That $(n - 1)$ is used to give an unbiased estimate of the *variance*.

It is important that we are aware of biases and have some idea of their magnitude. This is particularly true in the case of the selection of subsets of regression variables when the biases can be substantial when there are many subsets that are close competitors for selection.

## 6.2 Choice between two variables

To illustrate the bias resulting from selection, let us consider a simple example in which only two predictor variables, $X_1$ and $X_2$, are available and it has been decided *a priori* to select only one of them, the one that gives the smaller residual sum of squares when fitted to a set of data. For this case, it is feasible to derive mathematically the properties of the least-squares (or other) estimate of the regression coefficient for the selected variable, the residual sum

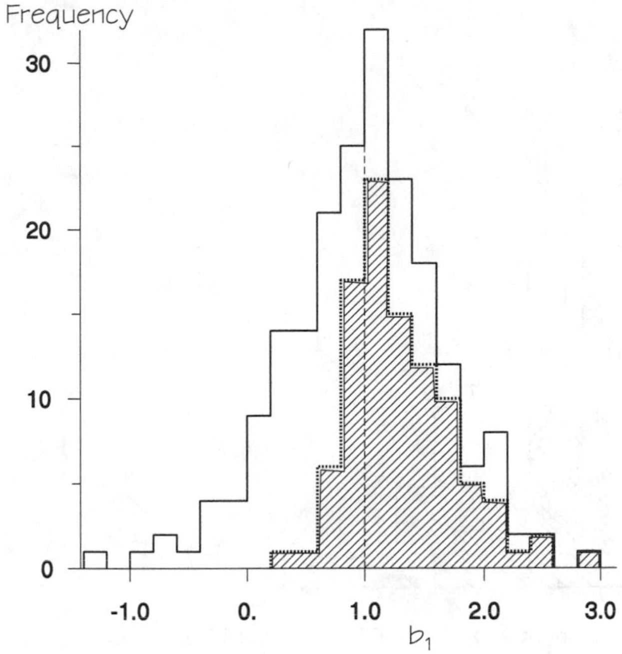


Figure 6.1 *Histogram of values of* $b_1$ *from 200 sets of artificial data. The outer histogram is for all data sets, the inner one is for those sets for which variable* $X_1$ *gave the smaller RSS. The thin vertical line is at the expected value of* $b_1$.

of squares, and other quantities of interest. However, before doing that, we present some simulation results.

The following example is constructed so that the expected residual sum of squares is the same, whichever of the two variables is selected. For added simplicity, the fitted models do not include a constant. Let us define

$$\begin{aligned} X_1 &= Z_1 \\ X_2 &= -3Z_1 + 4Z_2 \\ Y &= Z_1 + 2Z_2 + Z_3, \end{aligned}$$

where $Z_1$, $Z_2$ and $Z_3$ are independently sampled from the standard normal distribution. If we take a sample of $n$ independent observations and fit the two alternative models,

$$\begin{aligned} Y &= \gamma_1 X_1 + \text{residual} \\ Y &= \gamma_2 X_2 + \text{residual}, \end{aligned}$$

by least squares, then the expected values of the sample regression coefficients, $b_1$ and $b_2$, are 1.0 and 0.2 respectively, and the expected RSS is $5(n-1)$ for both models. Note, that the expected RSS with both variables in the model

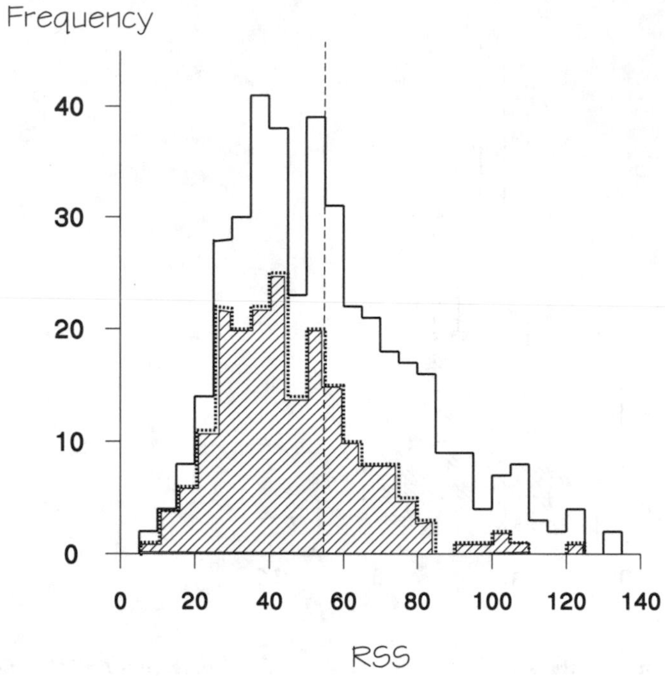


Figure 6.2 *Histograms of values of the RSS from 200 artificial data sets. The outer histogram is of all values (two per data set), while the inner one is for the model that gave the smaller RSS for each data set. The thin vertical line is at the expected value of the RSS.*

is $(n - 2)$, so that if a similar case arose in practice, both variables would probably be included in the selected model. The simulation could have been modified to give a much larger RSS with both variables in the model simply by increasing the variance of the residual variation, $Z_3$.

Using samples of size $n = 12$, 200 sets of artificial data were generated. The solid histogram shown in Figure 6.1 is that of the 200 sample values of $b_1$. These values averaged 0.991, which is close to the expected value; the average of the corresponding values of $b_2$ was 0.204. The inner, shaded histogram shown in Figure 6.1 is that of the values of $b_1$ when variable $X_1$ was selected. As can be seen, variable $X_1$ was usually selected when $b_1$ was above its expected value, but rarely selected when it was below. The corresponding histograms for $b_2$, which are not shown, look very similar. The average values of $b_1$ and $b_2$ for the data sets in which their corresponding variables were selected were 1.288 and 0.285. Variable $X_1$ was selected 99 times out of the 200 data sets.

Figure 6.2 shows the histograms of RSS's. The solid histogram is that for all 400 RSS's; that is, it includes two RSS's for each data set, one for each model. The inner, shaded histogram is that of RSS's for the selected models. The thin vertical line is at the expected value of the RSS (=55). The sample average of all the RSS's was 55.3, while the average for the selected models was 45.7.

Thus, for one highly artificial example, we have found the estimated regression coefficients of selected variables to be biased on the high side and the RSS to be too small. In practice, we often have many more variables or subsets of variables competing for selection, and in such cases, the biases are often far greater than here. When an exhaustive search has shown, say several hundred subsets of five, out of say twenty variables, which fit a set of data about equally well, perhaps so that Spj$\phi$tvoll's test finds no significant differences between them at the 5% level, then some of the regression coefficients for the best-fitting subset will probably be of the order of three standard errors from their expected values if they are estimated from the same data as were used to select that subset.

Let us now derive more general results for the two-variable case. Let us suppose that the true model is

$$Y = \beta_1 X_1 + \beta_2 X_2 + \epsilon, \tag{6.2}$$

where the residuals, $\epsilon$, have zero mean and variance $\sigma^2$. Later, we will also need to make assumptions about the shape of the distribution of the residuals. The least-squares estimate, $b_1$, of the regression coefficient for the simple regression of $Y$ upon $X_1$ is then

$$\begin{aligned} b_1 &= \boldsymbol{x}_1'\boldsymbol{y}/\boldsymbol{x}_1'\boldsymbol{x}_1 \\ &= (\beta_1\boldsymbol{x}_1'\boldsymbol{x}_1 + \beta_2\boldsymbol{x}_1'\boldsymbol{x}_2 + \boldsymbol{x}_1'\epsilon)/\boldsymbol{x}_1'\boldsymbol{x}_1, \end{aligned}$$

and hence,

$$\begin{aligned} E(b_1) &= \beta_1 + \beta_2\boldsymbol{x}_1'\boldsymbol{x}_2/\boldsymbol{x}_1'\boldsymbol{x}_1 \\ &= \gamma_1. \end{aligned}$$

Similarly,

$$\begin{aligned} E(b_2) &= \beta_2 + \beta_1\boldsymbol{x}_2'\boldsymbol{x}_1/\boldsymbol{x}_2'\boldsymbol{x}_2 \\ &= \gamma_2. \end{aligned}$$

Note that these are the expected values over all samples; no selection has been considered so far. The difference between $\gamma_1$ and $\beta_1$, and similarly that between $\gamma_2$ and $\beta_2$, is what we referred to earlier as the *omission* bias.

Now variable $X_1$ is selected when it gives the smaller RSS, or equivalently, when it gives the larger regression sum of squares. That is, $X_1$ is selected when

$$\boldsymbol{x}_1'\boldsymbol{x}_1 b_1^2 > \boldsymbol{x}_2'\boldsymbol{x}_2 b_2^2. \tag{6.3}$$

If we let $f(b_1,\ b_2)$ denote the joint probability density of $b_1$ and $b_2$, then the expected value of $b_1$ when variable $X_1$ is selected is

$$E(b_1 \mid X_1 \text{ selected}) = \frac{\int_R \int b_1 \, f(b_1, b_2) \, db_1 \, db_2}{\int_R \int f(b_1, b_2) \, db_1 \, db_2}, \qquad (6.4)$$

where the region $R$ in the $(b_1,\ b_2)$-space is that in which condition (6.3) is satisfied. The denominator of the right-hand side of (6.4) is the probability that variable $X_1$ is selected. The region $R$ can be reexpressed as that in which $|b_1| > C|b_2|$, where $C = (\boldsymbol{x}_2'\boldsymbol{x}_2/\boldsymbol{x}_1'\boldsymbol{x}_1)^{\frac{1}{2}}$. As the boundaries of this region are straight lines, it is relatively straightforward to evaluate (6.4) numerically for any assumed distribution of $b_1$ and $b_2$, given the sample values of $\boldsymbol{x}_1'\boldsymbol{x}_1$ and $\boldsymbol{x}_2'\boldsymbol{x}_2$. Similarly, by replacing the $b_1$ in the numerator of the right-hand side of (6.4) with $b_1^r$, we can calculate the $r^{th}$ moment of $b_1$ when $X_1$ is selected.

As $b_1$ and $b_2$ are both linear in the residuals, $\epsilon$, it is feasible to calculate $f(b_1,\ b_2)$ for any distribution of the residuals. However, if the distribution of the residuals departs drastically from the normal, we should be using some other method of fitting regressions than least squares, though in some cases, e.g. if the observations have a Poisson distribution or a gamma distribution with constant shape parameter, the maximum likelihood estimators of the regression coefficients are weighted least-squares estimators. If the residuals have a distribution which is close to normal then, by the Central Limit Theorem, we can expect the distribution of $b_1$ and $b_2$ to be closer to normal, particularly if the sample size is large. The results which follow are for the normal distribution.

Given the values of $X_1$ and $X_2$, the covariance matrix of $b_1$, $b_2$ is

$$\boldsymbol{V} = \sigma^2 \begin{bmatrix} (\boldsymbol{x}_1'\boldsymbol{x}_1)^{-1} & \boldsymbol{x}_1'\boldsymbol{x}_2(\boldsymbol{x}_1'\boldsymbol{x}_1)^{-1}(\boldsymbol{x}_2'\boldsymbol{x}_2)^{-1} \\ \boldsymbol{x}_1'\boldsymbol{x}_2(\boldsymbol{x}_1'\boldsymbol{x}_1)^{-1}(\boldsymbol{x}_2'\boldsymbol{x}_2)^{-1} & (\boldsymbol{x}_2'\boldsymbol{x}_2)^{-1} \end{bmatrix}.$$

It will simplify the mathematical expressions if we scale $X_1$ and $X_2$ so that $\boldsymbol{x}_1'\boldsymbol{x}_1 = \boldsymbol{x}_2'\boldsymbol{x}_2 = 1$, that is, if we replace $X_i$ with

$$\boldsymbol{x}_i^* = \boldsymbol{x}_i/(\boldsymbol{x}_i'\boldsymbol{x}_i)^{\frac{1}{2}}$$

and replace $b_i$ with

$$b_i^* = b_i(\boldsymbol{x}_i'\boldsymbol{x}_i)^{\frac{1}{2}}.$$

We will assume that such a scaling has been carried out and drop the use of the asterisks (*). The covariance matrix of the scaled $b$'s is then

$$\boldsymbol{V} = \sigma^2 \begin{bmatrix} 1 & \rho \\ \rho & 1 \end{bmatrix},$$

where $\rho = \boldsymbol{x}_1'\boldsymbol{x}_2$. The joint probability density of $b_1$ and $b_2$ is then

$$f(b_1, b_2) = \frac{exp\{-(\boldsymbol{b}-\boldsymbol{\gamma})'\boldsymbol{V}^{-1}(\boldsymbol{b}-\boldsymbol{\gamma})\}}{2\pi\sigma^2(1 - \rho^2)^{\frac{1}{2}}},$$

where $\boldsymbol{b}' = (b_1, b_2)$, $\boldsymbol{\gamma}' = (\gamma_1, \gamma_2)$, and

$$\boldsymbol{V}^{-1} = \frac{1}{\sigma^2(1 - \rho^2)} \begin{bmatrix} 1 & -\rho \\ -\rho & 1 \end{bmatrix}.$$

The argument of the exponential is

$$\begin{aligned} -\{(b_1-\gamma_1)^2 &- 2\rho(b_1-\gamma_1)(b_2-\gamma_2) + (b_2-\gamma_2)^2\}/\{2\sigma^2(1-\rho^2)\} \\ &= -\{[b_1-\mu(b_2)]^2 + (1-\rho^2)(b_2-\gamma_2)^2\}/\{2\sigma^2(1-\rho^2)\}, \end{aligned}$$

where $\mu(b_2) = \gamma_1 + \rho(b_2 - \gamma_2)$. Hence, we need to evaluate integrals of the form

$$\begin{aligned} I_r &= \int_{-\infty}^{\infty} \frac{exp[-(b_2-\gamma_2)^2/2\sigma^2]}{(2\pi\sigma^2)^{\frac{1}{2}}} \\ &\quad \times \int_{R(b_2)} \frac{b_1^r\, exp\{-[b_1-\mu(b_2)]^2/[2\sigma^2(1-\rho^2)]\}}{\{2\pi\sigma^2(1-\rho^2)\}^{\frac{1}{2}}}\, db_1\, db_2, \qquad (6.5) \end{aligned}$$

where the region of integration, $R(b_2)$, for the inner integral is $b_1 > |b_2|$ and $b_1 < -|b_2|$. The inner integral can be evaluated easily for low moments, $r$, giving for $r = 0$,

$$\Phi(z_1) + 1 - \Phi(z_2)$$

for $r = 1$

$$\sigma(1-\rho^2)^{\frac{1}{2}}[\phi(z_2) - \phi(z_1)] + \mu(b_2)[\Phi(z_1) + 1 - \Phi(z_2)];$$

for $r = 2$,

$$\begin{aligned} \sigma(1-\rho^2)^{\frac{1}{2}}\{[|b_2| + \mu(b_2)]\phi(z_2) + [|b_2| - \mu(b_2)]\phi(z_1)\} \\ + [\mu^2(b_2) + \sigma^2(1-\rho^2)][\Phi(z_1) + 1 - \Phi(z_2)], \end{aligned}$$

where $\phi$ and $\Phi$ are the probability density and distribution function of the standard normal distribution, and

$$\begin{aligned} z_1 &= [-|b_2| - \mu(b_2)]/[\sigma^2(1-\rho^2)]^{\frac{1}{2}} \\ z_2 &= [|b_2| - \mu(b_2)]/[\sigma^2(1-\rho^2)]^{\frac{1}{2}}. \end{aligned}$$

Numerical integration can then be used to determine $I_r$. Unfortunately, none of the derivatives of the kernel of (6.5) is continuous at $b_2 = 0$, so that Hermite integration cannot be used. However the kernel is well behaved on each side of $b_2 = 0$ so that integration in two parts presents no problems. This can be done using half-Hermite integration for which tables of the weights and ordinates have been given by Steen et al. (1969), and Kahaner et al. (1982).

Table 6.1 contains some values of the mean and standard deviation of $b_1$ when variable $X_1$ is selected. In this table, the expected value of $b_1$ over all cases, i.e. whether or not variable $X_1$ is selected, is held at 1.0. To apply the table when the expected value of $b_1$, i.e. $\gamma_1$, is not equal to 1.0, the X-variables should be scaled as described earlier, and the Y-variable should be scaled by dividing by $\gamma_1(\boldsymbol{x}_1'\boldsymbol{x}_1)^{\frac{1}{2}}$. The residual standard deviation after fitting both

Table 6.1 *Values of the expected value, $E(b_1|sel.)$, and standard deviation, std. devn.$(b_1|sel.)$, of $b_1$ when variable $X_1$ is selected, with $\gamma_1 = 1.0$*

| | | $\sigma = 0.3$ | | $\sigma = 0.5$ | |
|---|---|---|---|---|---|
| $\rho$ | $\gamma_2$ | $E(b_1\|sel.)$ | $Std.devn.(b_1\|sel.)$ | $E(b_1\|sel.)$ | $Std.devn.(b_1\|sel.)$ |
| -0.6 | 0.0 | 1.02 | 0.28 | 1.11 | 0.43 |
| | 0.5 | 1.08 | 0.25 | 1.21 | 0.39 |
| | 1.0 | 1.21 | 0.21 | 1.36 | 0.35 |
| | 1.5 | 1.39 | 0.18 | 1.53 | 0.32 |
| | 2.0 | 1.60 | 0.16 | 1.72 | 0.30 |
| 0.0 | 0.0 | 1.01 | 0.29 | 1.10 | 0.45 |
| | 0.5 | 1.05 | 0.28 | 1.15 | 0.44 |
| | 1.0 | 1.17 | 0.25 | 1.28 | 0.42 |
| | 1.5 | 1.35 | 0.23 | 1.46 | 0.40 |
| | 2.0 | 1.57 | 0.22 | 1.66 | 0.38 |
| +0.6 | 0.0 | 1.02 | 0.28 | 1.11 | 0.43 |
| | 0.5 | 1.01 | 0.29 | 1.09 | 0.46 |
| | 1.0 | 1.11 | 0.28 | 1.17 | 0.48 |
| | 1.5 | 1.30 | 0.27 | 1.34 | 0.51 |
| | 2.0 | 1.53 | 0.27 | 1.52 | 0.58 |

variables, $\sigma$, should be divided by the same quantity, $\gamma_2$ should be multiplied by $(\boldsymbol{x_2'x_2}/\boldsymbol{x_1'x_1})^{\frac{1}{2}}/\gamma_1$, and $\rho = \boldsymbol{x_1'x_2}/(\boldsymbol{x_1'x_1x_2'x_2})^{\frac{1}{2}}$. Thus, for the simulation at the start of this section, we had

$$\gamma_1 = 1.0,\ \gamma_2 = 0.2,\ x_1'x_1 = 12,\ x_2'x_2 = 300,\ x_1'x_2 = -3,\ \sigma = 1.$$

After scaling, and using asterisks as before to denote the scaled values, we have

$$\gamma_1^* = 1.0,\ \gamma_2^* = 1.0,\ \rho = -.05,\ \sigma^* = 1/\sqrt{12} = 0.289.$$

This is close to the entry in the table for $\rho = 0$, $\sigma = 0.3$, for which the expected value of $b_1$ is 1.17 when $X_1$ is selected. However, the average value in our simulations was 1.288, which is significantly larger than 1.17. The reason for the apparent discrepancy is that the theory above is for given values of $X_1$ and $X_2$, whereas $X_1$ and $X_2$ were random variables in our simulation, taking different values in each artificial data set. As a check, the simulations were repeated with fixed values of $X_1$ and $X_2$ such that $\boldsymbol{x_1'x_1} = 12$, $\boldsymbol{x_1'x_2} = 0$, $\boldsymbol{x_2'x_2} = 300$ and $\sigma = 1$. The average value of $b_1$ for the 106 cases out of 200 in which variable $X_1$ was selected was 1.202.

Figure 6.3 is intended to give a geometric interpretation of selection bias. The ellipses are for two different cases, and are ellipses of constant probability density in $(b_1,\ b_2)$ such that most pairs of values of $(b_1,\ b_2)$ are contained within them. For this figure, it is assumed that $X_1$ and $X_2$ have both been

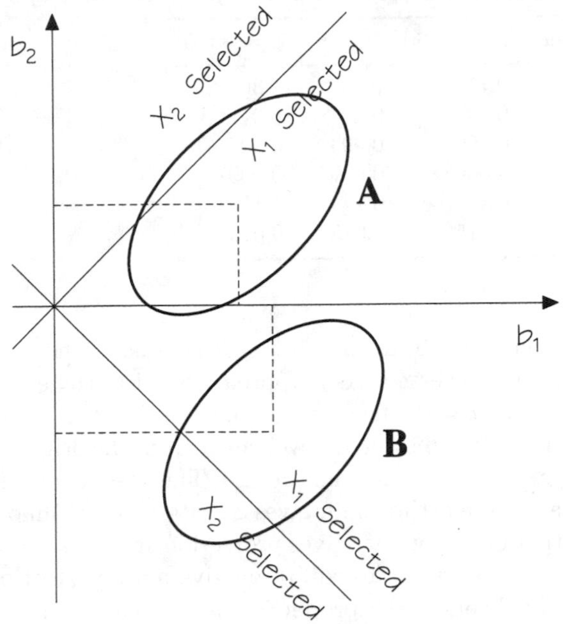


Figure 6.3 *A figure to illustrate the size and direction of selection bias.*

scaled to unit length so that the regions in which $X_1$ and $X_2$ are selected are bounded by lines at 45 degrees to the axes. Thus $X_1$ is selected in regions to the left and right of the origin, and $X_2$ is selected if $(b_1, b_2)$ is in the top or bottom regions.

Ellipse A represents a case in which $b_1$ is positive and $b_2$ is usually positive. The thin horizontal and vertical lines running from the centroid of the ellipse are at the unconditional expected values of $b_1$ and $b_2$. When $X_2$ is selected, $(b_1, b_2)$ is in the small sliver to the top left of the ellipse or just above it. Most of the sliver is above the expected value of $b_2$, so that $b_2$ is biased substantially in those rare cases in which it is selected. As the few cases in which $X_1$ is not selected give values of $b_1$ less than its expected value, $b_1$ is biased slightly on the high side when $X_1$ is selected.

Ellipse B represents a case in which the principal axis of the ellipse is perpendicular to the nearest selection boundary. In this case, far more of the ellipse is on the 'wrong' side of the boundary and the biases in both $b_1$ and $b_2$, when their corresponding variables are selected, are relatively large.

In both cases, A and B, the standard deviation of $b_1$ and $b_2$, when their variables are selected, are less than the unconditional standard deviations. This applies until the ellipses containing most of the joint distribution include the origin.

Table 6.2 *Coefficients in the rational polynomial (6.6) for three values of* $\rho$

| Parameter | $\rho = -0.6$ | $\rho = 0$ | $\rho = +0.6$ |
|---|---|---|---|
| $p_1$ | 0.71 | 0.56 | 0.36 |
| $p_2$ | 0.43 | 0.40 | 0.34 |
| $p_3$ | 0.090 | 0.094 | 0.103 |
| $p_4$ | 0.0067 | 0.0075 | 0.0097 |
| $p_5$ | 0.15 | 0.14 | 0.11 |
| $p_6$ | 0.015 | 0.019 | 0.029 |

Both ellipses shown here have $\rho > 0$; for $\rho < 0$ the directions of the major and minor axes of the ellipse are reversed. As $\rho$ approaches $\pm 1.0$, the ellipses become longer and narrower; for $\rho = 0$, the ellipses are circles. It can be seen that when $\gamma_1$ and $\gamma_2$ have the same signs and are well away from the origin, the biases are smallest when $\rho \gg 0$ and largest when $\rho \ll 0$. The case $\rho = 0$, that is, when the predictor variables are orthogonal, gives an intermediate amount of bias. The popular belief that orthogonality gives protection against selection bias is fallacious; highly correlated variables will often give more protection.

The above derivations have been of the properties of the regression coefficients. A similar exercise can be carried out using the joint distribution of the RSS's for the two variables, to find the distribution of the minimum RSS. This is somewhat simpler as both RSS's must be positive or zero, and the boundary is simply the straight line at which the two RSS's are equal.

In the two-variable case, it is possible to construct a function that approximates the selection bias, and then to use that function to eliminate much of the bias. From Table 6.1, it is obvious that the most important term in the function is that which measures the degree of competition between the variables. For $\sigma$ sufficiently small that only one boundary crossing needs to be considered, the bias in $b_1$ is well approximated by a rational polynomial of the form

$$E(b_1|X_1 \text{ selected}) = 1 + \sigma.\frac{p_1 + p_2x + p_3x^2 + p_4x^3}{1 + p_5x + p_6x^2}, \tag{6.6}$$

where $x = (\gamma_2 - 1)/\sigma$, and the coefficients $p_i$ are slowly changing functions of $\rho$. Using extensive tabulations, from which Table 6.1 was extracted, the (least-squares) fitted values of the parameters were as shown in Table 6.2 for three values of $\rho$, and for $x \geq 0$.

The discussion above applies to the case in which one and only one of the two predictors is selected. In a practical situation, both predictors or neither may be selected. This increases the number of alternative models to four. The regions of the $(b_1, b_2)$-space in which each model is selected are then more complex than those shown in Figure 6.3. This problem has been considered in detail by Sprevak (1976).

## 6.3 Selection bias in the general case and its reduction

The type of approach used in the previous section is not easily extended. In the general case, there are many boundaries between the regions in which different variables or subsets of variables are selected, so that numerical integration rapidly ceases to be feasible. Also, the selection bias is a function of the selection method which has been used, and of the stopping rule.

If the predictor variables are orthogonal, as, for instance, when the data are from a designed experiment or when the user has constructed orthogonal variables from which to select, then we can easily derive upper limits for the selection bias. If we scale all the predictor variables to have unit length, then the worst case is when all of the regression coefficients have the same expected value (or strictly that the absolute values of the expected values are the same). If all of the regression coefficients have expected value equal to $\pm\beta$ with sample standard deviation equal to $\sigma$ ($\sigma \ll \beta$), then if we pick just one variable, that with the largest regression coefficient in absolute value, the expected value of the absolute value of its regression coefficient is $\beta + \xi_1\sigma$, where $\xi_1$ is the first-order statistic for a random sample of $k$ values from the standard normal distribution, where $k$ is the number of variables available for selection. If we pick the three variables which give the best fit to a set of data, then the bias in the absolute values of the regression coefficients will have expected value equal to $\sigma(\xi_1 + \xi_2 + \xi_3)/3$, where $\xi_i$ is the $i^{th}$ order statistic. Thus, if we have say 25 available predictor variables, the bias in the regression coefficient of a single selected variable will be about 1.97 standard deviations.

The order-statistic argument gives only a rough guide to the likely size of the selection bias, though it does give an upper limit when the predictor variables are orthogonal. The selection bias can be higher than the order-statistic limit for correlated variables. In the author's experience, selection biases up to about three standard deviations are fairly common in real-life problems, particularly when an exhaustive search has been used to select the chosen subset of variables. In Chapter 6 we will see that the selection bias term is extremely important in estimating the magnitude of prediction errors, and in deciding upon a stopping rule.

To illustrate the extent of selection bias in practice, let us use the STEAM and POLLUTE data sets. We do not know the true population regression coefficients. What we can do is to split the data into two parts. The first part can be used to select a subset and to estimate LS regression coefficients for that subset. The second part can then be used as an independent data set to give unbiased estimates for the same subset al.ready selected.

The data sets were divided as nearly as possible into two equal parts. In the case of the STEAM data, which had 25 observations, 13 were used in the first part and 12 in the second. The two data sets were each split randomly into two parts, with the exercise repeated 100 times. An arbitrary decision was made to look at subsets of exactly three predictors plus a constant. Exhaustive searches were carried out to find the best-fitting subsets.

Table 6.3 *Subsets of 3 variables which gave best fits to random halves of the STEAM and POLLUTE data sets*

| STEAM | | | | POLLUTE | | | |
|---|---|---|---|---|---|---|---|
| Subset | | | Freq. | Subset | | | Freq. |
| 4 | 5 | 7 | 24 | 1 | 9 | 14 | 26 |
| 5 | 7 | 8 | 13 | 2 | 6 | 9 | 13 |
| 1 | 5 | 7 | 9 | 2 | 4 | 9 | 8 |
| 1 | 7 | 8 | 9 | 1 | 8 | 9 | 7 |
| 5 | 6 | 7 | 8 | 2 | 9 | 14 | 7 |
| 5 | 7 | 9 | 8 | 6 | 9 | 11 | 7 |
| 1 | 6 | 7 | 4 | 2 | 8 | 9 | 3 |
| 7 | 8 | 9 | 3 | 6 | 9 | 14 | 3 |
| 1 | 2 | 7 | 2 | 9 | 10 | 14 | 3 |
| 1 | 3 | 7 | 2 | 1 | 8 | 11 | 2 |
| 1 | 4 | 7 | 2 | 3 | 9 | 14 | 2 |
| 1 | 6 | 9 | 2 | 7 | 8 | 9 | 2 |
| 1 | 7 | 9 | 2 | 9 | 12 | 13 | 2 |
| 2 | 7 | 8 | 2 | | | | |
| Plus 10 others selected once | | | | Plus 15 others selected once | | | |

Table 6.4 *Results for the first split of the STEAM data*

| | First half of data | | Second half |
|---|---|---|---|
| | Regn. coeff. | Approx. s.e. | Regn. coeff. |
| Constant | 3.34 | 1.96 | 1.99 |
| Variable 4 | 0.19 | 0.05 | 0.60 |
| Variable 5 | 0.48 | 0.27 | –0.03 |
| Variable 7 | –0.080 | 0.011 | –0.082 |

Table 6.3 shows the different subsets that were selected for the two data sets. We note that variable number 7 (average temperature) was selected 94 times out of 100 for the STEAM data, while variable number 9 (% nonwhite in the population) was selected 91 times out of 100 for the POLLUTE data. These two will be considered 'dominant' variables.

For the first splitting of the STEAM data, the regression coefficients were as shown in Table 6.4.

The approximate standard errors shown above are the usual LS estimates applicable when the model has been chosen independently of the data.

Ignoring the constant in the model, which was not subject to selection, the regression coefficient for variable 4 has increased by about 8 standard errors from the data that selected it to the independent data, the regression

Table 6.5 *Frequency of shift of regression coefficients from the data used to select the model to independent data*

| Shift (z) in std. errors | STEAM data | POLLUTE data |
|---|---|---|
| < -5 | 10 | 9 |
| -5 to -4 | 19 | 7 |
| -4 to -3 | 29 | 22 |
| -3 to -2 | 46 | 44 |
| -2 to -1 | 59 | 71 |
| -1 to 0 | 43 | 58 |
| 0 to 1 | 33 | 37 |
| 1 to 2 | 15 | 30 |
| 2 to 3 | 9 | 13 |
| 3 to 4 | 4 | 6 |
| 4 to 5 | 15 | 2 |
| > 5 | 18 | 1 |
| Average shift | –0.71 | –0.90 |

coefficient for variable 5 has almost vanished, while that for variable 7 has remained steady.

The scale for each regression coefficient is different, so to combine the information on the shift of different regression coefficients from one half of the data to the other, the quantities

$$z_i \;=\; \frac{b_{2i} \;-\; b_{1i}}{s_{1i}}.sign(b_{1i})$$

were formed, where $b_{1i}$, $b_{2i}$ are the LS regression coefficients for variable number $i$ for each of the halves of the data, and $s_{1i}$ is the estimated standard error of the $i^{th}$ regression coefficient calculated from the first half. Thus $z_i$ is the shift, in standard errors, from the first half to the second half. The sign of $z_i$ is positive if $b_{2i}$ has the same sign as $b_{1i}$ and is larger in magnitude, and negative if the regression coefficient shrank or changed sign.

Table 6.5 shows the frequency of shifts of the regression coefficients, that is, of the $z_i$'s, for the two data sets. In the majority of cases, the unbiased regression coefficients were smaller, with an average shift of just under one standard error.

Let us separate out the two 'dominant' variables. The average shift for variable 7 for the STEAM data was –0.08 of a standard error, while that for variable 9 for the POLLUTE data was +0.04 of a standard error. Thus, in this case there appears to be very little overall average bias for the dominant variables, but an average bias of just over one standard error for the other variables.

Table 6.6 *Frequency of ratios of residual variance estimates for the data used for selection and for independent data*

| Variance ratio | STEAM data | POLLUTE data |
|---|---|---|
| 0.0 to 0.1 | 3 | 0 |
| 0.1 to 0.2 | 9 | 1 |
| 0.2 to 0.3 | 7 | 7 |
| 0.3 to 0.4 | 11 | 6 |
| 0.4 to 0.5 | 11 | 11 |
| 0.5 to 0.6 | 8 | 21 |
| 0.6 to 0.7 | 10 | 15 |
| 0.7 to 0.8 | 7 | 7 |
| 0.8 to 0.9 | 3 | 6 |
| 0.9 to 1.0 | 5 | 9 |
| 1.0 to 1.1 | 5 | 6 |
| 1.1 to 1.2 | 2 | 7 |
| 1.2 to 1.3 | 4 | 2 |
| 1.3 to 1.4 | 3 | 1 |
| > 1.4 | 12 | 1 |

Table 6.6 shows a histogram of the ratio $s_1^2/s_2^2$, where $s_1^2$, $s_2^2$ are the usual residual variance estimates for the two halves of the data for the subset of three variables which best fitted the first half of the data. We see that the average ratio was 0.76 for the STEAM data and 0.69 for the POLLUTE data. If $s_1^2$, $s_2^2$ had been independent estimates of the same variance, and the regression residuals have a normal distribution, then the expected value of this variance ratio is $\nu_2/(\nu_2 - 2)$, where $\nu_2$ is the number of degrees of freedom of $s_2^2$. These numbers of degrees of freedom are $12 - 4 = 8$ and $25 - 4 = 21$ for the STEAM and POLLUTE data sets, respectively. Thus, the expected values of $s_1^2/s_2^2$ are 1.33 and 1.11 for unbiased estimates of $s_1^2$. This gives a rough estimate of the extent of overfitting that has occurred.

In the last chapter, we looked at the use of bootstrapping residuals to estimate the bias due to selection. We now look at some other alternatives.

### *6.3.1 Monte Carlo estimation of bias in forward selection*

The simplest selection rule is forward selection; let us see if we can estimate selection biases in this case. Suppose that we have selected the first $(p - 1)$ variables and are considering which variable to add next. At this stage, the $(p - 1)$ selected variables will be in the first $(p - 1)$ rows of the triangular factorization. Let $r_{iy}$ be the $i^{th}$ projection of $Y$, that is the $i^{th}$ element in the vector $\boldsymbol{Q'y}$. Now if the true relationship between $Y$ and the complete set of

Table 6.7 *Least-squares projections for the STEAM data for a particular ordering of the predictors*

| Variable | Const. | 7 | 1 | 5 | 4 | 9 | 2 | 3 | 6 | 8 |
|---|---|---|---|---|---|---|---|---|---|---|
| Projn. | 47.12 | 6.75 | 3.05 | 1.12 | –0.94 | 0.59 | –0.56 | 0.46 | 0.42 | –0.94 |

predictor variables is

$$Y = X\beta + \epsilon,$$

where the residuals, $\epsilon$, are independently sampled from the normal distribution with zero mean and standard deviation $\sigma$, then

$$\begin{aligned} \boldsymbol{Q}'\boldsymbol{y} &= \boldsymbol{R}\boldsymbol{\beta} + \boldsymbol{Q}'\epsilon \\ &= \boldsymbol{R}\boldsymbol{\beta} + \eta \text{ say,} \end{aligned}$$

which means that the projections, $r_{iy}$, are normally distributed about their expected values, given by the appropriate element in $\boldsymbol{R}\boldsymbol{\beta}$, with standard deviation $\sigma$.

The reduction in RSS from adding next, the variable in row $p$, is $r_{py}^2$. Hence the variable in row $p$ is more likely to be selected next if its deviation from its expected value, $\eta_p$, is large, say greater than $\sigma$, and has the same sign as $\boldsymbol{R}\boldsymbol{\beta}$. We can then use a Monte Carlo type of method to estimate the bias in the projections when the corresponding variable is selected. The following is an algorithm for doing this.

1. Rotate the next selected variable into row $p$ if it is not already there.
2. Move the original projection, $r_{py}$, toward zero by a first guess of the bias, e.g. by $\hat{\sigma}$, where $\hat{\sigma}$ is an estimate of the residual standard deviation with all of the predictor variables in the model.
3. Generate pseudorandom normal vectors, $\eta_i$, with zero mean and standard deviation $\sigma$, and add these to the projections $r_{iy}$ for rows $p$, ..., $k$. Find whether the variable in row $p$ is still selected with these adjusted projections.
4. Repeat step 3 many times and average the values of $\eta_p$ for those cases in which the variable in row $p$ is selected. Take this average as the new estimate of the bias in $r_{py}$. Repeat steps 2 to 4 until the estimate of the bias stabilizes.

The above technique was applied to the STEAM and POLLUTE data sets used in earlier chapters. It was not appropriate to apply it to either the CLOUDS or DETROIT data sets as the first had no estimate of the residual standard deviation, and the DETROIT data set has only one degree of freedom for its residual.

For the STEAM data, the first five variables selected in forward selection are those numbered 7, 1, 5, 4 and 9 in that order. With these variables in that order, the projections in vector $\boldsymbol{Q}'\boldsymbol{y}$ are as in Table 6.7.

Table 6.8 *Estimating the bias in the projection for variable 1 for the STEAM data*

| Iteration | Starting bias estimate | Times out of 200 variable 1 selected | New bias estimate |
|---|---|---|---|
| 1 | 0.57 | 72 | 0.37 |
| 2 | 0.37 | 87 | 0.20 |
| 3 | 0.20 | 105 | 0.12 |
| 4 | 0.12 | 103 | 0.20 |

Table 6.9 *Estimating the bias in the projection for variable 5 for the STEAM data*

| Iteration | Starting bias estimate | Times out of 200 variable 5 selected | New bias estimate |
|---|---|---|---|
| 1 | 0.57 | 50 | 0.64 |
| 2 | 0.64 | 33 | 0.81 |
| 3 | 0.81 | 23 | 0.81 |

The sample estimate of the residual standard deviation is 0.57 with 15 degrees of freedom. Comparing the projections with this residual standard deviation suggests that we are only justified in including two or possibly three variables, plus the constant, in our model. Applying the above algorithm to the selection of the first variable, that is, variable number 7, after subtracting $\hat{\sigma} = 0.57$ from the projection in row 2, the variable was selected 200 times out of 200. The estimate of bias obtained by strictly applying the method above was the sum of 200 pseudorandom normal deviates with zero mean and standard deviation $\hat{\sigma}$, and this turned out to be +0.03 with the random number generator used. Clearly, there was no competition for selection in this case, and the bias in the projection is zero for all practical purposes.

There was more competition for the next position. Consecutive iterations are shown in Table 6.8. Using 0.20 as the bias estimate reduces the projection from 3.05 to 2.85.

For the third variable (row 4), the competition was greater. In this case the iterations gave the output shown in Table 6.9.

Using 0.81 as the bias estimate reduces the projection for variable 5 from 1.12 to 0.31.

Using the adjusted projections and back-substitution, the fitted 3-variable regression line changes from

$$Y = 8.57 - 0.0758X_7 + 0.488X_1 + 0.108X_5$$

to

$$Y = 9.48 - 0.0784X_7 + 0.637X_1 + 0.029X_5,$$

and the residual sum of squares for this model increases from 7.68 to 10.40.

Table 6.10 *Estimates of bias for the POLLUTE data*

| Variable number | Original projection | Times variable selected in last iteration | Bias estimate | Adjusted projection |
|---|---|---|---|---|
| Const. | 7284.0 | | | |
| 9 | 307.6 | 176 | 11.1 | 296.5 |
| 6 | –184.0 | 31 | –41.1 | –142.9 |
| 2 | –132.1 | 28 | –40.8 | –91.3 |

Notice that reducing the absolute size of the projections does not necessarily reduce the sizes of all the regression coefficients. It always reduces the size of the coefficient of the last variable selected, but not necessarily the sizes of the others.

For the POLLUTE data set, the residual standard deviation estimate is 34.9 with 44 degrees of freedom with all the variables in the model. The bias estimates again appeared to converge very rapidly. The results obtained are shown in Table 6.10.

Using these projections, the fitted three-variable regression line changes from

$$Y = 1208.1 + 5.03X_9 - 23.8X_6 - 1.96X_2$$

to

$$Y = 1138.7 + 4.65X_9 - 18.9X_6 - 1.35X_2$$

and the residual sum of squares increases from 82389 to 111642.

This simple "intuitive" method appears to produce good point estimates of regression coefficients, but has a number of shortcomings. First, when the first variable is selected, its sample projection may be appreciably larger in absolute value than its expected value. The method allows for that bias and for the fact that the variable may have been selected because some of the other projections deviated substantially from their expected values. However, any bias in these other projections was forgotten when we proceeded to select the next variable. This can easily be accommodated by estimating the biases of all the projections at each iteration instead of estimating only the bias for the projection of the selected variable. In most cases, this will make very little difference.

Another objection to the method is that it provides only point estimates without confidence limits or approximate standard errors.

Implicit in the method just described is the notion that if there is *apparently* competition amongst variables for selection, then the variable selected must have a sample projection that is above its expected value. Alternatively, the sample projection may have been smaller than its expected value so that a variable that was a clear superior to others appeared closer than it should. In this case, the above method reduces the projection even further and so introduces a bias that would not otherwise have arisen.

The above method will not be developed further, though similar methods can be developed for other selection procedures, and it is possible at great computational expense to obtain confidence limits. In the next section, a method based upon conditional likelihood will be described, but before proceeding to that, let us look briefly at other alternative ways of tackling the selection bias problem.

### *6.3.2 Shrinkage methods*

Figure 6.1 suggests that some kind of shrinkage should be applied. Two kinds of shrinkage have become popular; these are ridge regression and simple shrinkage of all the regression coefficients by the same factor.

The simplest form of shrinkage estimator is that suggested by James and Stein (1961). Let $T_i$, i = 1, 2, ..., k, be unbiased estimators of quantities $\mu_i$, each having the same variance $\sigma^2$. The $T_i$ are assumed to be uncorrelated. Consider the shrunken estimators

$$T_i^* = (1 - \alpha)T_i, \; 0 < \alpha < 1; \qquad (6.7)$$

that is, each estimate is shrunk toward zero by the same relative amount. The new estimates will be biased but have lower variances than the $T_i$'s. Suppose we trade off the bias against the reduced variance by minimizing a loss function, which is the expected squared error

$$\begin{aligned} \text{loss} &= E\sum_{i=1}^{k}(T_i^* - \mu_i)^2 \\ &= (\text{bias})^2 + \text{variance} \\ &= \alpha^2 \sum_{i=1}^{k} \mu_i^2 + k(1 - \alpha)^2\sigma^2. \end{aligned}$$

Then

$$\frac{d(\text{loss})}{d\alpha} = 2\alpha \sum_{i=1}^{k} \mu_i^2 - 2k(1 - \alpha)\sigma^2.$$

Setting this equal to zero gives

$$\alpha = \frac{k\sigma^2}{\sum \mu_i^2 + k\sigma^2}.$$

Unfortunately, this involves the unknown $\mu_i$'s. Now $\sum T_i^2$ is an unbiased estimator of $(\sum \mu_i^2 + k\sigma^2)$, and substituting this into (6.7) gives the estimator:-

$$T_i^* = \left(1 - \frac{k\sigma^2}{\sum T_i^2}\right) T_i.$$

Our derivation above assumed that $\alpha$ was not a random variable, yet we have now replaced it with a function of the $T_i$'s, thus invalidating the derivation.

If we allow for the variance of $\alpha$, it can be shown (see James and Stein (1961) for more details) that the estimator

$$T_i^* = \left(1 - \frac{(k-2)\sigma^2}{\sum T_i^2}\right) T_i \tag{6.8}$$

gives a smaller squared error for all $k > 2$. This is the James-Stein estimator.

In practice, $\sigma^2$ must be estimated. If the usual estimate

$$s^2 = \sum_{i=1}^{k} (T_i - \bar{T})^2/(k-1)$$

is used, where $\bar{T}$ is the sample mean of the $T_i$'s, then Stein (1962) shows that the $\sigma^2$ in the above estimators should be replaced with $s^2(k - 1)/(k + 1)$.

Lindley (pages 285-287 of Stein (1962)) suggested shrinkage toward the mean, rather than shrinkage toward zero. His estimator is

$$T_i^* = \bar{T} + \left(1 - \frac{(k-3)\sigma^2}{\sum(T - \bar{T})^2}\right)(T_i - \bar{T}).$$

The James-Stein estimator has been controversial. The following is a quote from Efron and Morris (1973):

> The James-Stein estimator seems to do the impossible. The estimator of each $\mu_i$ is made to depend not only on $T_i$ but on the other $T_j$, whose distributions seemingly are unrelated to $\mu_i$, and the result is an improvement over the maximum likelihood estimator no matter what the values of $\mu_1, \mu_2, ..., \mu_k$. Thus we have the 'speed of light' rhetorical question, 'Do you mean that if I want to estimate tea consumption in Taiwan I will do better to estimate simultaneously the speed of light and the weight of hogs in Montana?'

Of course, the other $T_i$'s (the speed of light and weight of hogs), are used to estimate the shrinkage factor $\alpha$, and a critical assumption is that all the $T_i$'s have the same variance. It seems improbable that these three disparate estimates would have the same variance, even after appropriate scaling say by measuring the speed of light in knots and the weight of hogs in carats.

James-Stein shrinkage cannot be applied directly to regression coefficients as they do not in general have the same variance and are usually correlated. However, the least-squares (LS) projections, from which we calculate the regression coefficients, do have these properties.

Let $t_i$ and $\tau_i$ denote the LS-projections and their expected values, and let $\boldsymbol{R}$ denote the upper-triangular Cholesky factor of the design matrix $\boldsymbol{X}$. Then as

$$\begin{aligned} \boldsymbol{Rb} &= \boldsymbol{t} \\ \boldsymbol{R\beta} &= \boldsymbol{\tau} \end{aligned}$$

where $\boldsymbol{b}$ and $\boldsymbol{\beta}$ are the LS-estimates of the regression coefficients and their expected values, then the loss function

$$\begin{aligned}\sum_{i=1}^{k}(t_i - \tau_i)^2 &= (\boldsymbol{t} - \boldsymbol{\tau})'(\boldsymbol{t} - \boldsymbol{\tau}) \\ &= (\boldsymbol{b} - \boldsymbol{\beta})'\boldsymbol{R}'\boldsymbol{R}(\boldsymbol{b} - \boldsymbol{\beta}) \\ &= (\boldsymbol{b} - \boldsymbol{\beta})'\boldsymbol{X}'\boldsymbol{X}(\boldsymbol{b} - \boldsymbol{\beta}).\end{aligned}$$

This is the sum of squares of the elements of $\boldsymbol{X}(\boldsymbol{b} - \boldsymbol{\beta})$. As the elements of $\boldsymbol{Xb}$ and $\boldsymbol{X\beta}$ are the LS-fitted and expected values, respectively, of the dependent variable, the sum is the sum of squares of differences between the fitted and expected values. Minimizing this sum is a reasonable objective function in many situations.

The factor $\alpha$ is then

$$\alpha = \frac{(k - 2)\sigma^2}{\sum t_i^2}.$$

The sum of squares of the LS-projections in the denominator is the regression sum of squares (regn. S.S.). Furthermore, if all the $t_i$'s are reduced by the same factor, so are the regression coefficients derived from them, so that our new estimates, $b_i^*$, of the regression coefficients are

$$b_i^* = \left(1 - \frac{(k - 2)\sigma^2}{regn.S.S.}\right) b_i.$$

This estimator is due to Sclove (1968). Normally, $\sigma^2$ will be replaced with an estimate $s^2\nu/(\nu + 2)$, where $s^2$ is the usual residual variance estimate and $\nu$ is its number of degrees of freedom.

Notice though, that we have derived the amount of shrinkage to apply to *unbiased* LS-regression coefficients. No allowance has been made for the bias introduced by selection. In most cases, the amount of shrinkage using James-Stein estimates will be much less than is needed to overcome selection bias. For instance, for the STEAM data, $k = 9$ (excluding the constant), $s^2 = 0.325$, and the regression sum of squares $= 58.9$. Using $s^2$ for $\sigma^2$ gives $b_i^* = 0.96 b_i$, i.e. the regression coefficients are shrunk by only 4%.

Ridge regression has been described in section 3.9. Much of the vast literature on ridge regression has focussed upon mean squared errors of the regression coefficients, i.e.

$$MSE(\hat{\boldsymbol{\beta}}) = E(\boldsymbol{\beta} - \hat{\boldsymbol{\beta}})'(\boldsymbol{\beta} - \hat{\boldsymbol{\beta}}),$$

where $\hat{\boldsymbol{\beta}}$ is an estimate of the vector of regression coefficients, $\boldsymbol{\beta}$. In most cases in practice, the mean squared errors of prediction are likely to be of more interest. Lawless and Wang (1976) have particularly emphasized this distinction, and have shown that while ridge regression can produce huge reductions in the $MSE(\hat{\boldsymbol{\beta}})$ when the $\boldsymbol{X}'\boldsymbol{X}$-matrix is ill-conditioned, it produces far less reduction in the $MSE(\boldsymbol{X}\hat{\boldsymbol{\beta}})$, that is, in the mean squared error of prediction, and sometimes produces a small increase.

Very little attention has been paid to the ridge trace idea of Hoel and Kennard (1970b) for variable selection, possibly because of its subjective nature. An explicit rule for deletion of variables has subsequently been given by Hoerl et al. (1986). They suggest using a modified t-statistic,

$$t = \hat{\beta}_i / s_i$$

where the estimates of the regression coefficients are given by

$$\hat{\boldsymbol{\beta}} = (\boldsymbol{X}'\boldsymbol{X} + d\boldsymbol{I})^{-1}\boldsymbol{X}'\boldsymbol{y}$$

after first shifting and scaling each $X$-predictor to have zero sample mean and unit sample standard deviation, and where the $s_i$'s are the square roots of the diagonal elements of

$$\hat{\sigma}^2(\boldsymbol{X}'\boldsymbol{X} + d\boldsymbol{I})^{-1}\boldsymbol{X}'\boldsymbol{X}(\boldsymbol{X}'\boldsymbol{X} + d\boldsymbol{I})^{-1}.$$

In their simulations, a range of significance levels was used, but those reported in their paper were for a nominal 20% level. These simulations showed that this ridge selection procedure gave good performance in terms of both $MSE(\hat{\boldsymbol{\beta}})$ and $MSE(\boldsymbol{X}\hat{\boldsymbol{\beta}})$ when the Lawless and Wang (1976) value for $d$ was used, i.e.

$$d = \frac{\text{residual mean square}}{\text{regression mean square}},$$

where the mean squares are evaluated using all the available predictors. The ridge regression estimator performed well with this value of $d$ in any case *without* subset selection. When some of the true $\beta_i$'s were zero, a moderate improvement was achieved using selection. In the case of the STEAM data, the value of $d = 0.325/58.9 = 0.0055$. That is, the diagonal elements of $\boldsymbol{X}'\boldsymbol{X}$ are incremented by about half of 1%.

Both ridge regression and James-Stein shrinkage require a knowledge of the size of the selection bias to make the best choice of the amount of shrinkage to apply. As each method has only one parameter controlling the amount of shrinkage, it cannot be controlled to eliminate or reduce the bias simultaneously in all parameters. We believe that the method of conditional likelihood to be described in section 5.4 is a more satisfactory method of achieving this.

### *6.3.3 Using the jackknife*

A popular method of bias reduction is the so-called jackknife (Quenouille (1956), Gray and Schucany (1972), Miller (1974)). Suppose that $T_n$ is a statistic based upon a sample of $n$ observations, and that

$$E(T_n) = \theta + \frac{a}{n} + \frac{b}{n^2} + o(n^{-2}),$$

where $\theta$ is a parameter or vector of parameters that we want to estimate. Then,

$$E\{nT_n - (n-1)T_{n-1}\} = \theta - \frac{b}{n(n-1)} + o(n^{-2}),$$

that is, the terms of order $n^{-1}$ in the bias are eliminated, while those of order $n^{-2}$ are reversed in sign and increased very slightly in magnitude.

The jackknife could be applied to the estimation of regression coefficients or the RSS for a model. Suppose that $T_n$ is the least-squares estimate of the regression coefficients for a subset of variables selected using a particular procedure and $n$ observations. As the bias is due to the fact that the regression coefficient is being estimated conditional upon a certain subset being selected, $T_{n-1}$ obtained from $(n-1)$ observations out of the $n$ must be subject to the same condition. A sample of $(n-1)$ observations can be obtained in $n$ different ways by deleting one of the $n$ observations. Consider all $n$ such samples and apply the same selection procedure to each. This may be quite feasible if the procedure is forward selection, sequential replacement, or one of the other "cheap" procedures, but will involve very substantial computational effort if the procedure is an exhaustive search. Suppose that in $m$ out of the $n$ cases the subset of interest is selected, then we can use the $m$ estimates of $T_{n-1}$ for these cases in the jackknife and average the results. In limited experiments by the author, the value of $m$ has usually been close to or equal to $n$ and rarely less than $n/2$, though there appears to be no reason why $m$ must be greater than zero. These experiments suggested that the jackknife may be fairly successful at removing bias, but the variance of the jackknife estimates was very large. There seems to be no reason for expecting selection bias to reduce roughly as $n^{-1}$, so that that part of the bias that is removed may be fairly small. Unless the term in $n^{-1}$ accounts for most of the bias, the increased variance in the resulting estimates is too high a price to pay.

The order-statistic argument used earlier in this section leads us to anticipate that selection bias may be roughly proportional to $n^{-\frac{1}{2}}$ when the predictor variables are orthogonal and are all equally good choices. Also, substitution in (6.6) gives the leading term in the bias as proportional to $n^{-\frac{1}{2}}$. To eliminate this type of bias, the jackknife statistic should be modified to

$$[n^{\frac{1}{2}}T_n \; - \; (n-1)^{\frac{1}{2}}T_{n-1}] \;/\; [n^{\frac{1}{2}} \; - \; (n-1)^{\frac{1}{2}}]. \tag{6.9}$$

If we write the jackknife estimator as

$$(f_nT_n \; - \; f_{n-1}T_{n-1})/(f_n \; - \; f_{n-1}),$$

where we have suggested that $f_n = \sqrt{n}$ is a suitable choice, then the estimator can be rewritten as

$$T_n \; + \; \frac{f_{n-1}}{f_n \; - \; f_{n-1}}(T_n \; - \; T_{n-1}).$$

Thus, the initial estimate $T_n$ is moved away from $T_{n-1}$ by a substantial multiple of the difference between them. The use of a Taylor series expansion shows that the square root jackknife adjusts the biased estimate by about twice as much as the choice $f_n = n$.

Freedman, Navidi and Peters (1988) have applied the jackknife to subset selection in regression, but not as described above. All $n$ sets of data, each

Table 6.11 *Model selection with two sets of data; hypothetical data*

| Rank | Selection data | Regn. coeff. data |
|---|---|---|
| Best | 3, 7, 14 | 3, 10, 11 |
| 2nd | 3, 7, 11 | 3, 10, 14 |
| 3rd | 3, 4, 8 | 3, 7, 11 |
| 4th | 3, 4, 7 | 3, 7, 10 |
| 5th | 3, 8, 11 | 3, 11, 14 |

with one case deleted, were used with the regression coefficients set to zero if a variable was not selected. They did not use the square root version of the jackknife. These authors, and Dijkstra and Veldkamp (1988) in the same volume of conference proceedings, have also used the 'bootstrap' technique with very little success. Platt (1982) had previously also advocated the use of the bootstrap after model selection.

### *6.3.4 Independent data sets*

Selection bias can be completely eliminated by using independent data sets for the selection of the model and for estimating the regression coefficients. It is rarely sensible to recommend this method in practice as it is inefficient in not using the information from the selection set of data in estimating the regression coefficients. In some cases, only the selected variables will have been measured for the second data set, though in other cases measurements of all variables will be available. If all of the variables have been measured for the second data set, then it is tempting to see if our selected subset is one of the best-fitting subsets of its size. Suppose that it is not; what do we do now? We may well find something like the following for the best-fitting subsets of three variables.

We notice that the best-fitting subset of three variables for the selection set of data does not appear among the best five for the set of data to be used for estimating the regression coefficients; let us suppose that it occurs much further down the list. We notice that the second-best subset from the first data set occurs quite high up on the other list. It looks like a good choice. This is a crude way of looking for the best-fitting subset for the combined data set, so why don't we do the search properly for the combined data set? But then we are back with the problem of selection bias if we use least-squares estimates of the regression coefficients for the best-fitting subset! Fortunately though, the selection bias should be a little smaller because of the larger sample size.

Roecker (1991) reports on a simulation study to investigate whether it is better to use separate subsets of the data for model selection and estimation. Generating 32 artificial sets of data for which the true models were known, and using half of the data for model selection and half for estimating the regression coefficients, she found that in all cases, the predictions were inferior to those in which all of the data were used for both selection and estimation. Using ordinary least-squares estimates of the regression coefficients, the mean squared errors ranged from 2% larger to 140% larger with 50/50 splitting.

## 6.4 Conditional likelihood estimation

A method that can usually be used to obtain parameter estimates, providing that we are prepared to make distributional assumptions, is maximum likelihood. We want to estimate parameters for a subset of variables after we have found that these variables give a better fit to a set of data than some other subsets. Taking the values of the $X$-variables as given, if we assume that the values of $Y$ are normally distributed about expected values given by $\boldsymbol{X\beta}$, where for the moment $X$ contains all of the predictor variables, with the same variance $\sigma^2$, and that the deviations of the $Y$'s from $\boldsymbol{X\beta}$ are independent, then the unconditional likelihood for a sample of $n$ observations is

$$\prod_{i=1}^{n} \phi\{(y_i - \sum \beta_j x_{ij})/\sigma\},$$

where $\phi$ is the standard normal probability density, that is $\phi(x) = e^{-\frac{1}{2}x^2}/\sqrt{2\pi}$.

Now given that a specific subset has been selected by some procedure (e.g. forward selection, sequential replacement, exhaustive search, etc.), many vectors of $\boldsymbol{y}$-vectors are impossible as they would not lead to the selection of that subset. The conditional likelihood is then proportional to the above likelihood for acceptable $y$-vectors and zero elsewhere. Hence, the likelihood of the sample values of $Y$, given $X$ and that a certain selection procedure has selected a subset of variables, $S$, is

$$\frac{\prod_{i=1}^{n} \phi\{(y_i - \sum \beta_j x_{ij})/\sigma\}}{\int ... \int (\text{the above density})\ dy_1\ ...\ dy_n}$$

in a region $R$ of the $Y$-space in which the procedure used selects subset $S$. The multidimensional integration is also over this region, and the value of the integral is the *a priori* probability that $S$ is selected given $X$. Substituting for $\phi$, the logarithm of the conditional likelihood (LCL) over region $R$ is then

$$\begin{aligned} \log_e(\text{conditional likelihood}) &= -(n/2)\log_e(2\pi\sigma^2) \\ &\quad -\sum_{i=1}^{n}\left(y_i - \sum \beta_j x_{ij}\right)^2/(2\sigma^2) \\ &\quad - \log_e(\text{Prob. S is selected}). \end{aligned} \tag{6.10}$$

The difficulty in using this conditional likelihood is clearly in evaluating the probability of selection of subset $S$, which is a function of the parameters $\boldsymbol{\beta}$ and $\sigma$. In simple cases, such as when there are only two $X$-variables or in forward selection when the $X$-variables are orthogonal, the probability that subset $S$ is selected can be evaluated explicitly. In general, we need to evaluate the probability that the regression sum of squares for subset $S$ is larger than those of others with which it was compared in the selection procedure. These regression sums of squares are quadratic forms in the $y_i$'s so that the region in the $Y$-space in which one of them exceeds another is quite complex.

The probability of selection, to be denoted by $P$, can be estimated by Monte Carlo methods in a manner similar to that used in section 5.3. The expected values of the projections, $\boldsymbol{Q}'\boldsymbol{y}$, are given by $\boldsymbol{R\beta}$. By adding vectors $\boldsymbol{\eta}$ to $\boldsymbol{R\beta}$, where the elements $\eta_i$ of $\boldsymbol{\eta}$ are sampled from the $N(0, \sigma^2)$ distribution, random vectors of projections can be obtained. These can be subjected to the selection procedure that found subset $S$. The proportion of times in which subset $S$ is selected then gives an estimate of $P$ for the vector $\boldsymbol{\beta}$ used. This is a feasible method if one of the "cheap" methods discussed in Chapter 3, such as forward selection, but it is not as practical in conjunction with an exhaustive search. An alternative method that it is feasible to use with an exhaustive search procedure is to consider only those subsets of variables found to be closely competitive with subset $S$. If, say the best 10 or 20 subsets of each size that were found during the search for subset $S$ were recorded, then these can be used. In the Monte Carlo simulations, the regression sum of squares for subset $S$ can then be compared with these other 9 or 19 subsets. The probability that subset $S$ fits better than these other subsets can then be used as an approximation to the required probability of selection.

Many ways of maximizing (6.10) are possible. For instance, a simplex method such as the Nelder and Mead (1965) algorithm could be used. Alternatively, the logarithm of $P$ could be approximated by a quadratic form in $\boldsymbol{\beta}$ by evaluating it at $k(k+1)/2$ points and fitting a quadratic surface. Either of these methods requires a fairly large number of estimates of $P$ and so requires a very substantial amount of computation.

We can write (6.10) as

$$LCL \;=\; const. \;-\; (RSS_k \;+\; \sum_j \delta_j^2)/(2\sigma^2) \;-\; \log_e P(\boldsymbol{\delta}), \qquad (6.11)$$

where $\boldsymbol{\delta} = \boldsymbol{R\beta} - \boldsymbol{Q}'\boldsymbol{y}$, and $P(\boldsymbol{\delta}) =$ the probability of selection for given $\boldsymbol{\delta}$. That is, $\boldsymbol{\delta}$ is the difference between the expected values of the projections and the sample values. We now maximize the LCL with respect to these deviations, $\boldsymbol{\delta}$, rather than with respect to $\boldsymbol{\beta}$. An alternative way of thinking of this method is as a transformation from the original $X$-variables to a set of orthogonal $Q$-variables. The regression coefficients with respect to which we are maximizing the LCL are the elements of $\boldsymbol{R\beta}$, which are the regression coefficients of $Y$ upon the columns of the $\boldsymbol{Q}$-matrix.

Differentiating (6.11) we obtain

$$\frac{d(LCL)}{d\delta_j} \;=\; -\;\delta_j\sigma^2 \;-\; \frac{dP/d\delta_j}{P}. \qquad (6.12)$$

By equating the left-hand side of (6.12) to zero, we obtain the following iter-

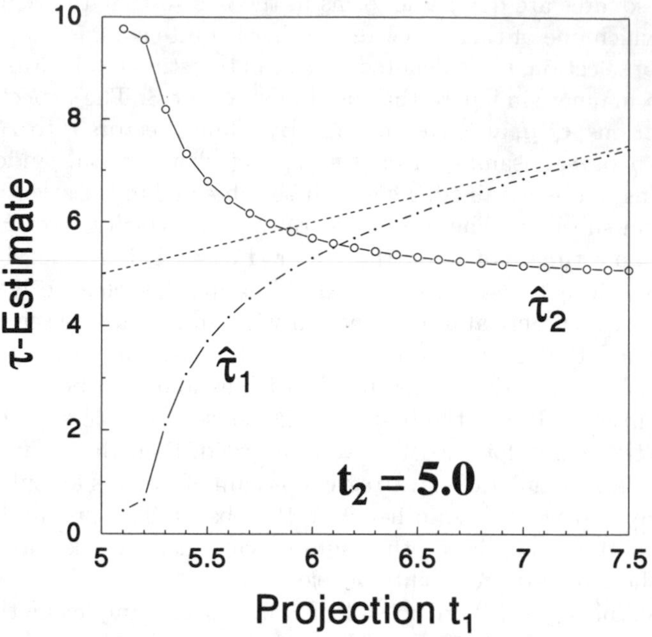


Figure 6.4 *Maximum likelihood estimation for the case of two orthogonal predictors when only one predictor is to be chosen.*

ative method for obtaining the $r$-th estimate, $\delta_j^{(r)}$, from the preceding one

$$\delta_j^{(r)} \quad = \quad - \ \sigma^2 \frac{dP/d\delta_j}{P},$$

where the right-hand side is evaluated at $\delta_j^{(r-1)}$. The least-squares solution, which corresponds to $\delta_j = 0$, can be used as a starting point.

This method seems attractive until it is applied. Let us look at the simple case in which we have to choose between two orthogonal predictors, $X_1$ and $X_2$. If we rewrite equation (6.11) in terms of the least-squares projections then we have

$$LCL \ = \ const. \ - \ \sum_{j=1}^{2} (t_j - \tau_j)^2/(2\sigma^2) \ - \ \log_e P,$$

where the constant, *const.*, has changed, and the $t_j$'s are the least-squares projections. If $|t_1| > |t_2|$, variable $X_1$ is chosen. As the $t_j$'s are normally distributed with means equal to $\tau_j$ and standard deviation equal to $\sigma$, the probability, $P$, of selection is

$$P = \int_{-\infty}^{\infty} \phi(\frac{t_1 - \tau_1}{\sigma}) \times \text{Prob}(-|t_1| < t_2 < |t_1|) dt_1.$$

This can easily be evaluated numerically and the maximum likelihood estimates of $\tau_1$ and $\tau_2$ determined. Figure 6.4 shows the solutions for the case $\sigma = 1$, $t_2 = 5$ and a range of values of $t_1 > t_2$. We see that if $t_1$ is at least one standard deviation larger than $t_2$, then the estimate of $\tau_1$ is a little less than $t_1$, while the estimate of $\tau_2$ is a little larger than $t_2$. However, when $t_1$ is only slightly larger than $t_2$, $\tau_1$ is shrunk to far below $t_1$, while $\tau_2$ is increased way above both $t_1$ and $t_2$.

In most practical situations, when $t_2 = 5$ standard deviations, and $t_1$ is larger (or $t_1 < -5$), both variables would be accepted. Similar curves were generated for smaller values of $t_2$, and these show similar behaviour, but they are a little more complex because there is an appreciable probability of $X_1$ being accepted when $t_1$ has the opposite sign to $\tau_1$.

With more competing variables, the behaviour is even more extreme. The amount by which the estimate of $\tau_1$ is moved away from $t_1$ is also very sensitive to the distribution of the least-squares projections when the $t_j$'s are close together.

## 6.5 Estimation of population means

It will be apparent by now that very little progress has been made on estimating regression coefficients after subset selection.

- We know that when there are competing models that fit almost as well as one another, the least-squares estimates of the regression coefficients for the best-fitting model are liable to be too large. If we think in terms of the least-squares projections rather than regression coefficients, then if there are 'dominant' variables, which are in all the competing models, and they are placed first in order, then there is little or no bias in their projections along the direction of the dependent variable, but the projections for the other variables are liable to be perhaps one or more standard deviations too large.
- We also know from the work of Stein, Sclove and Copas, that if our objective is prediction, then the regression coefficients should be shrunk *even if there are no competing models or there has been no model selection.* Breiman's garrote gives us an improved form of shrinkage that gives minimum squared *cross-validation* prediction errors, and simultaneously does the subset selection for us. However, it does not correct for selection bias. If by chance the complete random sample of data collected favours model $M_1$ slightly over its close competitors $M_2$, $M_3$, etc., then it will usually do so also if different samples of, say 80%, of the data are used for model selection and the corresponding remaining 20%'s are used for cross-validation. The bias is present in the complete data set; cross-validation cannot remove it.

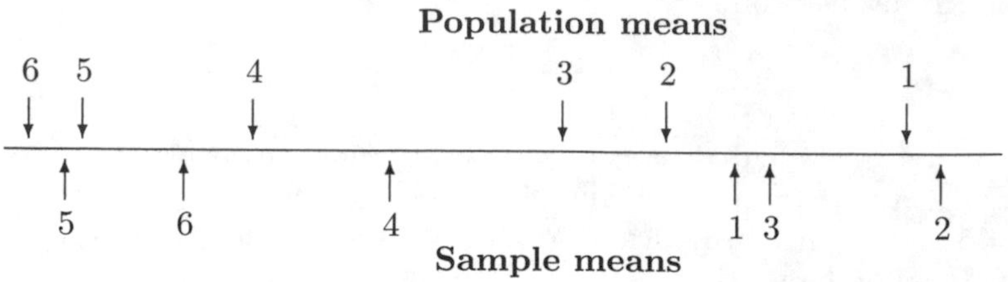


Figure 6.5 *A set of hypothetical positions of population and sample means.*

Can anything be done about the bias due to selecting the model that fits the best to a sample? Some progress has been made with a similar but simpler problem.

Suppose that we have $k$ populations and want to find the population with the largest mean. We take samples from each of the populations and choose that population for which the sample mean was the largest. Now we want to estimate the population mean for the chosen population. Notice that this is not the same problem as that of estimating the largest population mean. We may have chosen the wrong population, but having chosen one of the populations, it is the mean of this population which we want to estimate.

Interest in this problem may have originated with agricultural trials to determine, say the best yielding variety of potato, to grow in a particular environment. If there are say 20 varieties from which to choose, then the 20 varieties are grown and then the one with the largest sample yield is chosen. In practice, the experimentation is much more sophisticated. Perhaps after the first year, the 10 best-yielding varieties will be grown in the second year, then some of those will be eliminated for a third year. There is a considerable literature on the design of such experiments to maximize the chances of selecting the best variety. There is much less literature on the problem of estimating the yield for the variety that is finally selected.

Figure 6.5 illustrates the case in which we have samples from six different populations. By chance, the sample mean for population (1) is a little below its population mean, while the sample means for populations (2) and (3) came out above their population means and both exceeded the sample mean for population (1). In this case, we choose population (2) and want to estimate its population mean. In practice, we only have the lower half of Figure 6.5. It could be that the sample mean of population (2) is below its population mean, and that if we took another sample, then the mean from population might be much larger than any of the other sample means.

Miller (1996, 1997) shows that maximum likelihood performs badly in this case. Numerical solutions for the maximum likelihood estimator are presented for the case of just three populations. These show that as the sample means get closer, the estimate of the mean of the selected population rapidly decreases, as we found earlier.

There is some literature on the similar problem of estimating the largest population mean, which is of course a different problem. In some of this literature, maximum likelihood is used, but the maximum likelihood estimates of the population means are restricted to have the same order as the sample means.

Let us use square brackets to denote the populations when their sample means have been ordered, so that [1] denotes that population with the largest sample mean, [2] denotes the population with the second largest population mean, etc. We want to estimate the population mean, $\mu_{[1]}$. Let us denote the ordered sample means as $x_{[1]}$, $x_{[2]}$, ..., $x_{[k]}$.

The bias in using $x_{[1]}$ as the estimate of its population mean is greatest when all of the population means are identical. If we know that the population means are identical, then the average of the sample means is the maximum likelihood estimate of the common population mean. Hence it seems sensible to find an estimator that is a weighted average of those means that are close together but gives much less weight if there is a 'big' gap in the sample means.

If the $k$ population means are identical, and the sample means all have the same standard error, then the expected value of $x_{[1]}$ is equal to the first-order statistic from a sample of size $k$. This will be in the upper tail of the distribution of the means in the vicinity of the point in which the fraction in the tail is about $1/(k+1)$. This can be looked upon as an upper limit to the amount of correction to apply to $x_{[1]}$ to use it as an estimate of the population mean.

Cohen & Sackrowitz (1982) suggested estimators of the form

$$\hat{\mu}_{[1]} = \sum_{i=1}^{k} w_{i,k} x_{[1]}, \tag{6.13}$$

where the weights $w_{i,k}$, which are functions of the spacing of the sample means, have the following properties:

1. $w_{i,k} \geq 0$
2. The sum of the weights, $\sum_{i=1}^{k} w_{i,k} = 1$
3. $w_{i,k} \geq w_{i+1,k}$
4. $w_{i,k} = w_{i+1,k}$ if $x_{[i]} = x_{[i+1]}$
5. As any gap between two sample means increases, the weight for the sample mean after the gap tends to zero. If the gap is after the $m$-th sample mean, then $w_{i,k}$ tends to $w_{i,m}$ for $i = 1, ..., m$.

Cohen & Sackrowitz devised a family of weighting functions that satisfied the above criteria, and compared them.

Note, in the case $k = 2$, Cohen & Sackrowitz state that Stein showed in the discussion at a meeting in 1964, that $x_{[1]}$ is the minimax estimator of $\mu_{[1]}$, that is, over all possible pairs of population means, the maximum mean squared error of $x_{[1]}$, which occurs when $\mu_{[1]} = \mu_{[2]}$, is equalled or exceeded by all other estimators.

Various estimators have been suggested by a number of authors; see for instance Dahiya(1974), Vellaisamy (1992), Venter (1988), DuPreez & Venter (1989), and Venter & Steel (1991). In this last paper, the authors compared several estimators including the following:

$$\hat{\mu}_{[1]} = x_{[1]} - \frac{\sum_{i=2}^{k} w_i(x_{[1]} - x_{[i]})}{1 + \sum_{i=2}^{k} w_i}, \qquad (6.14)$$

where, with $w_1 = 1$,

$$\begin{aligned} t_i &= a.(x_{[1]} - x_{[i]})(x_{[i-1]} - x_{[i]}) \\ w_i &= \begin{cases} w_{i-1} & \text{if } t_i < 4 \\ w_{i-1}/(\sqrt{t_i} - 1) & \text{otherwise.} \end{cases} \end{aligned}$$

Here it is assumed that the sample means have been standardized so that they all have standard error = 1. Using simulation, they found values of $a$ for a range of $k$ to minimize the maximum mean squared error.

Cohen & Sackrowitz (1989) have derived an interesting two-stage estimator that is unbiased. An obvious two-stage procedure that yields an unbiased estimate is to use one sample to select the population, and then to estimate the mean from an independent sample from that population. While this gives an unbiased estimate, it does not use the information from the first sample. Let $x_{[1]}$ and $x_{[2]}$ be the largest two sample means (with the same standard errors = 1.0) from the first sampling, and let $y$ be the sample mean from an independent sample from population number [1], which also has the same standard error as the means from the first sampling. Cohen & Sackrowitz show that

$$z/2 - (1/\sqrt{2}) \frac{\phi(\sqrt{2}(z/2 - x_{[2]}))}{\Phi(\sqrt{2}(z/2 - x_{[2]}))}, \qquad (6.15)$$

where $z = x_{[1]} + y$, is a uniform minimum variance conditionally unbiased estimator (UMVCUE) of $\mu_{[1]}$. The proof of this result requires that the sample means have normal distributions with the same variance, though the population means may have any values. They also note that for other one-parameter cases with other distributions, the UMVCUE depends only upon $x_{[1]}$ and $x_{[2]}$, and the independent value $y$, and not upon the other $x_{[i]}$'s. Cohen & Sackrowitz comment 'We regard this as a negative result since we feel we would not recommend an estimator that did not use all the observations'.

The form of equation (6.15) is very appealing. If one population mean is very much larger than the others, so that the 'right' population is almost certain to be selected, then both $x_{[1]}$ and $y$ will be larger than $x_{[2]}$ and so the correction term will have $\Phi(..)$ close to one and hence be small. On the other hand, when the population means are very close together (relative to their standard errors), $y$ is likely to be much smaller than either $x_{[1]}$ or $x_{[2]}$ so that the argument of $\phi$ and $\Phi$ in the correction term will be negative. The ratio, $(1 - \Phi(x))/\phi(x)$ for positive $x$, is known as Mills' ratio. This ratio is often used in computational algorithms for the area in the tail of the normal distribution. For large $x$, it equals approximately $1/x - 1/x^3 + 3/x^5$. Using just the first term in Mills' ratio, the estimator (6.15) tends to $x_{[2]}$ as the argument of $\Phi$ tends to $-\infty$, which is as $y$ tends to $-\infty$. Thus, in the case in which all of the populations have the same mean, this method will give an estimate of that common mean that is close to the second-order statistic. We would expect this to be almost as bad as picking the first-order statistic.

A crude heuristic explanation of why the estimator (6.15) 'works' is that the independent observation $y$ is helping us discriminate between the two cases:

- $x_{[1]}$ is from the population with the largest mean, and it may be below its population mean, in which case we do not want to shrink it to estimate the population mean.
- Several population means are very close, and population [1] has been chosen only because its sample mean came out to be well above its population mean, in which case we would like to shrink it to estimate the population mean.

Figure 6.15 shows the mean squared errors and biases of three estimators of the mean of the population with the smallest sample mean. The sample means are from 10 populations with evenly spaced population means. Each sample mean has standard error = 1. The estimator [1] is the sample mean of the selected population. We see that it is by far the worst of the three estimators both in terms of mean squared error and bias. The second estimator, labelled [av.] is the average of the sample mean, $x_{[1]}$, and the mean, $y$, of an independent sample from the selected population. The third estimator, labelled [C&S], is from equation (6.15) above. We see that in terms of mean squared error, there is very little to choose, between the second and third estimators, though the Cohen & Sackrowitz estimator is unbiased.

Cohen & Sackrowitz also give the corresponding estimate when the sample $x_i$'s have the gamma distribution, $\Gamma(\nu, \theta_i)$, where $\nu$ is the shape parameter common to all of the $x_i$'s, but the scale parameters, $\theta_i$, differ. In this case, the estimator is

$$[z/(\nu+1)]\frac{I_{\beta(2,\nu)}(1 - x_{[2]}/z)}{1 - (x_{[2]}/z)^{\nu}}, \tag{6.16}$$

where $z = x_{[1]} + y$, $Y$ is distributed as $\Gamma(1, \theta_i)$, and $I_{\beta(2,\nu)}$ denotes the incomplete beta function with parameters $(2, \nu)$. Note that in this case, the distribution of the independent sample $Y$ has the same scale parameter as $x_{[1]}$ but shape parameter = 1.

## 6.6 Estimating least-squares projections

How can we apply the methods of the last section to estimating least-squares projections after selecting the model which is best in some sense?

The least-squares projections are approximately normally distributed. They

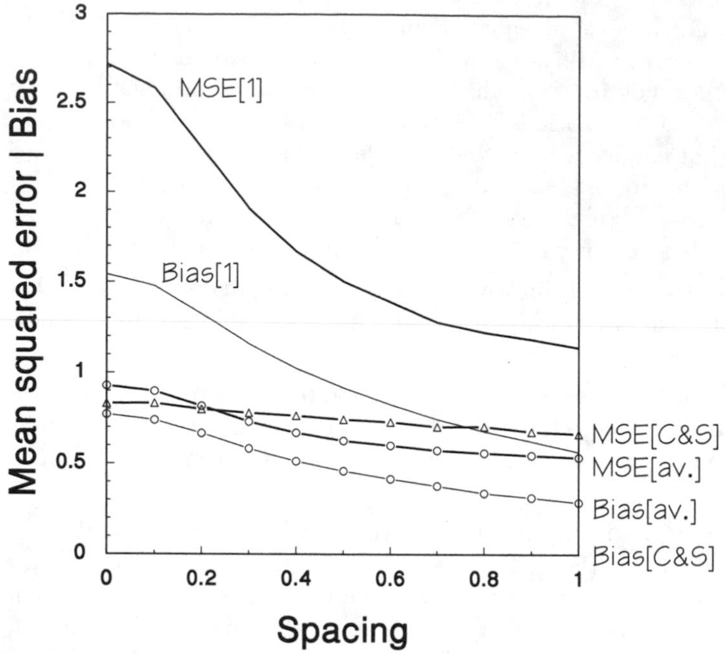


Figure 6.6 *Mean squared errors and biases for 3 estimators of the selected mean. [1] denotes the largest sample mean,* $x_{[1]}$*, [av] denotes the average* $(x_{[1]} + y)/2$*, and [C&S] denotes the estimator (6.15) for which the bias is zero. The spacing is that of the equally spaced population means.*

are projections upon a set of orthogonal directions, and they all have the same variance, equal to the residual variance. However, when we were looking at estimating the mean for a population, we had a scalar value, the sample mean, to work with. In the regression situation, we have a vector of projections for our subset of variables.

Let us look at a numeric example. Consider the POLLUTION data again. The three best-fitting subsets of three variables plus a constant were (2, 6, 9), (1, 9, 14), (6, 9, 14) and (2, 9, 14), with residual sums of squares of 82389, 83335, 85242, 88543 and 88920, respectively. With all of the variables in the model, the residual variance estimate is 1220 with 44 degrees of freedom. The standard deviation of the residuals, and hence of the projections, is therefore about 35.

Table 6.12 shows the least-squares projections for six of the variables, ordered so that variables 9 and 6, which featured most frequently in the best-fitting subsets, are at the start. We see that all of the projections except that for variable number 11 are large compared with the residual standard

Table 6.12 *Projections for the pollution data and minimum changes so that other subsets of 3 variables + intercept fit as well as subset (2, 6, 9) (Blanks indicate that changes to the projection effect both subsets equally)*

| Variable | Projctn. (2, 6, 9) | Changes to favour subset (1, 9, 14) | (6, 9, 14) | (2, 9, 14) | (6, 9, 11) |
|---|---|---|---|---|---|
| Constant | 7284.0 | | | | |
| 9 | 307.6 | | | | |
| 6 | –184.0 | +0.6 | | +11.1 | |
| 2 | –132.1 | +0.9 | +6.0 | –2.8 | +13.8 |
| 14 | 100.7 | +1.2 | +6.3 | +13.5 | +9.7 |
| 1 | 87.3 | +1.3 | | | –2.9 |
| 11 | –12.8 | | | | –22.7 |

deviation. The last three columns of Table 6.12 show the smallest changes to the projections, such that the second, third and fourth-best subsets give the same regression sums of squares as the best subset. We see that these changes are small compared with the residual standard deviation. If we had used Spj$\phi$tvoll's test, we would have found that none of these subsets, or several more, fit significantly worse than subset (2, 6, 9). Details of the derivation of the changes shown in this table are given in the appendix to this chapter.

Note that most of the projections in the best subset are reduced in magnitude while most of the variables competing to be in the best subset have their projections increased.

A weighted average of the changes to the projections seems an attractive way to reduce the selection bias. One method suggested in Miller (1996, 1997) was to use weights:

$$exp(-\alpha \sum (\delta_i^2/(\nu\sigma^2)),$$

where the $\delta_i$'s are the changes in the projections from the best subset, and $\nu$ is the number of projections changed. Much more work needs to be done on this. Perhaps the $\nu$ should be omitted. A small amount of simulation has been done by the author and this has suggested that the value of $\alpha$ should be approximately 1 without the $\nu$, or larger with it.

An alternative set of weights are the posterior probabilities from using Bayesian methods. For additional discussion see the next chapter.

# Appendix A
# Changing projections to equate sums of squares

Suppose we have two subsets containing $p_1$ and $p_2$ variables, respectively.

There may be some variables that are in both models. In this appendix, we derive the smallest changes to the least squares projections so that the regression sums of squares for the two models are equal. We actually generalize this to allow for a penalty term when the numbers of variables in the two models are different.

Let $t' = (t_1,\ t_2,\ ...,\ t_p)$ be a vector of projections such that the first $p_1$ are those of the variables in the first of two models, and they contain projections for the variables in the second model. The regression sum of squares for model 1 is $\sum_{i=1}^{p_1} t_i^2$.

Let us apply planar rotations to change the order of the variables so that the first $p_2$ become those for the variables in the second model. Let $A = \{a_{ij},\ i = 1,\ ...,\ p;\ j = 1,\ ...,\ p\}$ be the product of these planar rotations. The regression sum of squares for model 2 is the sum of squares of the first $p_2$ elements of $At$.

We want to find the smallest changes, $\delta_i$, to the projections such that

$$\sum_{i=1}^{p_1}(t_i + \delta_i)^2 = \sum_{i=1}^{p_2}[\sum_{j=1}^{p} a_{ij}(t_j + \delta_j)]^2 \ + \ P, \tag{A.1}$$

where $P$ is some penalty applied if the two models have different numbers of variables.

Suppose our objective function to be minimized is

$$S = \sum_{i=1}^{p} \delta_i^2. \tag{A.2}$$

Then, we have a quadratic objective function to minimize subject to a quadratic constraint. In general, such problems can be quite difficult to solve, but the simple form of the objective function makes this problem fairly simple, and the small sizes of the elements of $A$ ensures very rapid convergence.

Set $D$ equal to the difference between the left-hand and right-hand sides of (A.1). Let us eliminate one of the $\delta$'s, say $\delta_1$. That is, for any given values of $\delta_2,\ ...,\ \delta_p$, $\delta_1$ is chosen so that $D = 0$. Then differentiating $D$ we have, for $i \neq 1$,

$$\begin{aligned} \frac{dD}{d\delta_i} &= 0 \\ &= \frac{\partial D}{\partial \delta_i} + \frac{\partial D}{\partial \delta_1}.\frac{d\delta_1}{d\delta_i}. \end{aligned}$$

Differentiating (A.2) and setting it equal to zero, we need then to solve

$$\begin{aligned} \frac{dS}{d\delta_i} &= 0 \\ &= 2\delta_i \ + \ 2\delta_1.(d\delta_1/d\delta_i), \end{aligned}$$

Hence, from the derivative of (A.1),:

$$\delta_i = \delta_1.\frac{\partial D/\partial \delta_i}{\partial D/\partial \delta_1};$$

that is, the $\delta_i$'s should be proportional to the first derivatives, $\partial D/\partial \delta_i$.

Using a linear approximation, at each iteration we set

$$-D = \sum \Delta\delta_i . \partial D/\partial \delta_i,$$

where $\Delta\delta_i$ is the change in $\delta_i$. Setting $\Delta\delta_i = c.\partial D/\partial \delta_i$, and solving for the constant $c$, we arrive at

$$\Delta\delta_i = -D . \frac{\partial D/\partial \delta_i}{\sum(\partial D/\partial \delta_i)^2}.$$

CHAPTER 7

# Bayesian subset selection

## 7.1 Bayesian introduction

A large number of papers on the use of Bayesian methods to select models appeared in the 1990s. There were a few such papers prior to 1990, but the only one to attract much attention was probably that of Mitchell and Beauchamp (1988), though there was also discussion by Lindley of the paper by Miller (1984).

In classical or frequentist statistics, we specify a probability model for a random variable $Y$ as $p(y|\boldsymbol{\theta})$, where $\boldsymbol{\theta}$ is a parameter or vector of parameters, and $p$ is a probability, if $Y$ is a discrete variable, or a probability density if $Y$ is a continuous variable. $p(y|\theta)$ is then the likelihood of the value $y$ for variable $Y$ as a function of the parameter(s) $\theta$. If we have a sample of independent values of $Y$ then we multiply these probabilities or probability densities together to obtain the likelihood for the sample. One way of estimating $\boldsymbol{\theta}$ is to maximize the likelihood, or usually its logarithm, as a function of $\boldsymbol{\theta}$. Confidence limits can then be placed on the value of $\boldsymbol{\theta}$ by finding those values for which the vector of values of $Y$, $\boldsymbol{y}$, is reasonably plausible.

If we are prepared to specify a *prior* probability, $p(\boldsymbol{\theta})$ for $\boldsymbol{\theta}$, the joint probability of $\boldsymbol{y}$ and $\boldsymbol{\theta}$ is

$$p(\boldsymbol{y},\boldsymbol{\theta}) = p(\boldsymbol{y}|\boldsymbol{\theta}).p(\boldsymbol{\theta}).$$

Bayes' rule then reverses the roles of variables and parameters to give a *posterior* probability for the parameter(s) given the observed data

$$\begin{aligned} p(\boldsymbol{\theta}|\boldsymbol{y}) &= \frac{p(\boldsymbol{\theta},\boldsymbol{y})}{p(\boldsymbol{y})} \\ &= \frac{p(\boldsymbol{\theta}).p(\boldsymbol{y}|\boldsymbol{\theta})}{p(\boldsymbol{y})}, \end{aligned}$$

where

$$p(\boldsymbol{y}) = \int p(\boldsymbol{y}|\boldsymbol{\theta}).p(\boldsymbol{\theta}).d\boldsymbol{\theta}$$

(or $\sum p(\boldsymbol{y}|\boldsymbol{\theta})$ if $\boldsymbol{\theta}$ is discrete).

A good introductory text to Bayesian methods is that by Gelman et al. (1995).

The brief description above was of the use of Bayesian methods when only one model is being considered. Let us suppose that we have $K$ models, $M_1$, $M_2$,..., $M_K$. In subset selection, when we have $k$ variables and are prepared to consider any combination of those variables, $K = 2^k$; this is often a very large number.

Suppose that for model $M_j$, the variable $Y$ has density

$$p_j(Y|\boldsymbol{X}, \boldsymbol{\theta}_j),$$

where $\boldsymbol{X}$ consists of the values of the predictors in model $M_j$; for simplicity, it has not been given a subscript. At this stage, we are considering any kinds of models, not restricting our attention to linear models. In the linear models situation, with a normal distribution for the residuals, the form of $p_j(\boldsymbol{y}|\boldsymbol{X}, \boldsymbol{\theta}_j)$ would be

$$p_j(\boldsymbol{y}|X, \boldsymbol{\theta}_j) = (2\pi\sigma_j^2)^{-n/2} exp(-\sum_{i=1}^{n}(y_i - \boldsymbol{\beta}_j'\boldsymbol{x}_i)^2/\sigma_j^2),$$

where $\sigma_j^2$ is the residual variance and $\boldsymbol{\beta}_j'$ is the vector of regression coefficients for the $j^{th}$ model, $\boldsymbol{x}_i$ is the vector of values of the predictor variables for the $i^{th}$ case, and the vector of parameters, $\boldsymbol{\theta}_j$ comprises both $\boldsymbol{\beta}_j$ and $\sigma_j^2$; $n$ is the sample size.

Let $p(M_j)$ be the prior probability for model $M_j$, then its posterior probability is

$$p(M_j|\boldsymbol{y}, \boldsymbol{X}) = \frac{p_j(\boldsymbol{y}|\boldsymbol{X}).p(M_j)}{\sum_{j=1}^{K} p_j(\boldsymbol{y}|\boldsymbol{X}).p(M_j)}, \tag{7.1}$$

where

$$p_j(\boldsymbol{y}|\boldsymbol{X}) = \int p_j(\boldsymbol{y}|\boldsymbol{X}, \boldsymbol{\theta}_j).p_j(\boldsymbol{\theta}_j)\, d\boldsymbol{\theta}_j. \tag{7.2}$$

Thus, we must also specify a prior distribution for each $\boldsymbol{\theta}_j$, and then perform the multivariate integration over $\boldsymbol{\theta}_j$.

From (7.1) and (7.2), we see that $p(M_j|\boldsymbol{y}, \boldsymbol{X})$ can be small for a 'good' model if $p_j(\boldsymbol{\theta}_j)$ is badly chosen, e.g. if it is centered at the 'wrong' place or if it is given a very wide spread, so that very little weight is given in the integration in the region of the $\boldsymbol{\beta}_j$-space where the residual sum of squares is small.

The ratio of $p_j(\boldsymbol{y}|\boldsymbol{X})$ to the sum of the posterior probabilities for all of the models is known as the Bayes factor for model $M_j$, though some authors prefer to restrict the use of this term to the ratio of two posterior probabilities where it is the Bayes factor for one model relative to another.

Clearly, the posterior probabilities are strongly influenced by the two components of the prior probabilities, that is the prior for the model and the prior for the parameters in that model. Much of the work on the application of Bayes methods to model selection has been devoted to trying to reduce this influence and/or trying to find 'fair' priors.

## 7.2 'Spike and slab' prior

Mitchell and Beauchamp (1988) used 'spike and slab' priors for the selection of a linear model. For each variable, the prior probability that its regression coefficient was zero (the spike) was specified as

$$Prob(\beta_l = 0) \;=\; h_{l0},$$

while the slab, or density for nonzero values was spread uniformly between $-f_l$ and $+f_l$. Hence, the density for nonzero values of $\beta_l$ was $(1 - h_{l0})/(2f_l)$ for some 'large' positive $f_l$.

Obvious problems are starting to appear. How large is 'large'? Should we use the same value of $f_l$ for all variables? In any real problem, we usually expect the regression coefficients to be of different orders of magnitude. Perhaps $f_l$ for the various $X$-variables should be proportional to the reciprocal of the range (or some other measure of spread) of its corresponding $X$-variable. We shall leave such questions for the moment.

The other component of the specification of priors by Mitchell and Beauchamp was that the logarithm of the standard deviation of residuals, $\sigma$, should be uniformly distributed over the range $-\ln(\sigma_0)$ to $+\ln(\sigma_0)$.

This specification of priors is not the same as the two-part specification described at the end of the last section. We can easily derive the corresponding prior for model $M_j$. It is simply

$$p(M_j) \;=\; \prod_{l \in M_j} (1 - h_{l0}) \prod_{l \in \bar{M}_j} h_{l0},$$

where $l \in M_j$ denotes a variable in model $M_j$, and $l \in \bar{M}_j$ denotes an omitted variable.

With this specification of priors, the probability density of $\boldsymbol{y}$ given the model, $M_j$, the regression coefficients, $\boldsymbol{\beta_j}$, and $\sigma$ is

$$p(\boldsymbol{y}|M_j, \boldsymbol{\beta_j}, \sigma) \;=\; \frac{exp(-(RSS_j \;+\; (\boldsymbol{\beta_j} - \hat{\boldsymbol{\beta}}_j)'\boldsymbol{X}_j'\boldsymbol{X}_j(\boldsymbol{\beta_j} - \hat{\boldsymbol{\beta}}_j))/(2\sigma^2))}{(2\pi\sigma^2)^{n/2}}, \tag{7.3}$$

where $\hat{\boldsymbol{\beta}}_j$ is the vector of least-squares estimates of the regression coefficients, and $RSS_j$ is the residual sum of squares for model $M_j$.

If we now multiply (7.3) by $p(\boldsymbol{\beta_j}|M_j, \sigma)$, which is just $\prod_{l \in M_j} (2f_l)^{-1}$, and then integrate with respect to $\boldsymbol{\beta_j}$, we obtain

$$p(\boldsymbol{y}|M_j, \sigma) \;=\; \frac{exp(-RSS_j/(2\sigma^2))}{(\prod_{l \in M_j} (2f_l)).(2\pi\sigma^2)^{(n-p)/2}|\boldsymbol{X}_j'\boldsymbol{X}_j|^{1/2}}, \tag{7.4}$$

where $p$ is the number of variables in model $M_j$. This integral is only approximate as it has treated each integration from $-f_l$ to $+f_l$ as if it were from $-\infty$ to $+\infty$.

Now we multiply by $p(\sigma|M_j)$, which is proportional to $1/\sigma$, and integrate over $\sigma$ obtaining

$$p(\boldsymbol{y}|M_j) = \frac{(1/2)\Gamma((n-p)/2)}{2\ln(\sigma_0).(\prod_{l\in M_j}(2f_l)).\pi^{(n-p)/2}.|\boldsymbol{X}_j'\boldsymbol{X}_j|^{1/2}.RSS_j^{(n-p)/2}} \quad (7.5)$$

This integral is also approximate; it is the integral for $\sigma$ between 0 and $\infty$ instead of from $1/\sigma_0$ to $\sigma_0$.

Now we multiply (7.5) by $p(M_j)$ to obtain $p(\boldsymbol{y}, M_j)$. Summing over all the models gives $p(\boldsymbol{y})$. Then

$$\begin{aligned} p(M_j|\boldsymbol{y}) &= \frac{p(\boldsymbol{y}, M_j)}{p(\boldsymbol{y})} \\ &= \frac{g.\prod_{l\in \bar{M}_j}[2h_{l0}f_l/(1-h_{l0})].\pi^{(n-p)/2}}{|\boldsymbol{X}_j'\boldsymbol{X}_j|^{1/2}.RSS_j^{(n-p)/2}}, \end{aligned} \quad (7.6)$$

where $g$ is a normalizing constant.

The above derivation is essentially that in Mitchell and Beauchamp's paper, apart from some changes in notation to match more closely that in the remainder of this book. It is clear that the posterior probabilities given by (7.6) are very dependent upon the choice of $f_l$ and $h_{l0}$ for each variable.

If Mitchell and Beauchamp had used orthogonal projections instead of regression coefficients, the derivation would have been much simpler. Also, there is a very simple and natural answer to the problem of scaling. In fact, they probably used some kind of Cholesky factorization/orthogonalization in going from (7.3) to (7.4) above, though the details of the integration are not given in the paper.

Let us rewrite (7.3) as

$$p(\boldsymbol{y}|M_j, \boldsymbol{\beta_j}, \sigma) = \frac{exp(-(RSS_j + \sum_{i=1}^{p}(\tau_{ij}-t_{ij})^2)/(2\sigma^2))}{,}(2\pi\sigma^2)^{n/2} \quad (7.7)$$

where the $t_{ij}$'s are the least-squares projections for the $p$ variables in model $M_j$, and the $\tau_{ij}$'s are their expected values. Thus, in the notation of previous chapters, $\boldsymbol{R_j\beta_j} = \boldsymbol{\tau_j}$, $\boldsymbol{R_j\hat{\beta}_j} = \boldsymbol{t_j}$ where $\boldsymbol{R_j}$ is the Cholesky factorization of $\boldsymbol{X_j'X_j}$.

Recalling that the projections all have the same variance, $\sigma^2$, we can then set the prior for the spike as

$$\text{Prob.}(\tau_l = 0) = h_0,$$

where $h_0$ is the same for all predictors. Similarly, we can specify the prior for the slab as uniform over a range $(-f, f)$, where $f$ is the same for all predictors. In practice, we could look at the largest sample value of any projection for any model and add, say $4\sigma$, onto its absolute value.

Equation (7.4) is then replaced by

$$p(\boldsymbol{y}|M_j, \sigma) = \frac{exp(-RSS_j/(2\sigma^2))}{(2f)^p.(2\pi\sigma^2)^{(n-p)/2}}, \quad (7.8)$$

while equations (7.6) and (7.7) become

$$p(\boldsymbol{y}|M_j) = \frac{(1/2)\Gamma((n-p)/2)}{2\ln(\sigma_0).(2f)^p.\pi^{(n-p)/2}.RSS_j^{(n-p)/2}} \tag{7.9}$$

$$\begin{aligned} p(M_j|\boldsymbol{y}) &= \frac{p(\boldsymbol{y}, M_j)}{p(\boldsymbol{y})} \\ &= \frac{g.[2h_0 f/(1-h_0)]^{n-p}.\pi^{(n-p)/2}}{RSS_j^{(n-p)/2}}. \end{aligned} \tag{7.10}$$

If we now take logarithms of (7.10) and rearrange terms, then we find

$$\ln p(M_j|\boldsymbol{y}) = \text{const.} + \frac{n-p}{2}\left[\ln \pi[2h_0 f/(1-h_0)]^2 - \ln(RSS_j)\right],$$

which emphasizes the influence of $h_0$, and to a lesser extent, $f$, upon the choice of size of model. Notice that the posterior probability increases while the residual sum of squares $RSS_j$, decreases, provided the decrease is sufficiently large to exceed any change due to an increase in the number of variables, $p$, in the model.

## 7.3 Normal prior for regression coefficients

To overcome the dependence of the posterior probabilities on $f$, the half-range of the uniform prior for the regression coefficients, $\boldsymbol{\beta}$ or equivalently for the projections, $\boldsymbol{\tau}$, many authors now use a normal prior which of course has an infinite range.

For the $\tau_i$'s, let us assume the prior

$$p(\tau_i) = N(0, \sigma_0^2);$$

that is, that each $\tau_i$ is normally distributed with that normal distribution centered at zero and with variance $\sigma_0^2$. As the relationship between the regression coefficients and the expected values of the projections is $\boldsymbol{R\beta} = \boldsymbol{\tau}$ or $\boldsymbol{\beta} = \boldsymbol{R}^{-1}\boldsymbol{\tau}$, we have that

$$\begin{aligned} E(\boldsymbol{\beta\beta}') &= E(\boldsymbol{R}^{-1}\boldsymbol{\tau\tau}'\boldsymbol{R}^{-T}) \\ &= \boldsymbol{R}^{-1}\,\sigma_0^2\boldsymbol{I}\,\boldsymbol{R}^{-T} \\ &= \sigma_0^2(\boldsymbol{X}'\boldsymbol{X})^{-1}. \end{aligned}$$

It is in the latter form above that it is most often used in the literature.

One user of this prior was Atkinson (1978), who described it as the outcome of a fictitious experiment, and related the ratio $\sigma_0^2/\sigma^2$ to the ratio of the sample sizes for the false experiment and the actual one. The value of $\sigma_0^2$ is usually taken as much larger than $\sigma^2$; for instance, Smith and Kohn (1996) use a ratio of 100 but state that it does not make much difference for any ratio in the range 10–100.

Instead of assuming a uniform prior for $\log\sigma$ over a finite range, an inverse gamma distribution is sometimes assumed for $\sigma$. For instance, it may be assumed that $\nu\lambda/\sigma^2$ has a chi-squared distribution with $\nu$ degrees of freedom. Thus, $1/\lambda$ is the expected value of $1/\sigma^2$.

These priors, that is, the normal prior for the projections (or equivalently for the regression coefficients), and the inverse gamma for the distribution of $\sigma^2$ have been used by several authors including Garthwaite and Dickey (1992), and Clyde et al. (1996).

Using these priors instead of those of Mitchell and Beauchamp yields the probability density for $\boldsymbol{y}$ if $M_j$ is the correct model as

$$p(\boldsymbol{y}|M_j) = \frac{\Gamma((n+\nu)/2)(\lambda\nu)^{\nu/2}}{\Gamma(\nu/2)\pi^{n/2}\left(RSS_j + \frac{\sum t_{ij}^2}{C+1} + \lambda\nu\right)^{(\nu+n)/2}},$$

where $C = \sigma_0^2/\sigma^2$. The posterior probability for model $M_j$ is then

$$p(M_j|\boldsymbol{y}) = \frac{p(\boldsymbol{y}|M_j)p(M_j)}{\sum_j p(\boldsymbol{y}|M_j)p(M_j)}.$$

We still need a prior for $M_j$. If we assume a binomial model with independent probability $w$ that each variable is in the model, then

$$p(M_j) = w^p(1-w)^{k-p},$$

where $p$ is the number of variables in model $M_j$. Note that other authors, such as Fernandez et al. (2001a) have given each of the $2^k$ models equal prior probability.

Then,

$$\begin{aligned} \ln p(M_j|\boldsymbol{y}) &= \text{const.} + p\ln w + (k-p)\ln(1-w) \\ &\quad -\frac{\nu+n}{2}\ln\left(RSS_j + \frac{\sum t_{ij}^2}{C+1} + \lambda\nu\right). \end{aligned} \tag{7.11}$$

A similar expression to (7.11) has been obtained by Fernandez *et al* (2001a), though they do not use orthogonal projections and hence the Bayes factors look much more 'messy'. For instance, our regression sum of squares, $\sum t_{ij}^2$, becomes $\boldsymbol{y}'\boldsymbol{X}_j(\boldsymbol{X}_j'\boldsymbol{X}_j)^{-1}\boldsymbol{X}_j'\boldsymbol{y}$, and the residual sum of squares, $RSS_j$, is denoted as $\boldsymbol{y}'\boldsymbol{y}$ minus the above expression.

Chipman et al. (2001), and others, use the same priors but then look at fixing $\sigma^2$. For cases in which there is a moderate number of degrees of freedom for the estimate of the residual variance, this is quite reasonable and simplifies the mathematics substantially. This is equivalent to letting the number of degrees of freedom, $\nu$, tend to infinity and setting $\lambda = \sigma^2$. If we omit the integration over the distribution of $\sigma^2$ then we obtain

$$p(\boldsymbol{y}|M_j,\sigma^2) = \frac{\exp\left[-\left(RSS_j + \frac{\sum t_{ij}^2}{C+1}\right)/(2\sigma^2)\right]}{(C+1)^{p/2}.(2\pi\sigma^2)^{n/2}}$$

Table 7.1 *Bayes factors (B.F.) for the top 10 models selected using three values of $w$, and $C = 50$, for the POLLUTE data*

| $w = 0.25$ Model vars. | B.F. | $w = 0.5$ Model vars. | B.F. | $w = 0.75$ Model vars. | B.F. |
|---|---|---|---|---|---|
| 1 2 9 14 | 16.49% | 1 2 9 14 | 3.02% | 1 2 3 6 9 14 | 0.42% |
| 2 6 9 14 | 4.75% | 1 2 6 9 14 | 2.60% | 1 2 3 5 6 9 14 | 0.42% |
| 1 2 6 9 14 | 4,73% | 1 2 3 6 9 14 | 1.89% | 1 2 3 6 8 9 14 | 0.34% |
| 1 2 3 9 14 | 3.13% | 1 2 3 9 14 | 1.72% | 1 2 3 4 5 6 9 14 | 0.26% |
| 1 2 8 9 14 | 2.19% | 1 2 8 9 14 | 1.20% | 1 2 3 4 5 6 9 12 13 | 0.25% |
| 1 2 9 10 14 | 1.95% | 1 2 9 10 14 | 1.07% | 1 2 3 5 6 8 9 14 | 0.24% |
| 2 5 6 9 | 1.90% | 1 2 5 6 9 14 | 0.93% | 1 2 3 6 8 9 12 13 | 0.23% |
| 2 6 9 | 1.73% | 2 6 9 14 | 0.87% | 1 2 3 7 8 9 14 | 0.21% |
| 2 4 6 9 14 | 1.43% | 2 4 6 9 14 | 0.78% | 1 2 3 5 6 9 12 13 | 0.21% |
| 1 2 7 9 14 | 1.29% | 1 2 6 8 9 14 | 0.72% | 1 2 5 6 9 14 | 0.21% |

and hence,

$$\ln p(M_j|\boldsymbol{y},\sigma^2) = \text{const.} + p\log w + (k-p)\log(1-w) - \left(RSS_j + \frac{\sum t_{ij}^2}{C+1}\right)/(2\sigma^2) - \frac{p}{2}\log(C+1). \quad (7.12)$$

As the residual sum of squares for any model, $RSS_j$, plus the regression sum of squares, $\sum t_{ij}^2$, equals the total sum of squares, $\boldsymbol{y}'\boldsymbol{y}$, which is the same for all models, (7.12) can be simplified to

$$\ln p(M_j|\boldsymbol{y},\sigma^2) = \text{const.} + p\log w + (k-p)\log(1-w) - \frac{C}{C+1}.RSS_j/(2\sigma^2) - \frac{p}{2}\log(C+1), \quad (7.13)$$

where the $\boldsymbol{y}'\boldsymbol{y}$ has been absorbed into the constant term.

Chipman et al. then show that using different parameterizations of these priors can be chosen to give the Mallows' $C_p$, AIC and BIC stopping rules. There were other earlier derivations, see, for instance, Chow (1981).

Clearly, by the appropriate choice of parameters in the priors, particularly the choice of $w$, almost any size of subset can be forced to have the largest posterior probabilities. For any size of subset though, the residual sum of squares fixes the relative ordering of models of that size.

One way of overcoming this problem which has been sometimes been suggested (e.g. Berger and Pericchi (1996)) is to use part of the data as a training sample to give prior distributions. The posterior probabilities are then based upon the remainder of the data. In the linear regression situation with $k$ predictors plus a constant, the number of samples in the training sample would have to be at least $(k+1)$, plus a few more samples to give a crude estimate of $\sigma^2$. This seems to be a promising direction for research. Obviously work

Table 7.2 *Bayes factors (B.F.) for the top 10 models selected using three values of C, and w = 0.25, for the POLLUTE data*

| C = 5 | | C = 10 | | C = 50 | |
|---|---|---|---|---|---|
| Model vars. | B.F. | Model vars. | B.F. | Model vars. | B.F. |
| 1 2 9 14 | 3.53% | 1 2 9 14 | 6.12% | 1 2 9 14 | 16.49% |
| 1 2 6 9 14 | 2.25% | 1 2 6 9 14 | 3.31% | 2 6 9 14 | 4.75% |
| 1 2 3 9 14 | 1.58% | 1 2 3 9 14 | 2.26% | 1 2 6 9 14 | 4.73% |
| 1 2 3 6 9 14 | 1.24% | 2 6 9 14 | 1.93% | 1 2 3 9 14 | 3.13% |
| 2 6 9 14 | 1.23% | 1 2 8 9 14 | 1.62% | 1 2 8 9 14 | 2.19% |
| 1 2 8 9 14 | 1.17% | 1 2 3 6 9 14 | 1.53% | 1 2 9 10 14 | 1.95% |
| 1 2 9 10 14 | 1.06% | 1 2 9 10 14 | 1.46% | 2 5 6 9 | 1.80% |
| 2 4 6 9 14 | 0.81% | 2 4 6 9 14 | 1.09% | 2 6 9 | 1.73% |
| 1 2 7 9 14 | 0.74% | 1 2 7 9 14 | 0.99% | 2 4 6 9 14 | 1.43% |
| 1 2 4 9 14 | 0.71% | 1 2 4 9 14 | 0.94% | 1 2 7 9 14 | 1.29% |

needs to be done to decide how large the training sample should be, and to decide whether its size should be fixed or should be proportional to the total sample size. Then, should repeated sample sizes be used and the posterior probabilities averaged in some way?

To illustrate the use of the methods of this section, let us look at the POLLUTE data. Posterior probabilities were calculated for all $2^{15}$ models, using (7.12) for $w = 0.25$, 0.5 and 0.75, and $C = 5$, 10 and 50. Table 7.1 shows the 10 largest Bayes factors out of 32768 for $w = 0.25$, 0.5 and 0.75. These factors were scaled to sum to 100% over all 32768 models. For $w = 0.25$, the preferred models contain about 4-5 variables, while for $w = 0.5$ they contain about 5-6 variables, and for $w = 0.75$ they contain 6-8 variables. Thus, there is a moderate influence of the prior probabilities for the size of the model, on the posterior probabilities. Comparison with findings for this data set in previous chapters shows that most of these models have been seen before, though it appears that only about 3 or 4 variables had much predictive value. Using $w = 0.25$, the subset of variables numbered 1, 2, 9 and 14 looks strikingly superior to the others.

Table 7.2 shows the posterior probabilities for the three values of $C$ and for $w = 0.25$.

If we choose to give equal probability to each model rather to each variable, equation (7.13) is easily modified simply by leaving out the terms involving $w$, giving

$$\ln p(M_j|\boldsymbol{y}, \sigma^2) = \text{const.} - \frac{C}{C+1}.RSS_j/(2\sigma^2) - \frac{p}{2}\log(C+1), \quad (7.14)$$

Table (7.3) shows the 10 models giving the highest posterior probabilities for three values of $C$ using (7.14). The models chosen each have between 4 and 7 variables.

Table 7.3 *Bayes factors (B.F.) for the top 10 models selected using three values of C when equal prior probabilities are given to each possible model, for the POLLUTE data*

| $C = 5$ Model vars. | B.F. | $C = 10$ Model vars. | B.F. | $C = 50$ Model vars. | B.F. |
|---|---|---|---|---|---|
| 1 2 3 6 9 14 | 0.39% | 1 2 3 6 9 14 | 0.78% | 1 2 9 14 | 3.02% |
| 1 2 3 5 6 9 14 | 0.33% | 1 2 6 9 14 | 0.56% | 1 2 6 9 14 | 2.60% |
| 1 2 3 6 8 9 14 | 0.28% | 1 2 3 5 6 9 14 | 0.52% | 1 2 3 6 9 14 | 1.89% |
| 1 2 6 9 14 | 0.23% | 1 2 3 6 8 9 14 | 0.44% | 1 2 3 9 14 | 1.72% |
| 1 2 5 6 9 14 | 0.21% | 1 2 5 6 9 14 | 0.40% | 1 2 8 9 14 | 1.20% |
| 1 2 3 4 5 6 9 14 | 0.19% | 1 2 3 9 14 | 0.38% | 1 2 9 10 14 | 1.07% |
| 1 2 3 7 8 9 14 | 0.19% | 1 2 9 14 | 0.34% | 1 2 5 6 9 14 | 0.93% |
| 1 2 3 5 6 8 9 14 | 0.18% | 1 2 6 8 9 14 | 0.32% | 2 6 9 14 | 0.87% |
| 1 2 6 8 9 14 | 0.17% | 1 2 3 8 9 14 | 0.31% | 2 4 6 9 14 | 0.78% |
| 1 2 3 6 8 9 12 13 | 0.17% | 1 2 6 9 14 15 | 0.29% | 1 2 6 8 9 14 | 0.72% |

For the POLLUTE data, we evaluated all possible subsets. For data sets with larger numbers of predictors, this is not feasible. Chapter 3 discusses various alternative algorithms. Two further methods have been introduced in the context of Bayesian methods, namely Markov Chain Monte Carlo model composition ($MC^3$), and Gibbs sampling, or stochastic search variable selection (SSVS).

$MC^3$ was introduced into Bayesian subset selection by Madigan and York (1995) and used since then in Raftery et al. (1997), Hoeting et al. (1997), Fernandez et al. (2001a) and by others. The algorithm starts with a random subset of the variables. A Bayes factor, apart from the scaling factor, is calculated for it. The Markov chain has 3 possible states from the current model. One is to add a variable, another is to delete a variable, and the third is to remain in the same state. In the computer program of Fernandez et al., a random integer between 0 and $k$ is chosen. If it is zero, then the chain stays with the same model. Otherwise, if the random integer is $j$, then the status of the variable with that number is reversed. That is, the variable is added to the subset if it was not in it, or it is removed if it was in it. The Bayes factor is evaluated for the new model. If it exceeds the Bayes factor for the previous model, it is accepted; otherwise, a random decision is made to either accept the new model or stay with the old one. The algorithm is run for a prespecified number of steps and can then be restarted from a new random start.

This method is an extension on the Efroymson stepwise regression algorithm. It will usually find any subset which the Efroymson algorithm will find, but the stochastic nature of the algorithm means that it may sometimes find better-fitting subsets. It could of course be extended to exchange two or more variables at a time, and this would give it a better chance of finding the subsets which fit well in fields such as meteorology and near-infrared spectroscopy.

The Gibbs sampling method as applied to Bayesian sampling is described in George and McCulloch (1993), and Diebolt and Robert (1994). The latter paper is not an application to subset selection but to the estimation of missing values.

A recent paper by Han and Carlin (2001) reviews eight different ways of doing Monte Carlo sampling to estimate posterior probabilities.

Both $MC^3$ and Gibbs sampling are very inefficient algorithms for attempting to find the best subsets. Some authors have reported that certain models have been 'visited' thousands of times in their searches. While it is difficult to limit the number of times that different models are tried, an efficient search algorithm should be able to keep the number of revisits down to, say single digit numbers. Some authors have even suggested that the number of visits to each model should be used to derive the posterior probabilities for the models. The numeric algorithms used in many of the papers on Bayesian model selection could be substantially improved both in accuracy (using orthogonal reduction methods as opposed to normal equation methods), and in respect to the search algorithms used. The papers by Smith and Kohn (1996), and Kohn et al. (2001) are notable exceptions to this statement.

The paper by Fernandez et al. (2001b), which applies the methods of Fernandez et al. (2001a) to the analysis of gross domestic product for different countries using 41 predictors, ran for nearly 3 hours on a fast computer (fast in 2001!). They state (footnote on page 567) that the 'largest computational burden lies in the evaluation of the marginal likelihood of each model'. As they use linear models with normally distributed residuals, the evaluation of the likelihood essentially means evaluating the residual sum of squares (RSS) for each model. As they only add or delete one variable at a time, it requires very little computation to calculate this from the RSS for the previous model, if orthogonal reduction methods are being used.

The George and McCulloch paper uses a different prior for the regression coefficients than most others. Instead of a finite prior probability of a zero value (the OUT variables), and a wide normal distribution for the nonzero coefficients (the IN variables), they are assumed to have been sampled from one of two normal distributions, one with a widespread, corresponding to the nonzero coefficients of other authors, and the other distribution having a very narrow spread. Both distributions are centered at zero. The algorithm starts with all of the variables in the model. The initial estimates of the regression coefficients are the least-squares estimates, and the residual variance is used as the first estimate of the variance, $\sigma^2$, for the least-squares projections; though they actually use the equivalent of this applied to the small regression coefficients. The ratio of the variance for the large regression coefficients and the small coefficients is fixed. They actually use a range of ratios, 10, 100, 1000, 10000 and 100000.

The iterative cycle then is as follows for iteration $j$:

1. Sample $\boldsymbol{\beta_j}$ from $f(\boldsymbol{\beta_j}|\boldsymbol{y}, \sigma_{j-1}, M_{j-1})$, the latest posterior probability for each $\boldsymbol{\beta_j}$.
2. Sample $\sigma_j$ from $f(\sigma_j|\boldsymbol{y}, \boldsymbol{\beta_j}, M_{j-1})$, the latest posterior distribution for $\sigma$.
3. For each variable, sample $\gamma = 0$ (out) or 1 (in) for each variable, taking the variables in a random order, from the posterior probability that the variable is from the IN or OUT set of variables. This posterior probability is calculated from its current regression coefficient. If it is large compared with its standard error based upon $\sigma$, then the probability is high that it is from the IN set of variables.

Full details are given in the paper. This algorithm effectively starts with something like the final subset from a stepwise regression algorithm, and then only visits a fairly small number of subsets close to the subset at the end of the first iteration. However, it can add or drop more than one variable at each iteration, so it is more likely to find good subsets in fields such as meteorology and near-infrared spectroscopy where differences between predictor variables need to be found.

With both of these search procedures, the evaluation of Bayes factors is limited to those subsets visited. Thus, whereas Table 7.1 is based upon all 32768 subsets, the equivalent table using one of these search procedures might relate to, say a few hundred subsets, and the posterior probabilities would be much larger. Thus, for instance, in Table 12 of Fernandez et al. (2001a), the posterior probabilities range from 3.61% to 2.02% for the top 8 models for the analysis of a set of crime data.

## 7.4 Model averaging

If we use some combination of the priors used in the previous section then we end up, not with a single regression equation, but with a mixture of a large number of regressions. It is usually not practical to use all of the possible models, and the number is truncated at some point. For instance, if we look at only the first 10 models list in Table 7.1 using $w = 0.25$, then if the models in the table are numbered from 1 to 10, we take as our final equation:

$$\begin{aligned} Y &= 0.418 \times \text{Regression equation for } M_1 \\ &\quad +0.120 \times \text{Regression equation for } M_2 \\ &\quad +0.120 \times \text{Regression equation for } M_3 \\ &\quad +... \\ &\quad +0.033 \times \text{Regression equation for } M_{10} \end{aligned}$$

where the fractions 0.418, 0.120, 0.120, ..., 0.033 are in the same ratios as the posterior probabilities for the models, and sum to 1.0.

The regression coefficients in general will be different in each of these equations and will be slightly smaller than the least squares coefficients for the same models because the prior distribution for the $\boldsymbol{\beta}$'s (or equivalently for

Table 7.4 *Least-squares projections for the POLLUTE data for a specified ordering of the variables*

| Variable | Projection |
|---|---|
| 9 | 307.6 |
| 14 | 156.5 |
| 2 | –143.7 |
| 1 | 139.2 |
| 6 | –67.2 |
| 3 | –64.0 |
| 4 | 8.6 |
| 5 | –55.6 |
| 7 | –16.9 |
| 8 | 30.2 |
| 10 | –14.1 |
| 11 | 2.1 |
| 12 | –11.3 |
| 13 | 46.5 |
| 15 | 3.2 |

the expected least squares projections, the $\boldsymbol{\tau}$'s) tends to shrink them slightly. Recalling that

$$p(\boldsymbol{y}|\boldsymbol{\beta_j}, \sigma, M_j) \;=\; \text{const.}\exp\left(-\frac{RSS_j \;+\; \sum(\tau_{ij} - t_{ij})^2}{2\sigma^2}\right).\exp\left(-\frac{\sum \tau_{ij}^2}{2C\sigma^2}\right),$$

where $C$ = say 100. If we collect together only those terms in the exponent which involve $\tau_{ij}$, we find that we have

$$\frac{1}{2\sigma^2}\left[\sum \frac{C+1}{C}\left(\tau_{ij} - \frac{C}{C+1}t_{ij}\right)^2\right];$$

that is, compared with the least-squares estimates, the distribution of $\tau_{ij}$ is centered about $\frac{C}{C+1}t_{ij}$ rather than $t_{ij}$, but the variance is reduced from $\sigma^2$ to $\frac{C}{C+1}\sigma^2$. The values for the $\boldsymbol{\beta}$'s are simply obtained by solving $\boldsymbol{R_j\beta_j} = \boldsymbol{\tau_j}$, where $\boldsymbol{R_j}$ is the Cholesky factorization of $\boldsymbol{X}_j'\boldsymbol{X}_j$, and the elements of $\boldsymbol{\beta_j}$ will be shrunk from the least-squares values by the same factor.

Notice that our resulting equation, or mixture of equations, in this case uses variables 1 to 10 and 14, that is it uses 11 of the 15 available predictors. In some situations, this will defeat the purpose of subset selection, if the objective were to reduce the cost of measuring all of the variables in the future.

However, by not focussing upon the single best-fitting model, or the one with the highest posterior probability, but by averaging over a number of models, we substantially reduce the bias in the regression coefficients.

Let us look at the models in Table 7.3 for $C$ = 50, that is, on the right-

Table 7.5 *Least-squares projections for the POLLUTE data for three more specified orderings of the variables*

| | | Model 8 | | Model 9 | |
|---|---|---|---|---|---|
| Variable | Projn. | Variable | Projn. | Variable | Projn. |
| 9 | 307.6 | 9 | 307.6 | 9 | 307.6 |
| 14 | 156.5 | 14 | 156.5 | 14 | 156.5 |
| 2 | -143.7 | 2 | -143.7 | 2 | -143.7 |
| 6 | -127.6 | 1 | 63.6 | 1 | 92.1 |
| 4 | 46.4 | (6) | -110.7 | (6) | -94.3 |
| 1 | 74.1 | others | 0.0 | (4) | 59.5 |
| | | | | others | 0.0 |

hand side of the table. All of the models contain the variables numbered 2, 9 and 14, and all except models 8 and 9 also contain variable number 1. Let us reorder the variables so that these ones come first, followed by numbers 6 and 3, which also feature frequently. The least-squares projections for this ordering (omitting the projection for the constant) are as shown in Table 7.4. The residual variance with all of the variables in the model is 1220.0 for this data set. The square root of this value is just under 35, and this is the standard deviation of the projections. We see that the first four projections are much larger than this in magnitude, while the remainder are of this magnitude or smaller.

Now let us reorder the variables again in an order suitable for estimating the parameters for models 8 and 9. The left-hand pair of columns in Table 7.5 shows such an ordering. To fit model 8, we put the projections for variable number 4 and all subsequent variables equal to zero; to fit model 9, we put the projections for model 1 and all following variables equal to zero. If we do this, and then reorder so that variables numbered 9, 14, 2 and 1 come first (after the constant), then the resulting projections are those shown in the second and third pairs of columns in Table 7.5. Note that the projection for variable number 1 is not zero because it was not orthogonal to either variable 4 or variable 6. The solution of $\boldsymbol{R}\hat{\boldsymbol{\beta}} = \boldsymbol{t}$ with the appropriate ordering of the Cholesky factor $\boldsymbol{R}$ will give $\hat{\beta}_1 = 0$ with either of these two columns of projections used for $\boldsymbol{t}$.

Using the projections for variables 9, 14, 2 and 1 from Table 7.5, the posterior density for $\tau_1$, that is, the expected value for the projection for variable 1 after including a constant and variables 9, 14 and 2 in the model, is as shown in Figure 7.1, based upon only the 10 models with highest posterior probability. Both the mean and variance have been reduced by the factor $C/(C+1)$ $= 50/51$, or a reduction of about 2%, as discussed earlier. The means of the normal densities shown in this figure are at 62.3 (model 8), 90.3 (model 9) and 136.5 for the other models. The heights are scaled by the posterior probabilities for the 10 models, with models 1 to 7 and 10 combined.

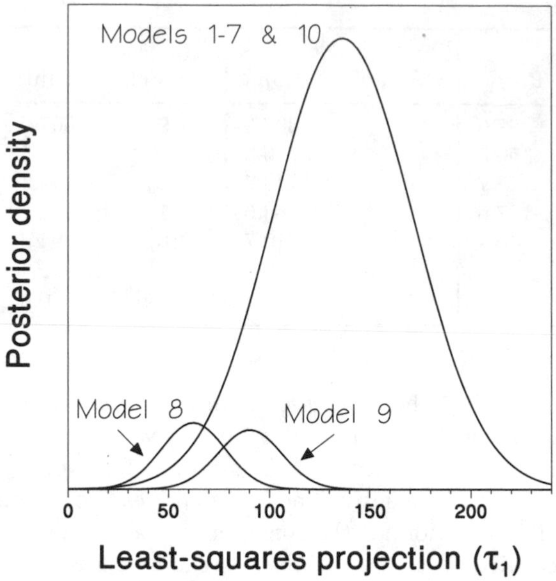


Figure 7.1 *Posterior density for the projection for variable $X_1$ for the POLLUTE data.*

In Figure 7.1, the variance for the estimate of $\tau_1$ is much smaller for models 8 and 9, which do not contain variable $X_1$, than for the other models. To obtain the projections in Table 7.5 for model 8, the projections for variables $X_1$ and $X_4$ in column 2 (and all following variables) were changed to zero exactly. That is, deterministic values with zero variance were used for these projections instead of stochastic ones with variance equal to 1220. For model 9, only the value of the projection variable $X_1$, not that for variable $X_4$, was changed to zero.

The overall mean of the mixture of these distributions is 129.7, compared with the least-squares projection of 139.2 for this variable (see Table 7.4). This drop of 9.5 is about 0.27 of a standard deviation. The 10 models used above had total posterior probability of just under 15%. In practice, it would be sensible to use far more than this, and it can be anticipated that the reduction in the estimated projections would have been larger, with some reductions also for variables 9, 14 and 2 which were in all of the first 10 models, which would not have been in all of, say the first 1,000.

We have thus a computational method for estimating the projections, and hence the regression coefficients, for just one model. But can there be any justification for its use? This is analogous to the question of whether it is valid to use the models selected in earlier chapters using the Efroymson stepwise procedure, or some other search procedure with a stopping rule such as Mallows' $C_p$ or the AIC. In the next section, this question is discussed with particular reference to its use with Bayesian methods.

## 7.5 Picking the best model

Can we just use the model giving the highest posterior probability? There has been very little discussion of this approach in the Bayesian literature until now. It was discussed by Dennis Lindley and me privately after the meeting at which Miller (1984) was presented, but that discussion did not appear in the printed proceedings of the meeting. More recently, Draper (1995) has discussed the idea. He expresses a model in two parts $(S, \theta)$, where $S$ denotes 'structure', and $\theta$ denotes the parameters within that structure. 'Structure' in general will include not only the inclusion or exclusion of variables, but also other common components of data analysis such as families of transformations (e.g. to transform a nonlinear model to a linear one, or to make residuals approximately normally distributed), a model for outliers, modelling correlation or nonconstant variances, etc. He uses the notation $S^*$ to denote the 'best-choice' model out of those examined. He refers to a number of authors who have emphasized the 'degree of overconfidence' generated by basing inferences and predictions on the same data set on which the search for structure occurred'. He makes particular reference to the paper by Adams (1990), who reported on a large computer simulation, which compared the effects of various decisions on transformations, variable selection, and outlier rejection, on the validity of inferences and the degree of overfitting which resulted.

Let $j*$ be the number of our 'best' model. In the argument which follows, it does not matter what criterion we used to arrive at our choice of this model, provided that some objective and reproducible method was used. Thus, model $M_{j*}$ may be the one which gave the largest posterior probability out of those examined, or it may have been selected using Mallows' $C_p$, or the AIC, BIC, RIC, Efroymson's stepwise regression, etc.

We will use the notation $M_j > M_i$ to indicate that model $M_j$ is a better model than model $M_i$ using whatever criterion and search method we have used to choose our 'best' model.

We want to find a posterior distribution for $\boldsymbol{\beta_{j*}}$, given that model $M_{j*}$ came out best. Conceptually, this is not difficult. Distributions of extremes are not that difficult to find as is evident from the huge amount of work which was done in this area since the 1930s, culminating in a set of 3 asymptotic distributions that are often associated with the name of Gumbel, see e.g. Gumbel (1958).

We want to be able to say

$$p(\boldsymbol{\beta_{j*}}|\boldsymbol{y}, M_{j*} > \text{ other } M_j, \sigma, M_{j*}) = \frac{p(\boldsymbol{\beta_{j*}}, \boldsymbol{y}, M_{j*} > \text{ other } M_j|\sigma, M_{j*})}{p(\boldsymbol{y}, M_{j*} > \text{ other } M_j|\sigma, M_{j*})}. \tag{7.15}$$

We will simplify this slightly by treating $\sigma$ as known. It can easily be re-introduced if desired, and then the expressions integrated with respect to the prior distribution assumed for $\sigma$.

The numerator of (7.15) can be written as

$$\begin{aligned}
& p(\boldsymbol{\beta_{j*}}, \boldsymbol{y}, M_{j*} > \text{ other } M_j | M_{j*}) \\
= \; & p(\boldsymbol{y}|\boldsymbol{\beta_{j*}}, M_{j*} > \text{ other } M_j, M_{j*}). \\
& p(\boldsymbol{\beta_{j*}}|M_{j*} > \text{ other } M_j, M_{j*}).p(M_{j*} > \text{ other } M_j|M_{j*}) \\
= \; & \frac{\exp\left(-\frac{RSS_j \; + \; \sum(\tau_{ij}-t_{ij})^2}{2\sigma^2}\right)}{(2\pi\sigma^2)^{n/2}} \\
& \times \frac{\exp\left(-\frac{\sum\tau_{ij}^2}{2C\sigma^2}\right)}{(2\pi C\sigma^2)^{p/2}} \\
& \times \prod_{j \neq j*} Prob.(M_{j*} > M_j), \qquad (7.16)
\end{aligned}$$

where $p$ is the number of variables, including the intercept, in model $M_{j*}$. The denominator of (7.15) is obtained then by integrating with respect to $\boldsymbol{\beta}$.

Apart from the middle term in (7.16), this is exactly the same as the likelihood equation derived in Chapter 6. The last term is the one which utilizes the criterion used to choose the best model. For some criterion, it will require numerical integration, or perhaps estimation using Monte Carlo methods. In most cases in practice, it will only be feasible to evaluate the final term in (7.16) for a small number of the most important alternative models.

CHAPTER 8

# Conclusions and some recommendations

Let us conclude by posing a number of questions and examining how far they can be answered.

**Question 1.**
**How can we test whether there is any relationship between the predictors and the predictand?**

This is a frequent question in the social and biological sciences. Data have been collected, on say 20 or 50 predictors, and this may have been augmented with constructed variables such as reciprocals, logarithms, squares, and interactions of the original variables. An automatic computer package may have selected 5 or 10 of these predictors, and has probably output an $R^2$ value for the selected subset. Could we have done as well if the predictors had been replaced with random numbers or columns from the telephone directory?

If the package used the Efroymson stepwise algorithm, sometimes simply called stepwise regression, then Table 4.4 or formula (4.1) can be used to test whether the value of $R^2$ could reasonably have arisen by chance if there is no real relationship between the $Y$-variable and any of the $X$-variables. Clearly, the more exhaustive the search procedure used, the higher the $R^2$ value that can be achieved. References are given in section 4.1 to tables for other search algorithms, though there is scope for the extension of these tables. Some of these tables allow for nonorthogonality of the predictors, others do not. In fact, the degree of correlation among the predictors does not make much difference to the distribution of $R^2$.

Alternatively, if the number of observations exceeds the number of available predictors, the Spj$\phi$tvoll test described in section 4.2 can be used to test whether the selected subset fits significantly better than just a constant.

**Question 2.**
**Does one subset fit significantly better than another?**

If the number of observations exceeds the number of available predictors, then the Spj$\phi$tvoll test described in section 4.2 provides a satisfactory answer to this question. An attractive feature of the Spj$\phi$tvoll test is that it does not require the assumption that either model is the true model. The test is that one model fits better than another over the range of the $X$-variables in the available data. It is possible to modify the test to apply for extrapolated $X$'s, though this has not been described in detail here.

The Spj$\phi$tvoll test is fairly conservative, that is, it tends to say that subsets do not differ significantly unless one is strikingly better than the other.

In the case in which the number of available predictors equals or exceeds the number of observations, there is no general test available or possible. The situation is akin to that in the analysis of designed experiments when there is no replication. If the experimenter is prepared to take the gamble that high-order interactions can be used as a measure of the residual variation, then an analysis can proceed. Similarly, if the researcher gambles on some variables having no effect, or he thinks that he has a reasonable estimate of residual variation from other data sources, then some kind of risky analysis can proceed. Of course, if the judgment that certain variables have no effect is taken after a preliminary analysis of the data, the resulting estimate of residual variance is likely to be artificially small.

**Question 3.**
**How do we find subsets that fit well?**

Many automatic procedures have been described in Chapter 3. Exhaustive search, using a branch-and-bound algorithm, for the best-fitting subsets of all sizes is typically feasible if we have not more than approximately 25 available predictors. It is often sensible to try one of the cheap methods first, perhaps sequential replacement. This will usually show up the 'dominant' variables and give an idea of the likely size of the final subset. At this stage, it is often wise to use some of the standard regression diagnostic tools (see e.g. Belsley et al. (1980), Gunst and Mason (1980) or Cook and Weisberg (1982)). These could show up a nonlinear relationship with one of the dominant variables, or outliers that may be errors in the data, or very influential observations.

If the cheap method has shown that there is very little reduction in the residual sum of squares between fitting, say 8 variables, and fitting all of them, then an exhaustive search can be restricted to subsets of 8 or fewer variables. Such a search may be feasible when it is not feasible to search for the best-fitting subsets of all sizes.

As a rough rule, the feasible number of subsets that can be searched is of the order of $10^7$. There are $2^{25} = 3.3 \times 10^7$ possible subsets out of 25 predictors, including the empty subset and the complete set of 25, so that this is close to the limit of feasibility for subsets of all sizes. If we have 50 available predictors, then an exhaustive search for best-fitting subsets of all sizes will usually not be feasible, but it will be feasible to search for the best-fitting subset of 6 or fewer variables.

If an exhaustive search is not feasible, then a sequential procedure that adds or removes two variables at a time will sometimes find much better-fitting subsets than one-at-a-time algorithms.

**Question 4.**
**How many variables should be included in the final subset, assuming that it is required for prediction?**

Before that question can be answered, we need to know what method is to be used to estimate the regression coefficients. If ordinary least squares, or one of the robust alternatives, is to be used with no attempt to correct for selection bias, then using all the available predictors will often yield predictions with a smaller $MSEP$ than any subset.

If the conditional maximum likelihood method of section 6.4, or some other method, is used to partially adjust for selection bias, then minimizing the (falsely) estimated $MSEP$ given by (5.26) is often a reasonable stopping rule with random predictors, while minimizing Mallows' $C_p$ is the equivalent for fixed predictors. However, it is always possible to construct examples using fixed orthogonal predictors for which any given stopping rule will perform badly. This follows directly from Mallows (1973). In most practical cases, the stopping rule is not critical, provided that there is a correction for selection bias in the regression coefficients. The use of Mallows' $C_p$, even when the predictors are random, or of Akaike's Information Criterion, or an $F$-to-enter of just under 2.0, or of minimizing the $PRESS$ statistic, can all be expected to give about the same result as using the true $MSEP$, for up to about 20-30 predictors. For larger numbers of predictors, the Risk Inflation Factor (RIC) should be considered. In choosing a subset of variables, we risk both including variables which have no predictive power, and of leaving out variables which should be included. The RIC makes it progressively more difficult for variables to be included as the number of available predictors is increased.

It is often instructive to add a few extra predictors generated using a random number generator, as was done in chapter 4, to discover when any of these are selected.

If the cost of measuring the variables is an important consideration then a stopping rule that selects a smaller subset should be used such as using a higher $F$-to-enter. At the moment, there are no accurate formulae for the true $MSEP$ after subset selection, using either least squares regression coefficients, or bias-corrected coefficients.

Notice that for the purpose of *prediction*, we are looking at $F$-to-enter's of the order of 1.5 to 2.0. If the Spj$\phi$tvoll test is being used for *hypothesis testing,* a test at the 5% level may be equivalent to using an $F$-to-enter of 8 to 15, depending upon the numbers of predictors and observations, and the structure of the sample correlations among the predictors.

All of the above assumes that future predictions will be for $X$-variables that span the same space as those used for the model selection and calibration.

It is extremely hazardous to extrapolate beyond this region. The emphasis throughout this monograph has been upon finding models that fit and describe relationships within the space of the $X$-predictors. Unless there is established theory to justify a particular form of model, there is no reason to believe it will fit well outside of the calibration region.

**Question 5.**
**How should we estimate regression coefficients?**

A conditional maximum likelihood method was described in section 5.4, and also in Chapter 7. It requires simulation, is very slow, neglects the bias due to the stopping rule, and appears to overcorrect for the bias. However, the cost of the computer time is now usually very small compared with the cost of the data collection, and to the value that can be attached to the predictions in some cases. A simple alternative, which has not been investigated here, is the jackknife suggested near the end of section 5.3, using the square root of the sample size.

The 'off-the-peg' alternatives, such as James-Stein/Sclove shrinkage and ridge regression are not designed to reduce selection bias. They are intended primarily to reduce the variance of the regression coefficients at the expense of adding a small amount of bias. The variance of least-squares regression coefficients of best-fitting subsets are often very much smaller than those for models chosen independently of the data, so that subset selection has already done what these shrinkage estimators are designed to do. The simulation results in Table 5.16 show that the use of Sclove shrinkage of the regression coefficients of the selected variables always give a small improvement in $MSEP$ over the use of least-squares estimates, while the use of ridge regression can sometimes give a larger improvement and can sometimes be disastrous.

If you use ordinary least squares, or any other method which does not allow for selection bias, such as one of the robust regression methods, then you must expect some of the estimated regression coefficients to be too large. It is a very useful exercise to do some simulations. Generate 100 data sets, using the same values of the $X$-variables, and the same number of cases. Then use your estimated regression coefficients, with zeroes for excluded variables. Calculate the fitted $Y$-values, then add noise. Repeat the subset selection procedure and the estimation method that you used on each of the 100 artificial data sets. You will probably find that subsets other than the original one will be selected in most cases. Look at the estimated regression coefficients for the original subset for each replicate, whether or not the subset was the selected one for that replicate. You will usually find that the estimated regression coefficients will have shrunk, sometimes quite considerably.

This method is very similar to the method of bootstrapping standardized residuals suggested in Chapter 5.

**Question 6.**
**Can the use of subset regression techniques for prediction be justified?**

There are many practical situations in which the cost of measuring the $X$-predictors is a major consideration. If there is no cost associated with obtaining future $X$-predictors, then the use of ridge regression using the Lawless-Wang ridge parameter, or the Sclove estimator, will often be preferable alternatives to subset selection. They have the important advantage that their properties are known. If cost of measurement of the $X$-predictors is a consideration, then the loss due to poor predictions should be traded off against it in deciding how many predictors to use. True cross-validation, as described in section 5.3, can be used to obtain a realistic estimate of the $MSEP$ provided that no extrapolation outside of the space of the $X$-predictors is required.

**Question 7.**
**What alternatives are there to subset selection?**

In many cases, for example, in the social or biological sciences, relationships between variables are monotonic and a simple linear regression (perhaps after using a transformation such as taking logarithms) is an adequate empirical approximation. In the physical sciences, the shape of the regression curve must often be approximated with more precision. One way to do this is by augmenting the predictor variables with polynomial or cross-product terms. This often gives rise to situations in which the cheap 'one-at-a-time' selection procedures pick poor subsets, while an exhaustive search procedure is not feasible. One alternative for this situation is to use projection pursuit (see e.g. Huber (1985), Friedman (1987)). There is often some prior knowledge of the system being modelled in the physical sciences that enables perhaps partial differential equations to be formulated, which partially describe the system and leave only part of it to be modelled empirically. This will often be preferable to the use of black-box techniques.

Bayesian Model Averaging (BMA) is an important alternative that has emerged in the last few years. This uses a weighted average of all of the models considered, where the weight given to each model is the posterior prior for that model. Note that the weights are given to the models, not to the individual variables. Thus, the final model includes all of the predictors, though usually some of them are given so little weight that they are effectively excluded.

# References


Abramowitz, M. and Stegun, I. (1964), *Handbook of mathematical functions*, U.S. Govt. Printing Office: Washington, D.C.

Adams, J.L. (1990), *A computer experiment to evaluate regression strategies*, Proc. Statist. Computing Section, Amer. Statist. Assoc., 55–62.

Aitkin, M.A. (1974), *Simultaneous inference and the choice of variable subsets in multiple regression*, Technometrics, **16**, 221–227.

Akaike, H. (1969), *Fitting autoregressive models for prediction*, Ann. Inst. Statist. Math., **21**, 243–247.

______ (1973), *Maximum likelihood identification of Gaussian autoregressive moving average models*, Biometrika, **60**, 255–265.

______ (1977), *On entropy maximisation principle*, Applications of statistics (Krishnaiah, P.R., ed.), North Holland: Amsterdam, pp. 27–41.

Allen, D.M. (1974), *The relationship between variable selection and data augmentation and a method for prediction*, Technometrics, **16**, 125–127.

Altman, N. and Leger, C. (1997), *On the optimality of prediction-based selection criteria and the convergence rates of estimators*, J. Roy. Statist. Soc., Series B, **59**, 205–216.

Armstrong, R.D. and Kung, M.T. (1982), *An algorithm to select the best subset for a least absolute value regression problem*, TIMS Studies of Management Sci., **33**, 931–936.

Armstrong, R.D., Beck, P.O. and Kung, M.T. (1984), *Algorithm 615: The best subset of parameters in least absolute value regression*, ACM Trans. on Math. Software (TOMS), **10**, 202–206.

Atkinson, A.C. (1978), *Posterior probabilities for choosing a regresion model*, Biometrika, **65**, 39–48.

______ (1985), *Plots, transformations and regression*, Oxford Univ. Press: Oxford.

Banachiewicz, T. (1938), *Methode de resolution numerique des equations lineaires, du calcul des determinants et des inverses, et de reduction des formes quadratiques*, Comptes-rendus mensuels des sciences mathematiques et naturelles, Acad. Polonaise des Sci. et des Lettres, 393–404.

Bancroft, T.A. and Han, C-P. (1977), *Inference based on conditional specification: a note and a bibliography*, Internat. Statist. Rev., **45**, 117–127.

Barnett, V.D. and Lewis, T. (1978), *Outliers in statistical data*, Wiley: New York.

Bartlett, M.S. (1951), *An inverse matrix adjustment arising in discriminant analysis*, Ann. Math. Statist., **22**, 107–111.

Baskerville, J.C. and Toogood, J.H. (1982), *Guided regression modeling for prediction and exploration of structure with many explanatory variables*, Technometrics, **24**, 9–17.

Beale, E.M.L. (1970), *Note on procedures for variable selection in multiple regression*, Technometrics, **12**, 909–914.

Beale, E.M.L., Kendall, M.G. and Mann, D.W. (1967), *The discarding of variables in multivariate analysis*, Biometrika, **54**, 357–366.

Belsley, D.A., Kuh, E. and Welsch, R.E. (1980), *Regression diagnostics: identifying influential data and sources of collinearity*, Wiley: New York.

Bendel, R.B. (1973), *Stopping rules in forward stepwise-regression*, Ph.D. thesis, Biostatistics Dept., Univ. of California at Los Angeles.

Bendel, R.B. and Afifi, A.A. (1977), *Comparison of stopping rules in forward "stepwise" regression*, J. Amer. Statist. Assoc., **72**, 46–53.

Benedetti, J.K. and Brown, M.B. (1978), *Strategies for the selection of log-linear models*, Biometrics, **34**, 680–686.

Berger, J.O. and Pericchi, L.R. (1996), *The intrinsic Bayes factor for model selection and prediction*, J. Amer. Statist. Assoc., **91**, 109–122.

Berk, K.N. (1978a), *Gauss-Jordan v. Choleski*, Comput. Science and Statist.: 11th Annual Symposium on the Interface, Inst. of Statist., N. Carolina State Univ.

——— (1978b), *Comparing subset regression procedures*, Technometrics, **20**, 1–6.

Biondini, R., Simpson, J. and Woodley, W. (1977), *Empirical predictors for natural and seeded rainfalls in the Florida Area Cumulus Experiment (FACE), 1970-1975*, J. Appl. Meteor., **16**, 585–594.

Borowiak, D. (1981), *A procedure for selecting between two regression models*, Commun. in Statist., **A10**, 1197–1203.

Boyce, D.E., Farhi, A. and Weischedel, R. (1974), *Optimal subset selection: Multiple regression, interdependence and optimal network algorithms*, Lecture Notes in Economics and Mathematical Systems, vol. 103, Springer-Verlag: Berlin.

Breiman, L. (1995), *Better subset regression using the nonnegative garrote*, Technometrics, **37**, 373–384.

——— (1996), *Bagging predictors*, Machine Learning, **26**, 123–140.

Breiman, L. and Spector, P. (1992), *Submodel selection and evaluation in regression: the X-random case*, Int. Statist. Review, **60**, 291–319.

Breiman, L., Friedman, J.H., Olshen, R.A. and Stone, C.J. (1984), *Classification and regression trees*, Wadsworth: Belmont, CA.

Broersen, P.M.T. (1986), *Subset regression with stepwise directed search*, Applied Statist., **35**, 168–177.

Brown, M.B. (1976), *Screening effects in multidimensional contingency tables*, Appl. Statist., **25**, 37–46.

Brown, P.F., Fearn, T. and Vannucci, M. (1999), *The choice of variables in multivariate regression: A non-conjugate Bayesian decision theory approach*, Biometrika, **86**, 635–648.

Brown, R.L., Durbin, J. and Evans, J.M. (1975), *Techniques for testing the constancy of regression relationships over time*, J. Roy. Statist. Soc., **B, 37**, 149–163.

Bryant, P.G. and Cordero-Brana, O.I. (2000), *Model selection using the minimum description length principle*, The Amer. Statist., **54**, 257–268.

Burnham, K.P. and Anderson, D.R. (1998), *Model Selection and Inference: A Practical Information-Theoretic Approach*, Springer-Verlag: New York.

Butler, R.W. (1982), *Bounds on the significance attained by the best-fitting regressor variable*, Appl. Statist., **31**, 290–292.

——— (1984), *The significance attained by the best-fitting regressor variable*, J. Amer. Statist. Assoc., **79**, 341–348.

Chan, T.F. (1982), *Algorithm 581: an improved algorithm for computing the singular value decomposition*, ACM Trans. Math. Software (TOMS), **8**, 84–88.

Chan, T.F., Golub, G.H. and LeVeque, R.J. (1983), *Algorithms for computing the sample variance: analysis and recommendations*, The Amer. Statistician, **37**, 242–247.

Chatfield, C. (1995), *Model uncertainty, data mining and statistical inference* (with discussion), J. Roy. Statist. Soc., A, **158**, 419–466.

Chipman, H., George, E.I. and McCulloch, R.E. (2001), *The practical implementation of Bayesian model selection*, Model Selection (Lahiri, P., ed.), Lecture notes - monographs, vol. 38, Inst. Math. Statist., pp. 65–116.

Clarke, M.R.B. (1980), *Choice of algorithm for a model-fitting system*, Compstat 1980, Physica-Verlag: Vienna, pp. 530–536.

______ (1981), *Algorithm AS163: A Givens algorithm for moving from one linear model to another without going back to the data*, Appl. Statist., **30**, 198–203.

Clyde, M., Desimone, H. and Parmigiani, G. (1996), *Prediction via orthogonalized model mixing*, J. Amer. Statist. Assoc., **91**, 1197–1208.

Cohen, A. and Sackrowitz, H.B. (1982), *Estimating the mean of the selected population*, Statistical Decision Theory and Related Topics, III (Gupta, S.S. and Berger, J., ed.), vol. 1, Academic Press: New York, pp. 243–270.

______ (1989), *Two stage conditionally unbiased estimators of the selected mean*, Statist. and Prob. Letters, **8**, 273–278.

Cook, R.D. and Weisberg, S. (1982), *Residuals and influence in regression*, Chapman and Hall: London.

Copas, J.B. (1983), *Regression, prediction and shrinkage*, J. Roy. Statist. Soc., Series B, **45**, 311–354, incl. discussion.

Cox, D.R. and Hinkley, D.V. (1974), *Theoretical statistics*, Chapman and Hall: London.

Cox, D.R. and Snell, E.J. (1974), *The choice of variables in observational studies*, Appl. Statist., **23**, 51–59.

______ (1981), *Applied statistics: principles and examples*, Chapman and Hall: London.

Dahiya, R.C. (1974), *Estimation of the mean of the selected population*, J. Amer. Statist. Assoc., **69**, 226–230.

Dempster, A.P., Schatzoff, M. and Wermuth, N. (1977), *A simulation study of alternatives to ordinary least squares*, J. Amer. Statist. Assoc., **72**, 77–106, (incl. discussion by Hoerl, Allen, Smith, Bingham and Larntz, Thisted, and rejoinder by the authors).

Derflinger, G. and Stappler, H. (1976), *A correct test for the stepwise regression analysis*, COMPSTAT 1976: Proceedings in Computational Statistics (Gordesch, J. and Naeve, P., ed.), Physica-Verlag: Wien, pp. 131–138.

Diebolt, J. and Robert, C.P. (1994), *Estimation of finite mixture distributions through Bayesian sampling*, J. Roy. Statist. Soc., Series B, **56**, 363–375.

Diehr, G. and Hoflin, D.R. (1974), *Approximating the distribution of the sample $R^2$ in best subset regressions*, Technometrics, **16**, 317–320.

Diggle, P.J. and Gratton, R.J. (1984), *Monte Carlo methods of inference for implicit statistical models*, J. Roy. Statist. Soc., Series B, **46**, 193–227, incl. discussion.

Dijkstra, D.A. and Veldkamp, J.H. (1988), *Data-driven selection of regressors and the bootstrap*, On Model Uncertainty and its Statistical Implications (T.K. Dijkstra, ed.), Springer-Verlag: Berlin, pp. 17–38.

Dongarra, J.J., Bunch, J.R., Moler, C.B. and Stewart, G.W. (1979), *LINPACK Users Guide*, Soc. for Industrial and Appl. Math.: Philadelphia.

Donoho, D.L. and Johnstone, I.M. (1994), *Ideal spatial adaptation by wavelet shrinkage*, Biometrika, **81**, 425–456.

Draper, D. (1995), *Assessment and propagation of model uncertainty* (with discussion), J. Roy. Statist. Soc., B, **57**, 45–97.

Draper, N.R. and Smith, H. (1981), *Applied regression analysis, 2nd edition*, Wiley: New York.

Draper, N.R. and van Nostrand, R.C. (1979), *Ridge regression and James-Stein estimation: review and comments*, Technometrics, **21**, 451–466.

Draper, N.R., Guttman, I. and Kanemasu, H. (1971), *The distribution of certain regression statistics*, Biometrika, **58**, 295–298.

Draper, N.R., Guttman, I. and Lapczak, L. (1979), *Actual rejection levels in a certain stepwise test*, Commun. in Statist., **A8**, 99–105.

DuPreez, J.P. and Venter, J.H. (1989), *A group adaptive empirical Bayes estimator of normal means*, J. Statist. Planning Inference, **22**, 29–41.

Edwards, D. and Havranek, T. (1987), *A fast model selection procedure for large families of models*, J. Amer. Statist. Assoc., **82**, 205–213.

Efron, B. (1982), *The jackknife, the bootstrap and other resampling plans*, Soc. for Industrial and Appl. Math.: Philadelphia.

Efron, B. and Morris, C. (1973), *Combining possibly related estimation problems*, J. Roy. Statist. Soc., **B, 35**, 379–421, (incl. discussion).

Efron, B. and Tibshirani, R.J. (1993), *An introduction to the bootstrap*, Chapman and Hall: New York.

Efroymson, M.A. (1960), *Multiple regression analysis*, Mathematical Methods for Digital Computers (Ralston, A. and Wilf, H.S., ed.), vol. 1, Wiley: New York, pp. 191–203.

Elden, L. (1972), *Stepwise regression analysis with orthogonal transformations*, Master's thesis, Unpubl. report, Mathematics Dept., Linkoping Univ., Sweden.

Everitt, B. (1974), *Cluster analysis*, Heinemann: London.

Farebrother, R.W. (1974), *Algorithm AS79: Gram-Schmidt regression*, Appl. Statist., **23**, 470–476.

——— (1978), *An historical note on recursive residuals*, J. Roy. Statist. Soc., **B, 40**, 373–375.

——— (1988), *Linear least squares computations*, Marcel Dekker: New York.

Fearn, T. (1983), *A misuse of ridge regression in the calibration of a near infrared reflectance instrument*, Appl. Statist., **32**, 73–79.

Feiveson, A.H. (1994), *Finding the best regression subset by reduction in nonfull-rank cases*, SIAM J. Matrix Anal. Applic., **15**, 194–204.

Fernandez, C., Ley, E. and Steel, M.F.J. (2001), *Benchmark priors for Bayesian model averaging*, J. Econometrics, **100**, 381–427.

——— (2001), *Model uncertainty in cross-country growth regressions*, J. Appl. Econometrics, **16**, 563–576.

Fisher, J.C. (1976), *Homicide in Detroit: the role of firearms*, Criminology, **14**, 387–400.

Forsythe, A.B., Engelman, L., Jennrich, R. and May, P.R.A. (1973), *A stopping rule for variable selection in multiple regression*, J. Amer. Statist. Assoc., **68**, 75–77.

Forsythe, G.E. and Golub, G.H. (1965), *On the stationary values of a second-degree polynomial on the unit sphere*, SIAM J., **13**, 1050–1068.

Foster, D.P. and George, E.I. (1994), *The risk inflation criterion for multiple regression*, Ann. Statist., **22**, 1947–1975.

Frank, I. and Friedman, J. (1993), *A statistical view of some chemometrics tools (with discussion)*, Technometrics, **35**, 109–148.

Freedman, D.A., Navidi, W. and Peters, S.C. (1988), *On the impact of variable selection in fitting regression equations*, On Model Uncertainty and its Statistical Implications (T.K. Dijkstra, ed.), Springer-Verlag: Berlin, pp. 1–16.

Friedman, J.H. (1987), *Exploratory projection pursuit*, J. Amer. Statist. Assoc., **82**, 249–266.

Fu, W.J. (1998), *Penalized regressions: the Bridge versus the Lasso*, J. Comput. Graphical Statist., **7**, 397–416.

Furnival, G.M. (1971), *All possible regressions with less computation*, Technometrics, **13**, 403–408.

Furnival, G.M. and Wilson, R.W. (1974), *Regression by leaps and bounds*, Technometrics, **16**, 499–511.

Gabriel, K.R. and Pun, F.C. (1979), *Binary prediction of weather events with several predictors*, 6th Conference on Prob. and Statist. in Atmos. Sci., Amer. Meteor. Soc., pp. 248–253.

Galpin, J.S. and Hawkins, D.M. (1982), *Selecting a subset of regression variables so as to maximize the prediction accuracy at a specified point*, Tech. report, Nat. Res. Inst. for Math. Sciences, CSIR, P.O. Box 395, Pretoria, South Africa.

______ (1986), *Selecting a subset of regression variables so as to maximize the prediction accuracy at a specified point*, J. Appl. Statist., **13**, 187–198.

Garside, M.J. (1965), *The best subset in multiple regression analysis*, Appl. Statist., **14**, 196–200.

______ (1971a), *Algorithm AS 37: Inversion of a symmetric matrix*, Appl. Statist., **20**, 111–112.

______ (1971b), *Some computational procedures for the best subset problem*, Appl. Statist., **20**, 8–15.

______ (1971c), *Algorithm AS38: Best subset search*, Appl. Statist., **20**, 112–115.

Garthwaite, P.H. and Dickey, J.M. (1992), *Elicitation of prior distributions for variable-selectiom problems in regression*, Ann. Statist., **20**, 1697–1719.

Gelman, A., Carlin, J.B., Stern, H.S. and Rubin, D.B. (1995), *Bayesian data analysis*, Chapman and Hall: London.

Gentle, J.E. and Hanson, T.A. (1977), *Variable selection under $L_1$*, Proc. Statist. Comput. Section, Amer. Statist. Assoc., pp. 228–230.

Gentle, J.E. and Kennedy, W.J. (1978), *Best subsets regression under the minimax criterion*, Comput. Science and Statist.: 11th Annual Symposium on the Interface., Inst. of Statist., N. Carolina State Univ., pp. 215–217.

Gentleman, J.F. (1975), *Algorithm AS88: Generation of all ${}^{N}C_R$ combinations by simulating nested Fortran DO-loops*, Appl. Statist., **24**, 374–376.

Gentleman, W.M. (1973), *Least squares computations by Givens transformations without square roots*, J. Inst. Maths. Applics., **12**, 329–336.

______ (1974), *Algorithm AS75: Basic procedures for large, sparse or weighted linear least squares problems*, Appl. Statist., **23**, 448–454.

______ (1975), *Error analysis of QR decompositions by Givens transformations*, Linear Algebra and its Applics., **10**, 189–197.

George, E.I. (2000), *The variable selection problem*, J. Amer. Statist. Assoc., **95**, 1304–1308.

George, E.I. and Foster, D.P. (2000), *Calibration and empirical Bayes variable selection*, Biometrika, **87**, 731–747.

George, E.I. and McCulloch, R.E. (1993), *Variable selection via Gibbs sampling*, J. Amer. Statist. Assoc., **88**, 881–89.

Gilmour, S.G. (1996), *The interpretation of Mallows's $C_p$-statistic*, The Statistician, **45**, 49–56.

Golub, G.H. (1969), *Matrix decompositions and statistical calculations*, Statistical Computation (Milton, R.C. and Nelder, J.A., ed.), Academic Press: New York.

Golub, G.H. and Styan, G.P.H. (1973), *Numerical computations for univariate linear models*, J. Statist. Comput. Simul., **2**, 253–274.

Goodman, L.A. (1971), *The analysis of multidimensional contingency tables: Stepwise procedures and direct estimation methods for building models for multiple classifications*, Technometrics, **13**, 33–61.

Gray, H.L. and Schucany, W.R. (1972), *The generalized jackknife statistic*, Marcel Dekker: New York.

Grossman, S.I. and Styan, G.P.H. (1973), *Optimality properties of Theil's BLUS residuals*, J. Amer. Statist. Assoc., **67**, 672–673.

Gunst, R.F. and Mason, R.L. (1980), *Regression analysis and its application*, Marcel Dekker: New York.

Hall, P. (1989), *On projection pursuit regression*, Ann. Statist., **17**, 573–588.

Hammarling, S. (1974), *A note on modifications to the Givens plane rotation*, J. Inst. Maths. Applics., **13**, 215–218.

Han, C. and Carlin, B.P. (2001), *Markov chain Monte Carlo methods for computing Bayes factors: a comparative review*, J. Amer. Statist. Assoc., **96**, 1122–1132.

Hannan, E.J. (1970), *Multiple time series*, Wiley: New York.

Hannan, E.J. and Quinn, B.G. (1979), *The determination of the order of an autoregression*, J. Roy. Statist. Soc., **B, 41**, 190–195.

Hansen, M.H. and Yu, B. (2001), *Model selection and the principle of minimum description length*, J. Amer. Statist. Assoc., **96**, 746–774.

Hartigan, J.A. (1975), *Clustering algorithms*, Wiley: New York.

Hawkins, D.M. (1980), *Identification of outliers*, Chapman and Hall: London.

Healy, M.J.R. (1968a), *Algorithm AS6: Triangular decomposition of a symmetric matrix*, Appl. Statist., **17**, 195–197.

——— (1968b), *Algorithm AS7: Inversion of a positive semi-definite symmetric matrix*, Appl. Statist., **17**, 198–199.

Hemmerle, W.J. (1975), *An explicit solution for generalized ridge regression*, Technometrics, **17**, 309–314.

Hemmerle, W.J. and Brantle, T.F. (1978), *Explicit and constrained generalized ridge regression*, Technometrics, **20**, 109–120.

Hjorth, U. (1982), *Model selection and forward validation*, Scand. J. Statist., **9**, 95–105.

——— (1994), *Computer intensive statistical methods: Validation model selection and bootstrap*, Chapman and Hall: London.

Hocking, R.R. (1976), *The analysis and selection of variables in linear regression*, Biometrics, **32**, 1–49.

Hocking, R.R. and Leslie, R.N. (1967), *Selection of the best subset in regression analysis*, Technometrics, **9**, 531–540.

Hocking, R.R., Speed, F.M. and Lynn, M.J. (1976), *A class of biased estimators in linear regression*, Technometrics, **18**, 425–437.

Hoerl, A.E. and Kennard, R.W. (1970a), *Ridge regression: biased estimation for nonorthogonal problems*, Technometrics, **12**, 55–67.

———(1970b), *Ridge regression: applications to nonorthogonal problems*, Technometrics, **12**, 69–82.

Hoerl, A.E., Kennard, R.W. and Baldwin, K.F. (1975), *Ridge regression: some simulations*, Commun. in Statist., **4**, 105–123.

Hoerl, A.E., Kennard, R.W. and Hoerl, R.W. (1985), *Practical use of ridge regression: a challenge met*, Appl. Statist., **34**, 114–120.

Hoerl, R.W., Schuenemeyer, J.H. and Hoerl, A.E. (1986), *A simulation of biased estimation and subset selection regression techniques*, Technometrics, **28**, 369–380.

Hoeting, J.A., Madigan, D., Raftery, A.E. and Volinsky, C.T. (1997), *Bayesian model averaging: A tutorial*, Statist. Sci., **14**, 382–417, (Corrected version from http://www.stat.washington.edu/www/research/online/hoeting1999.pdf).

Huber, P.J. (1985), *Projection pursuit*, Ann. Stat., **13**, 435–525, incl. discussion.

Hurvich, C.M. and Tsai, C-L. (1990), *The impact of model selection on inference in linear regression*, The Amer. Statistician, **44**, 214–217.

James, W. and Stein, C. (1961), *Estimation with quadratic loss*, Proc. 4th Berkeley Symposium on Probability and Statistics (Neyman, J. and Le Cam L., ed.), vol. 1, pp. 362–379.

Jeffers, J.N.R. (1967), *Two case studies in the application of principal component analysis*, Appl. Statist., **16**, 225–236.

Jennings, L.S. and Osborne, M.R. (1974), *A direct error analysis for least squares*, Numer. Math., **22**, 325–332.

Jennrich, R.I. (1977), *Stepwise regression*, Statistical Methods for Digital Computers (Enslein,K., Ralston, A. and Wilf, H.S., ed.), Wiley: New York, pp. 58–75.

Jolliffe, I.T. (1982), *A note on the use of principal components in regression*, Appl. Statist., **31**, 300–303.

Jones, M.C. and Sibson, R. (1987), *What is projection pursuit?*, J. Roy. Statist. Soc., Series A, **150**, 1–36, incl. discussion.

Judge, G.G. and Bock, M.E. (1978), *The statistical implications of pre-test and Stein-rule estimators in econometrics*, North Holland: Amsterdam.

Kahaner, D., Tietjen, G. and Beckmann, R. (1982), *Gaussian quadrature formulas for* $\int_0^\infty e^{-x^2} g(x)dx$, J. Statist. Comput. Simul., **15**, 155–160.

Kahn, H. (1956), *Use of different Monte Carlo sampling techniques*, Symposium on Monte Carlo Methods (Meyer, H.A., ed.), Wiley: New York, pp. 146–190.

Kailath, T. (1974), *A view of three decades of linear filtering theory*, I.E.E.E. Trans. on Inf. Theory, **IT-20**, 145–181.

Kaiser, J.H. (1972), *The JK method: a procedure for finding the eigenvectors and eigenvalues of a real symmetric matrix*, The Computer J., **15**, 271–273.

Kendall, M.G. and Stuart, A. (1961), *The advanced theory of statistics*, vol. 2, Griffin: London.

Kennedy, W.J. and Bancroft, T.A. (1971), *Model building for prediction in regression based upon repeated significance tests*, Ann. Math. Statist., **42**, 1273–1284.

Kohn, R., Smith, M. and Chan, D. (2001), *Nonparametric regression using linear combinations of basis functions*, Statist. and Computing, **11(4)**, 313–322.

Kudo, A. and Tarumi, T. (1974), *An algorithm related to all possible regression and discriminant analysis*, J. Japan. Statist. Soc., **4**, 47–56.

Kullback, S. (1959), *Information theory and statistics*, Wiley: New York.

Kullback, S. and Leibler, R.A. (1951), *On information and sufficiency*, Ann. Math.

Statist., **22**, 79–86.

LaMotte, L.R. and Hocking, R.R. (1970), *Computational efficiency in the selection of regression variables*, Technometrics, **12**, 83–93.

Lane, L.J. and Dietrich, D.L. (1976), *Bias of selected coefficients in stepwise regression*, Proc. Statist. Comput. Section, Amer. Statist. Assoc., pp. 196–200.

Lawless, J.F. (1978), *Ridge and related estimation procedures: theory and practice*, Commun. in Statist., **A7**, 139–164.

——— (1981), *Mean squared error properties of generalized ridge estimators*, J. Amer. Statist. Assoc., **76**, 462–466.

Lawless, J.F. and Wang, P. (1976), *A simulation study of ridge and other regression estimators*, Commun. in Statist., **A5**, 307–323.

Lawrence, M.B., Neumann, C.J. and Caso, E.L. (1975), *Monte Carlo significance testing as applied to the development of statistical prediction of tropical cyclone motion*, 4th Conf. on Prob. and Statist. in Atmos. Sci., Amer. Meteor. Soc., pp. 21–24.

Lawson, C.L. and Hanson, R.J. (1974), *Solving least squares problems*, Prentice-Hall: New Jersey, (reprinted 1995, Society for Industrial and Applied Mathematics: Philadelphia as vol. 15 in its Classics series).

Lawson, C.L., Hanson, R.J., Kincaid, D.R. and Krogh, F.T. (1979), *Basic linear algebra subprograms for FORTRAN usage*, ACM Trans. on Math. Software (TOMS), **5**, 308–323.

Lee, T.-S. (1987), *Algorithm AS223: Optimum ridge parameter selection*, Appl. Statist., **36**, 112–118.

Linhart, H. and Zucchini, W. (1986), *Model selection*, Wiley: New York.

Longley, J.W. (1967), *An appraisal of least squares programs for the electronic computer from the point of view of use*, J. Amer. Statist. Assoc., **62**, 819–841.

——— (1981), *Modified Gram-Schmidt process vs. classical Gram-Schmidt*, Commun. in Statist., **B10**, 517–527.

Lovell, M.C. (1983), *Data mining*, The Rev. of Econ. and Statist., **65**, 1–12.

Madigan, D. and York, J. (1995), *Bayesian graphical models for discrete data*, Intern. Statist. Rev., **63**, 215–232.

Maindonald, J.H. (1984), *Statistical computation*, Wiley: New York.

Mallows, C.L. (1973), *Some comments on $C_p$*, Technometrics, **15**, 661–675.

——— (1995), *More comments on $C_p$*, Technometrics, **37**, 362–372.

Mantel, N. (1970), *Why stepdown procedures in variable selection*, Technometrics, **12**, 621–625.

Mason, R.L. and Gunst, R.F. (1985), *Selecting principal components in regression*, Statist. and Prob. Letters, **3**, 299–301.

McCabe, G.P., Jr. (1978), *Evaluation of regression coefficient estimates using $\alpha$-acceptability*, Technometrics, **20**, 131–139.

McDonald, G.C. and Galarneau, D.I. (1975), *A Monte Carlo evaluation of some ridge-type estimators*, J. Amer. Statist. Assoc., **70**, 407–416.

McDonald, G.C. and Schwing, R.C. (1973), *Instabilities of regression estimates relating air pollution to mortality*, Technometrics, **15**, 463–482.

McIntyre, S.H., Montgomery, D.B., Srinavasan, V. and Weitz, B.A. (1983), *Evaluating the statistical significance of models developed by stepwise regression*, J. Marketing Res., **20**, 1–11.

McKay, R.J. (1979), *The adequacy of variable subsets in multivariate regression*, Technometrics, **21**, 475–479.

Miller, A.J. (1984), *Selection of subsets of regression variables*, J. Roy. Statist. Soc., Series A, **147**, 389–425, incl. discussion.

——— (1989), *Updating means and variances*, J. Comput. Phys., **85**, 500–501.

——— (1992), *As 273: Comparing subsets of regressor variables*, Appl. Statist., **41**, 443–457.

——— (1992), *AS274. Least squares routines to supplement those of Gentleman*, Appl. Statist., **41**, 458–478.

——— (1996), *Estimation after model building: A first step*, COMPSTAT 1996, Proceedings in Computational Statistics (Prat, A., ed.), Physica-Verlag: Heidelberg, pp. 367–372.

——— (1997), *Estimation after stepwise regression using the same data*, Computing Science and Statistics (Interface '96) (Billard, L. and Fisher, N.I., ed.), vol. 28, Interface Foundation of N. America: Fairfax Station, pp. 539–544.

——— (2000), *Another look at subset selection using linear least squares*, Commun. in Statist., Theory and Methods, **29**, 2005–2018.

Miller, R.G. (1962), *Statistical prediction by discriminant analysis*, Meteor. Monographs, vol. 4, no. 25, Amer. Meteor. Soc.

——— (1974), *The jackknife - a review*, Biometrika, **61**, 1–15.

Mitchell, T.J. and Beauchamp, J.J. (1988), *Bayesian variable selection in linear regression* (with discussion), J. Amer. Statist. Assoc., **83**, 1023–1036.

Morgan, J.A. and Tatar, J.F. (1972), *Calculation of the residual sum of squares for all possible regressions*, Technometrics, **14**, 317–325.

Morrison, D.F. (1967), *Multivariate statistical methods*, McGraw-Hill: New York.

Naes, T., Irgens, C. and Martens, H. (1986), *Comparison of linear statistical methods for calibration of NIR instruments*, Appl. Statist., **35**, 195–206.

Narendra, P.M. and Fukunaga, K. (1977), *A branch and bound algorithm for feature subset selection*, IEEE Trans. Comput., C, **26**, 917–922.

Narula, S.C. and Wellington, J.F. (1977a), *Prediction, linear regression and minimum sum of relative errors*, Technometrics, **19**, 185–190.

——— (1977b), *An algorithm for the minimum sum of weighted absolute errors regression*, Commun. in Statist., **B6**, 341–352.

——— (1979), *Selection of variables in linear regression using the sum of weighted absolute errors criterion*, Technometrics, **21**, 299–306.

Nash, J.C. (1979), *Compact numerical methods for computers: linear algebra and function minimisation*, Adam Hilger (Inst. of Physics): Bristol.

Nelder, J.A. and Mead, R. (1965), *A simplex method for function minimization*, Comput. J., **7**, 308–313.

Obenchain, R.L. (1975), *Ridge analysis following a preliminary test of the shrunken hypothesis*, Technometrics, **17**, 431–441.

Oliker, V.I. (1978), *On the relationship between the sample size and the number of variables in a linear regression model*, Commun. in Statist., **A7**, 509–516.

Osborne, M.R. (1976), *On the computation of stepwise regressions*, Aust. Comput. J., **8**, 61–68.

Osborne, M.R., Presnell, B. and Turlach, B.A. (1998), *Knot selection for regression splines via the Lasso*, Computing Science and Statistics (Interface '98) (Weisberg, S., ed.), vol. 30, Interface Foundation of N. America: Fairfax Station, pp. 44–49.

——— (2000), *On the Lasso and its dual*, J. Comput. Graphical Statist., **9**, 319–337.

Piessens, R. de D.-K., Uberhuber, C. and Kahaner, D. (1983), *QUADPACK, A quadrature subroutine package*, Springer-Verlag: Berlin.

Plackett, R.L. (1950), *Some theorems in least squares*, Biometrika, **37**, 149–157.

Platt, C.A. (1982), *Bootstrap stepwise regression*, Proc. Bus. and Econ. Sect., Amer. Statist. Assoc., Amer. Statist. Assoc.: Washington, D.C., pp. 586–589.

Pope, P.T. and Webster, J.T. (1972), *The use of an F-statistic in stepwise regression procedures*, Technometrics, **14**, 327–340.

Press, S.J. (1972), *Applied multivariate analysis*, Holt, Rinehart and Winston: New York.

Quenouille, M.H. (1956), *Notes on bias in estimation*, Biometrika, **43**, 353–360.

Raftery, A.E., Madigan, D. and Hoeting, J.A. (1997), *Bayesian model averaging for linear regression models*, J. Amer. Statist. Assoc., **92**, 179–191.

Rencher, A.C. and Pun, F.C. (1980), *Inflation of $R^2$ in best subset regression*, Technometrics, **22**, 49–53.

Ridout, M.S. (1982), *An improved branch and bound algorithm for feature subset selection*, Appl. Statist., **37**, 139–147.

Rissanen, J. (1978), *Modeling by shortest data description*, Automatica, **14**, 465–471.

——— (1987), *Stochastic complexity*, J. Roy. Statist. Soc., Series B, **49**, 223–239.

Roecker, E.B. (1991), *Prediction error and its estimation for subset-selected models*, Technometrics, **33**, 459–468.

Ronchetti, E. (1985), *Robust model selection in regression*, Statist. and Prob. Letters, **3**, 21–23.

Ronchetti, E. and Staudte, R.G. (1994), *A robust version of Mallows's $C_p$*, J. Amer. Statist. Assoc., **89**, 550–559.

Ronchetti, E., Field, C. and Blanchard, W. (1997), *Robust linear model selection by cross-validation*, J. Amer. Statist. Assoc., **92**, 1017–1023.

Roodman, G. (1974), *A procedure for optimal stepwise MSAE regression analysis*, Operat. Res., **22**, 393–399.

Rothman, D. (1968), *Letter to the editor*, Technometrics, **10**, 432.

Rushton, S. (1951), *On least squares fitting of orthonormal polynomials using the Choleski method*, J. Roy. Statist. Soc., **B, 13**, 92–99.

Savin, N.E. and White, K.J. (1978), *Testing for autocorrelations with missing observations*, Econometrika, **46**, 59–67.

Schatzoff, M., Tsao, R., and Fienberg, S. (1968), *Efficient calculation of all possible regressions*, Technometrics, **10**, 769–779.

Scheffe, H. (1959), *The analysis of variance*, Wiley: New York.

Schwarz, G. (1978), *Estimating the dimension of a model*, Ann. Statist., **6**, 461–464.

Sclove, S.L. (1968), *Improved estimators for coefficients in linear regression*, J. Amer. Statist. Assoc., **63**, 596–606.

Seber, G.A.F. (1977), *Linear regression analysis*, Wiley: New York.

——— (1984), *Multivariate observations*, Wiley: New York.

Shao, J. (1993), *Linear model selection by cross-validation*, J. Amer. Statist. Assoc., **88**, 486–494.

——— (1997), *An asymptotic theory for linear model selection*, Statistica Sinica, **7**, 221–264.

Silvey, S.D. (1975), *Statistical inference*, Chapman and Hall: London.

Smith, B.T., Boyle, J.M., Dongarra, J.J., Garbow, B.S., Ikebe, Y., Klema, V.C. and Moler, C.B. (1976), *Matrix eigensystem routines - EISPACK Guide*, Springer-Verlag: Berlin.

Smith, G. and Campbell, F. (1980), *A critique of some ridge regression methods*, J. Amer. Statist. Assoc., **75**, 74–103, (incl. discussion by Thisted, Marquardt, van

Nostrand, Lindley, Oberchain, Peele and Ryan, Vinod, and Gunst, and authors' rejoinder).

Smith, M. and Kohn, R. (1996), *Nonparametric regresion using Bayesian variable selection*, J. Econometrics, **75**, 317–343.

Sparks, R.S., Zucchini, W. and Coutsourides, D. (1985), *On variable selection in multivariate regression*, Commun. in Statist., **A14**, 1569–1587.

Spath, H. (1992), *Mathematical algorithms for linear regression*, Academic Press.

Spjøtvoll, E. (1972a), *Multiple comparison of regression functions*, Ann. Math. Statist., **43**, 1076–1088.

______(1972b), *A note on a theorem of Forsythe and Golub*, SIAM. J. Appl. Math., **23**, 307–311.

Sprevak, D. (1976), *Statistical properties of estimates of linear models*, Technometrics, **18**, 283–289.

Steen, N.M., Byrne, G.D. and Gelbard, E.M. (1969), *Gaussian quadratures for the integrals* $\int_0^\infty exp(-x^2)f(x)dx$ *and* $\int_0^b exp(-x^2)f(x)dx$, Math. of Comput., **23**, 661–671.

Stein, C. (1960), *Multiple regression*, Contributions to Probability and Statistics (Olkin, I. et al., ed.), Stanford Univ. Press: Stanford.

Stein, C.M. (1962), *Confidence sets for the mean of a multivariate normal distribution*, J. Roy. Statist. Soc., **B, 24**, 265–296, (incl. discussion).

Stewart, G.W. (1973), *Introduction to matrix computations*, Academic Press: New York.

______(1977), *Perturbation bounds for the QR factorization of a matrix*, SIAM J. Numer. Anal., **14**, 509–518.

Stone, M. (1977), *An asymptotic equivalence of choice of model by cross-validation and Akaike's criterion*, J. Roy. Statist. Soc., Series B, **39**, 44–47.

______(1978), *Cross-validation: A review*, Math. Operationsforsch., Ser. Statist., **9**, 127–139.

______(1979), *Comments on model selection criteria of Akaike and Schwarz*, J. Roy. Statist. Soc., Series B, **41**, 276–278.

Tarone, R.E. (1976), *Simultaneous confidence ellipsoids in the general linear model*, Technometrics, **18**, 85–87.

Thall, P.F., Simon, R. and Grier, D.A. (1992), *Test-based variable selection via cross-validation*, J. Comp. Graph. Statist., **1**, 41–61.

Thompson, M.L. (1978), *Selection of variables in multiple regression: Part I. A review and evaluation. Part II. Chosen procedures, computations and examples*, Internat. Statist. Rev., **46**, 1–19 and 129–146.

Tibshirani, R. (1996), *Regression shrinkage and selection via the lasso*, J. Roy. Statist. Soc., **B, 58**, 267–288.

Vellaisamy, P. (1992), *Average worth and simultaneous estimation of the selected subset*, Ann. Inst. Statist. Math., **44(3)**, 551–562.

Venter, J.H. (1988), *Estimation of the mean of the selected population*, Commun. Statist.-Theory Meth., **17(3)**, 791–805.

Venter, J.H. and Steel, S.J. (1991), *Estimation of the mean of the population selected from k populations*, J. Statist. Comput. Simul., **38**, 1–14.

Vinod, H.D. and Ullah, A. (1981), *Recent advances in regression*, Marcel Dekker: New York.

Wallace, C.S. and Freeman, P.R. (1987), *Estimation and inference by compact cod-*

*ing*, J. Roy. Statist. Soc., Series B, **49**, 240–252.

Wallace, T.D. (1977), *Pretest estimation in regression: a survey*, Amer. J. Agric. Econ., **59**, 431–443.

Wampler, R.H. (1970), *A report on the accuracy of some widely used least squares programs*, J. Amer. Statist. Assoc., **65**, 549–565.

——— (1979a), *Solutions to weighted least squares problems by modified Gram-Schmidt with iterative refinement*, ACM Trans. on Math. Software (TOMS), **5**, 457–465.

——— (1979b), *Algorithm 544. L2A and L2B, weighted least squares solutions by modified Gram-Schmidt with iterative refinement*, ACM Trans. on Math. Software (TOMS), **5**, 494–499.

Ward, L.L. (1973), *Is uncorrelating the residuals worth it?*, Master's thesis, Unpubl. M.A. thesis, Mathematics Dept., McGill Univ., Montreal, Canada.

Weisberg, S. (1980), *Applied linear regression*, Wiley: New York.

Wellington, J.F. and Narula, S.C. (1981), *Variable selection in multiple linear regression using the minimum sum of weighted absolute errors criterion*, Commun. in Statist., **B10**, 641–648.

Wilkinson, J.H. (1965), *The algebraic eigenvalue problem*, Oxford Univ. Press: Oxford.

Wilkinson, J.H. and Reinsch, C. (1971), *Handbook for automatic computation. Vol.II. Linear algebra*, Springer-Verlag: Berlin.

Wilkinson, L. and Dallal, G.E. (1981), *Tests of significance in forward selection regression with an F-to-enter stopping rule*, Technometrics, **23**, 377–380.

Zhang, P. (1992), *Inference after variable selection in linear regression models*, Biometrika, **79**, 741–746.

Zirphile, J. (1975), *Letter to the editor*, Technometrics, **17**, 145.

Zurndorfer, E.A. and Glahn, H.R. (1977), *Significance testing of regression equations developed by screening regression*, 5th Conf. on Prob. and Statist. in Atmos. Sci., Amer. Meteor. Soc., pp. 95–100.

# Index